Intelligent and Sustainable Engineering Systems for Industry 4.0 and Beyond

The Fourth Industrial Revolution, Industry 4.0, aims to significantly improve the flexibility, versatility, usability and efficiency of future smart factories. However, the concept of Industry 4.0 is not only limited to the factory but also encompasses the entire life cycle of the product, that is, from production and suppliers, to end users. Industry 4.0 delivers seamless vertical and horizontal integration down the entire value chain and across all layers of the automation pyramid.

Industry 4.0 has its roots in a project for the high-tech strategy of the German Government back in 2011, which led to the progression of cyber-physical systems into cyber-physical production systems (CPPS). CPPS can make intelligent decisions through real-time communication and cooperation between manufacturing entities. Smart Factory, which is based on CPPS and artificial intelligence (AI), is one of the key associated initiatives of Industry 4.0. This enables flexible production of high-quality personalized products with mass efficiency. Another important aspect of Industry 4.0 is sustainable engineering systems that can help make its processes align with the United Nations Sustainable Development Goals (UN SDGs). Sustainable and intelligent engineering systems such as 5G, Industrial IoT, robotics and automation, renewable energy, logistics and even intelligent waste management can be the main enablers of Industry 4.0.

This is a multidisciplinary book and is meant for anyone with a basic engineering background interested in acquiring a solid foundation in the fundamental concepts and state-of-the-art research trends in Industry 4.0. It explores the application of AI and machine learning as well as sustainable engineering systems, which can be the main drivers for Industry 4.0 and beyond and have a significant impact on the UN SDGs.

Intelligent and Sustainable Engineering Systems for Industry 4.0 and Beyond

Edited by
Tulsi Pawan Fowdur, Dragorad A. Milovanovic,
and Zoran S. Bojkovic

CRC Press is an imprint of the
Taylor & Francis Group, an **informa** business

Designed cover image: © Shutterstock, Iurii Motov

First edition published 2025
by CRC Press
2385 NW Executive Center Drive, Suite 320, Boca Raton FL 33431

and by CRC Press
4 Park Square, Milton Park, Abingdon, Oxon, OX14 4RN

CRC Press is an imprint of Taylor & Francis Group, LLC

ISBN: 9781032840697 (hbk)
ISBN: 9781032841205 (pbk)
ISBN: 9781003511298 (ebk)

DOI: 10.1201/9781003511298

Typeset in Times
by codeMantra

Contents

Preface

Industry 4.0 is based on advanced manufacturing and operations techniques as well as the integration of digital technologies in smart manufacturing ecosystems. The Industry 4.0 paradigm in fact encompasses the whole life cycle of the product, from production and supplier to the end user, and is not restricted to the factory. Industry 4.0 has its roots in a project on the high-tech strategy of the German Government back in 2011, which led to the progression of the concept of cyber-physical systems (CPSs) to cyber-physical production systems (CPPSs). CPPS makes us of real-time communication and cooperation between manufacturing entities to enable intelligent decisions. Smart Factory, which is based on CPPS and AI, is one of the main initiatives of Industry 4.0.

One of the main drivers of Industry 4.0 is the fifth generation of mobile communications known as 5G, which will support the ongoing developments by providing powerful and pervasive connectivity between machines, people and objects. With several high-end connectivity features such as ultra-high reliability, low latency, 100% coverage, availability, high data rates and support for massive IoT deployments, 5G has positioned itself as a key enabler of industry 4.0. Several Industry 4.0 applications such as smart logistics, connected production robots, predictive maintenance, product prototyping and equipment visualization are benefitting immensely from the adoption of 5G. Each of these application use cases has different requirements in terms of throughput, latency, reliability and device density, which can be met by the different 5G service classes. Additionally, by providing advanced support to AR and VR, 5G will streamline the adoption of these technologies, which will be extremely beneficial to Industry 4.0 and beyond.

Another important aspect of Industry 4.0 is sustainable engineering systems that can help make its processes align with the United Nations Sustainable Development Goals (UN SDGs). Since Industry 4.0 is not only about the integration and application of digital technologies but also about sustainable practices in industry, it aligns with the SDGs. The UN SDGs are empowered through the application of resource-efficient and sustainable manufacturing processes. There are 17 goals, and many of them can be achieved through Industry 4.0. For instance, technologies such as IoT sensors, AI algorithms and big data analytics are the core of Industry 4.0, and these new advances can help reduce energy consumption, decrease the level of waste and minimize environmental impacts. In general sustainable and intelligent engineering systems such as 5G, Industrial IoT, robotics and automation, renewable energy, logistics and intelligent waste management can be the main enablers of Industry 4.0, which drives the UN SDGs simultaneously.

The purpose of this book therefore is to explore the application of AI and machine learning as well as sustainable engineering systems in multidisciplinary fields such as renewable energy, climate change, power transmission, transportation, telecommunications and waste management, which can be the main drivers for Industry 4.0 and beyond as well as have a significant impact on the UN SDGs. The book comprises 15 chapters, and a brief overview of these chapters is as follows.

Chapter 1 gives an overview of the basic principles of Industry 4.0, along with its contributions towards the UN SDGs. Chapter 2 of the book provides an overview of 5G/6G technology for smart logistics, predictive maintenance and virtual reality/augmented reality in manufacturing. Chapter 3 surveys applications of the Industrial Internet of Things. Chapter 4 seeks to provide a contribution to a better understanding of emerging trends and research directions of robotics and automation. Chapter 5 presents a paradigm shift towards sustainable production with additive manufacturing. Chapter 6 summarizes the push for predictive maintenance. Chapter 7 focuses on enabling technologies with Industry 4.0 for renewable energy. Chapter 8 presents a case study of solid waste management. Chapter 9 concentrates on logistics management versus supply chain management. Chapter 10 deals with the challenges and obstacles of drone technology in urban environments. Chapter 11 presents smart digitalization technologies for future resilient and sustainable energy systems. Chapter 12 investigates the challenges and potential of blockchain technology integration. Chapter 13 explores 5G/6G energy efficiency in mobile communications for a sustainable future. The goal of Chapter 14 is AI-powered agile project management. The book ends with Chapter 15, which is devoted to Industry 4.0 intelligence on Edge computing.

Editors

Tulsi Pawan Fowdur received his BEng (Hons) degree in Electronic and Communication Engineering with first-class honours from the University of Mauritius in 2004. He was also the recipient of a gold medal for producing the best degree project at the Faculty of Engineering in 2004. In 2005 he obtained a full-time PhD scholarship from the Tertiary Education Commission of Mauritius and was awarded his PhD degree in Electrical and Electronic Engineering in 2010 from the University of Mauritius. He is also a registered chartered engineer of the Engineering Council of the UK, a fellow of the Institute of Telecommunications Professionals of the UK and a senior member of the IEEE. He joined the University of Mauritius as an academic in June 2009 and is presently an associate professor at the Department of Electrical and Electronic Engineering of the University of Mauritius. His research interests include Mobile and Wireless Communications, Multimedia Communications, Networking and Security, Telecommunications Applications Development, Internet of Things and AI. He has published several papers in these areas and is actively involved in research supervision, reviewing papers and organizing international conferences.

Dragorad A. Milovanovic received the Diploma Electrical Engineering and Magister degree from the University of Belgrade, Serbia. He was a research assistant from 1987 to 1991 and PhD researcher from 1991 to 2001 at the Department of Electrical Engineering, where his interests include the design of digital communications systems. He has been working as an R&D engineer for DSP software development in the digital television industry. He is also serving as an ICT lecturer and consultant for digital media and medicine/sports/business/arts informatics in the development of innovative solutions. He has participated in research projects and published more than 300 papers in international journals and conference proceedings. He also coauthored textbooks and chapters in multimedia communications published by several publishers. Present projects include adaptive coding of volumetric 3D Video and PointCloud formats and 5G/6G massive/immersive communication.

Zoran S. Bojkovic is a full professor of Electrical Engineering at the University of Belgrade, Serbia, a life senior member of IEEE, a full member of the Engineering Academy of Serbia and a member of the Scientific Society of Serbia. He was and still is a visiting professor worldwide. He is the author and co-editor of more than 500 publications: monographs, books, book chapters, peer-reviewed journals and conference and symposium papers. Some books have been translated in China, India, Canada and Singapore. He has been a consultant to industry, research, institutes and academia worldwide. His research focuses on computer networks, multimedia communications, 3D video coding, smart grids, green communications, 5G and beyond. He had plenary keynotes and invited talks at international conferences. He is a highly regarded expert in the IEEE, contributing to the growth of the communication industry and society reviewing processes in many books and journals, as well as organizing special sessions and workshops, being general chair and TPC member at numerous conferences all over the world.

Contributors

Robert T. F. Ah King
Department of Electrical and Electronic Engineering
University of Mauritius
Moka, Mauritius

Iswaree Aubeeluck-Ragoonauth
Department of Chemical and Environmental Engineering
University of Mauritius
Moka, Mauritius

Cristina Barrado
Computer Architecture Department
Universitat Politècnica de Catalunya BarcelonaTech (UPC)
Barcelona, Spain

Vandana Bassoo
Department of Electrical and Electronic Engineering
University of Mauritius
Moka, Mauritius

Natasa Bojkovic
Department of Economy, Management and Organization in Traffic and Transport
University of Belgrade
Belgrade, Serbia

Zoran S. Bojkovic
University of Belgrade
Belgrade, Serbia

Milan Djordjevic
PM Guru LLC
Phoenix, Arizona, USA

Bhuvaneshwar Doorgakant
University of Mauritius
Moka, Mauritius

Tulsi Pawan Fowdur
Department of Electrical and Electronic Engineering
University of Mauritius
Moka, Mauritius

Mahendra Gooroochurn
Mechanical and Production Engineering Department
University of Mauritius
Moka, Mauritius

Mussawir Ahmad Hosany
Department of Electrical and Electronic Engineering
University of Mauritius
Moka, Mauritius

Visham Hurbungs
Faculty of Information, Communication & Digital Technologies, Department of Software & Information Systems
University of Mauritius
Moka, Mauritius

Jovana Kuljanin
Physics Department - Aeronautical Division
Universitat Politècnica de Catalunya BarcelonaTech (UPC)
Barcelona, Spain

Dragorad A. Milovanovic
University of Belgrade
Belgrade, Serbia

Vladan Pantovic
University Educons
Belgrade, Serbia

Doorgha Ragoobur
Department of Chemical and Environmental Engineering
University of Mauritius
Moka, Mauritius

Bhimsen Rajkumarsingh
Department of Electrical and Electronic Engineering
University of Mauritius
Moka, Mauritius

Ramful Raviduth
Mechanical & Production Engineering Department
University of Mauritius
Moka, Mauritius

Satyadev Rosunee
Department of Applied Sustainability and Enterprise Development
University of Mauritius
Moka, Mauritius

Asish Seeboo
Civil Engineering Department
University of Mauritius
Moka, Mauritius

Milan Smigic
Centar za upravljanje projektima - CPM
Belgrade, Serbia

Geeta D. Somaroo
Department of Chemical and Environmental Engineering
University of Mauritius
Moka, Mauritius

Vladimir Terzija
School of Engineering
Newcastle University
Newcastle upon Tyne, United Kingdom

Roshan Unmar
Department of Applied Sustainability and Enterprise Development
University of Mauritius
Moka, Mauritius

Tanja Zivojinović
Department of Economy, Management and Organization in Traffic and Transport
University of Belgrade
Belgrade, Serbia

1 Industry 4.0, Artificial Intelligence and the UN SDGs

Tulsi Pawan Fowdur, Dragorad A. Milovanovic, and Zoran S. Bojkovic

1.1 INTRODUCTION

The industrial revolution denotes a pivotal shift from agrarian and craft-based economies to ones dominated by manufacturing industries. Arnold [1] outlines that it signifies not just a historical period but a profound economic transformation. Key facts include technological advancements, cultural shifts and socioeconomic changes. The first industrial revolution saw the adoption of steam and water-powered manufacturing processes, while the second revolution introduced electrical technology to boost productivity. The third revolution, marked by IT and electronics, prioritised production automation. Finally, the fourth industrial revolution aimed to offer personalised customisation through the utilisation of artificial intelligence (AI) [2]. The fifth industrial revolution focuses on introducing personalisation personalised and human involvement into manufacturing techniques. Industry 5.0 is occasionally perceived as a counter to traditional industrial approaches due to its emphasis on personalisation involving human interaction.

Figure 1.1 illustrates the industrial revolution from I1.0 to I4.0. A more detailed review of the industrial revolution is given in the following subsections.

1.1.1 INDUSTRY 1.0

Early in the 18th century, Industry 1.0 was the era which changed the manufacturing and transportation sectors. This revolution is characterised by mechanisation and steam power, which transformed the social and economic environment of the world. Steam power replaced human muscular force during this revolution, raising production across a wide range of industrial businesses. For example, this led to an increase in the productivity of the manufacturing sector, including the spinning industry where weaving looms were employed [3]. Additionally, steam power paved the way for the development of locomotives and ships. Thus, long-distance transportation became more feasible. Globalisation therefore made trade and interconnections between far-off places easier. Small enterprises therefore evolved to cater to wider markets [4].

DOI: 10.1201/9781003511298-1

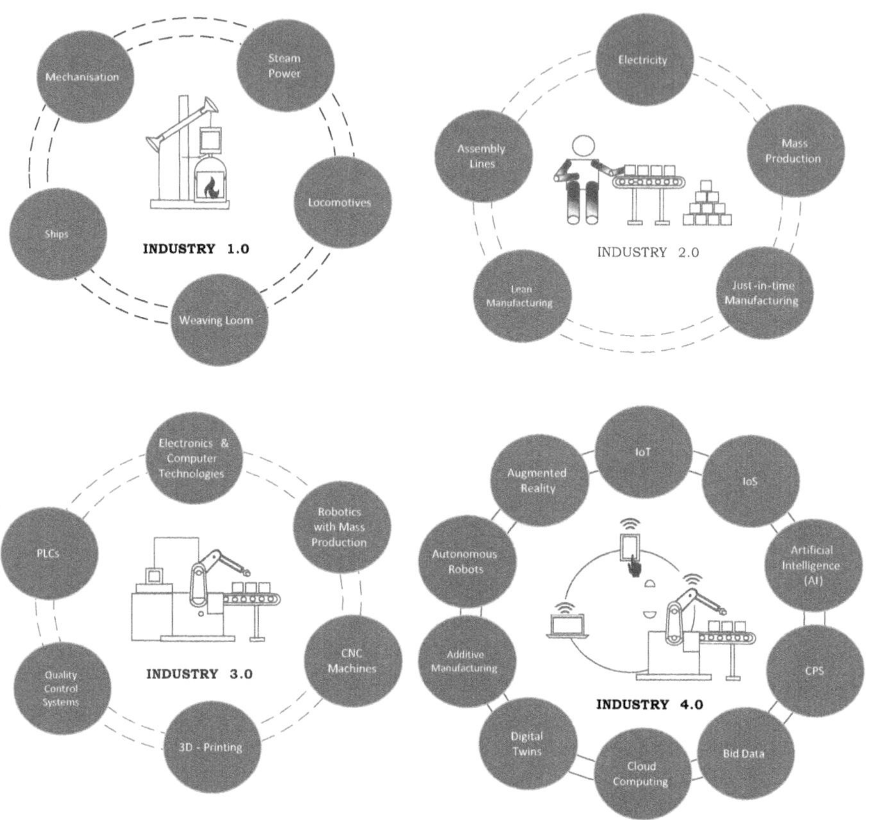

FIGURE 1.1 Stages of the industrial revolution.

This transformative period shifted society's focus from agrarian to industrial, introducing machinery capable of mass-producing goods previously crafted by hand. Factories, along with advancements in transportation, facilitated a more efficient distribution of goods over longer distances and at reduced costs. Examples include water-powered looms revolutionising cloth weaving in mills and steam trains facilitating transportation of people and merchandise to factories and markets. Concurrently, agriculture underwent mechanisation, enhancing food production efficiency and enabling rural workers to migrate to urban centres for factory employment. Unlike a military revolution, this was a societal revolution altering the way people lived, worked and conducted business [5].

Textile industries have emerged as one of the primary beneficiaries of these transformations. The key technological advancements of the era are closely associated with the Watt steam engine, renowned for its superior power compared to the Newcomen engine, leading to numerous subsequent innovations and uses. In addition to surpassing the water wheel in strength, the steam engine offered increased versatility in manufacturing capabilities [6].

1.1.2 Industry 2.0

The era of mass production, sometimes referred to as the Second Industrial Revolution (I2.0), emerged in the 19th century. This is the phase where electricity became the predominant energy source. Unlike water and steam in Industry 1.0, electricity could be manipulated and directed to machines, which increased their portability.

Moreover, improvements were made in management tools to increase industrial efficiency. During this period, the division of labour was spread, and each worker became more centred and specialised in their respective task. The implementation of assembly lines and other mass production techniques proliferated mass production. Furthermore, American mechanical engineer Frederick Taylor developed analysis job functions techniques to maximise employee productivity [7].

During Industry 2.0, the prosthetics sector underwent profound changes due to the impact of two wars, which led to the emergence of prosthetics as a specialised field. The aftermath of the American Civil War resulted in a significant number of amputees, prompting the recognition of the necessity to provide prosthetic devices to veterans without financial burden. This commitment by the US government, known as the Great Civil War Benefaction, sparked a surge in prosthetic technology development, resulting in numerous patented designs. Notably, this period marked the first application of industrialised thinking to systematically tackle challenges in prosthetic design and production [8].

1.1.3 Industry 3.0

The third industrial revolution became prevalent in the 20th century. The advent of electronics and computer technologies made manufacturing automation feasible in factories, thereby relieving workers of labour-intensive duties. The goal was to enhance manufacturing capabilities by offering increased versatility, shorter production times, personalised goods, rapid response to changing consumer preferences, improved process management and enhanced accuracy. Initially, manual loading systems were employed for tasks such as loading granules into injection moulding machines. However, in the late 1990s, automated injection moulding machines were introduced, boosting efficiency in part production and overall output volumes. This transition occurred during a challenging economic period characterised by factors such as the oil crisis of the 1970s, which dampened demand and fuelled inflation. To remain competitive amidst economic pressures, businesses and countries had to focus on cost savings and sales growth, necessitating the adoption of new organisational strategies [9].

Moreover, robotics with mass production was marked by this period [7]. Industry 3.0 saw the introduction of computer numeric control (CNC) machines, which enabled semi-automatic production based on 3D models and schematics which are programmatically operated by workers using specialised software. Moreover, complex-shaped products were realisable with the 3D printers, which were incorporated with CNC machines. Likewise, quality control was facilitated with the advent of special measuring systems, like optical stations and X-ray machines, which increased industrial productivity [10]. Furthermore, programmable logic controllers

(PLCs) that provided control and automation abilities were introduced in this phase. Manufacturing processes were enhanced with these controllers, which monitored inputs, made decisions and controlled output. Complex logic and networked systems also became feasible through developments in networking technologies and programming [11].

1.1.4 Industry 4.0

The fourth industrial revolution, i.e. Industry 4.0, originated in 2011 from a project in the high-tech strategy of the German government. It advanced the concept of cyber-physical systems (CPSs) into cyber-physical production systems (CPPSs) [12–14]. CPPS can make intelligent decisions through real-time communication and cooperation between "manufacturing entities" [6]. This evolution is reshaping competition dynamics, industry structures and customer expectations. It alters competition dynamics as companies adapt their business models through Internet of Things (IoT) adoption and factory digitisation. From a market standpoint, digital technologies enable the provision of innovative digital solutions, such as Internet-based services integrated into products. Operationally, digital technologies like CPS aim to decrease setup times, labour and material expenses and processing durations, ultimately boosting productivity in production processes [15].

Industry 4.0 improves operational and strategic decision-making for businesses by introducing real-time data analysis and flexible manufacturing processes. The widespread installation of sensors in physical objects due to their greater affordability and the integration of Information and Communication Technologies (ICTs) into industrial settings have made this change possible. These systems monitor and manage goods and other machinery through ongoing feedback cycles. They produce enormous volumes of data, which are used to update virtual representations of actual operations, ultimately giving rise to the idea of a "smart factory." As a result, numerous technologies have come into existence and been deployed in production systems since the birth of digital manufacturing in the 1980s. One such technology is cloud computing, which enables on-demand production services [5].

1.1.5 Beyond Industry 4.0

Beyond Industry 4.0, mainly four features, including the Internet of Senses (IoS), Wireless Power Transfer (WPT), AI in Collaborative IoT (CIoT) and Industrial Metaverse, are envisioned. IoS would back applications like emotion monitoring, event management, weather forecasting and emission detection. In addition, the wireless technology, WPT, would be able to transmit electromagnetic power to IoT devices over the air. AI's role would remain significant in Industry 4.0+ since it fosters CIoT collaboration and addresses model-free optimisation issues in a complex industrial networking environment. Moreover, data-driven CIoT (DCIoT) collaboration would also be required for higher-level intelligence. Furthermore, Industry 4.0+ would explore the concept of an industrial metaverse, which is an industrial ecological system that merges real-world industrial operations with the digital IIoT. Through technologies like Extended Reality (XR), AI, IoT, cloud computing, blockchain and 5G/6G, seamless connections between machines and humans would be encouraged [16].

1.2 DESIGN PRINCIPLES OF INDUSTRY 4.0

The basic principles of Industry 4.0 lie in connecting systems, machines and work divisions to form intelligent networks along the value chain that can function independently and autonomously. In this regard, there are mainly six established design principles such as interoperability, virtualisation, decentralisation, real-time capability, service orientation and modularity, which are considered by manufacturers in Industry 4.0.

1.2.1 Interoperability

Interoperability refers to the capability of various manufacturing systems including machines, devices, sensors and robots to seamlessly connect, communicate and function together in real time through the IoT. This encompasses the utilising technologies, smart factories as well as labourers which improved coordination and efficiency [17]. Additionally, it englobes CPS, IoT, IoS and smart factories. The integration of computational, networking and physical components makes real-time monitoring and control feasible by CPS, which increases productivity. Likewise, IoT allows the transmission, collection and analysis of data by connecting various devices and sensors to the Internet, while IoS allows cloud-based services to those connected devices. Furthermore, IoT, CPS and IoS technologies are used by smart factories to improve operational effectiveness and optimise manufacturing procedures [18].

Moreover, communication standards remain crucial enablers of Industry 4.0, which allowed the connection of CPS and human resources via the Internet in enterprises. Modern protocols like Message Queuing Telemetry Transport (MQTT) and Open Platform Communications Unified Architecture (OPC-UA) are being introduced to improve interoperability and expandability. Indeed, MQTT is widely spread in the IoT environment, which makes the integration of PLCs easier into IoT settings, while OPC-UA is identified as the communication standard for Industry 4.0 in RAMI 4.0 architecture. Seamless data transmission is also made possible through common data formats like JavaScript Object Notation (JSON) or XML. Middleware solutions serve as intermediates, while edge computing technologies process data closer to the source, decreasing bandwidth requirements [12]. The main components of interoperability are shown in Figure 1.2.

1.2.2 Virtualisation

Virtualisation entails the creation of digital models of production processes with virtual testing and predictive analysis features without neither physical prototypes nor disruption in production. Through this process, CPS monitors physical processes while sensor data are connected to virtual or simulation models. Consequently, a virtual replica of the physical world is produced [17].

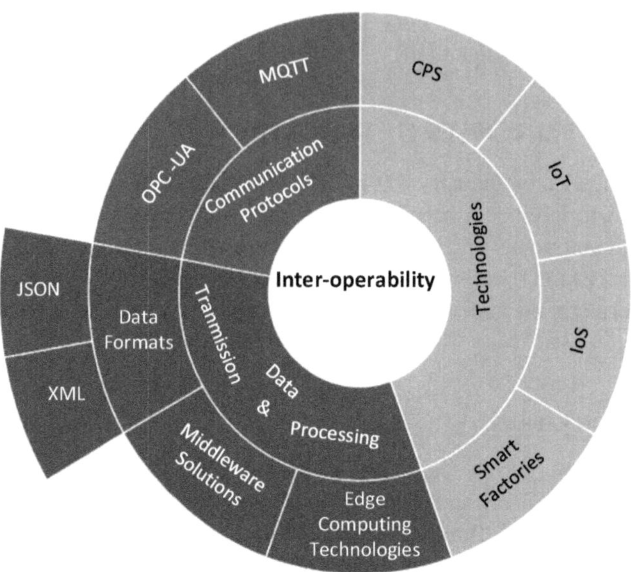

FIGURE 1.2 Components of interoperability in I4.0.

Industry 4.0 benefits from virtual manufacturing since capitalising on advanced technologies such as Big Data Analytics, AI and IoT for the designing and testing of products before their physical creation helps manufacturers to identify potential issues and adjust accordingly, which improves their efficiency and productivity. Virtualisation allows the company to respond to market dynamics accordingly by shifting their production strategies towards customised products as well as reducing costs associated with materials, labour and fast-to-market time delivery, allowing them to better meet their customer preferences and have a competitive advantage in the market. The transition into an integrated network fosters collaboration among manufacturing enterprises, allowing them to leverage each other core competencies which enhance overall productivity. Furthermore, virtualisation supports sustainable manufacturing practices by optimising resource utilisation, reducing waste and minimising their ecological footprint.

Figure 1.3 shows some of the main virtualisation technologies and the corresponding benefits derived from them.

1.2.3 Decentralisation

Centralised system control is becoming increasingly challenging due to the rising demand for mass production. Thus, decentralisation is essential where embedded computers allow CPS to operate autonomously and make independent decisions. Those physical systems have networked sensors fuelling automation dependent on performance data. Efficiency and productivity are boosted since only in the event of system failure that human intervention is needed [17,18].

Moreover, the authors of [17] illustrate decentralisation as dividing a complex system into smaller parts, thereby promoting independent decision-making and

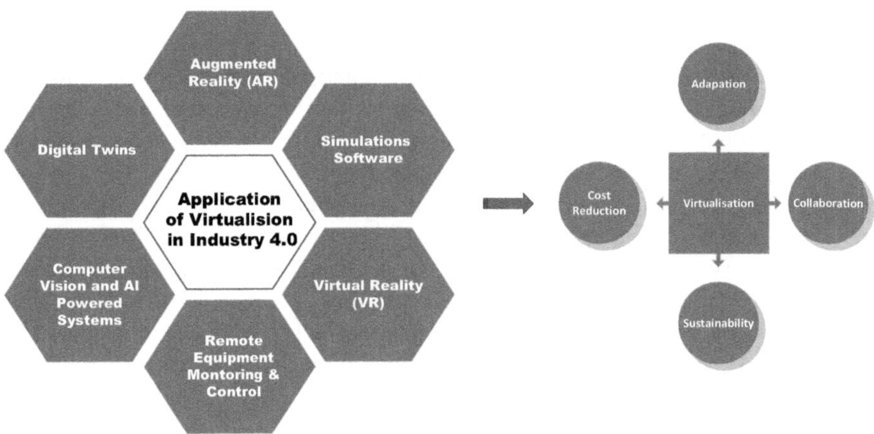

FIGURE 1.3 Virtualisation technologies and benefits.

improving organisational flexibility. Decentralisation is permissible in Industry 4.0 due to its unparalleled technologies like IoT, cobotics, augmented reality and AI, which boost production system autonomy by facilitating autonomous problem-solving and decision-making operations. Hence, autonomous intelligent factories benefit from these through scheduling, planning and process control [19].

Decentralisation is leveraged by Industry 4.0 using advanced technologies including blockchain, edge computing and direct peer-to-peer communication protocols like MQTT and Constrained Application Protocol (CoAP), for data management and decision-making. This boosts security and adaptability in autonomous and networked manufacturing settings [12].

Overall, decentralisation promotes agility, innovation and accountability within an organisation by empowering lower levels of the hierarchy to make decisions and take action. It allows organisations to respond more effectively to dynamic environments, leverage the expertise and creativity of individual agents and adapt to changing market conditions. Figure 1.4 summarises the benefits of decentralisation.

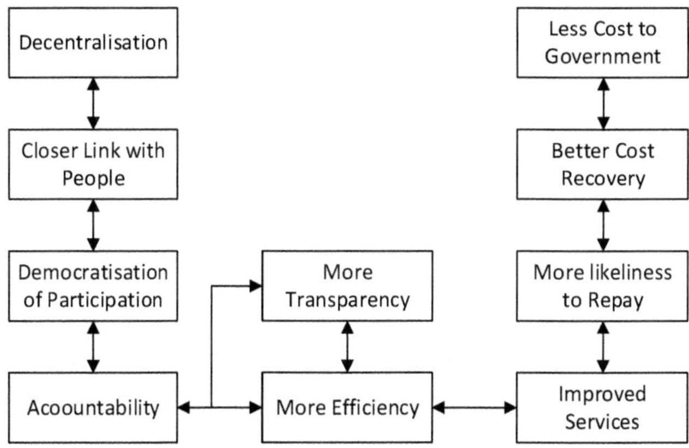

FIGURE 1.4 Benefits of decentralisation.

1.2.4 Modularity

Modularity is about the establishment of a standard protocol among products and communications networks to create systems that interact with each other and require minimal rectifications. Issues being faced in the industry require modern solutions and continuous innovations to tackle them, and thus modularity provides the flexibility to amend these changes and issues. Modular systems are highly scalable, which makes them most appropriate to react rapidly to modifications without compromising current investments. One of the advantages of this principle is its capability to handle mass customisation. It allows products to be manufactured according to the preferences of clients without sacrificing sales and efficiency. This is all possible through the use of various standardised modules to create multiple unique products [17,18].

1.2.5 Service Orientation

Service orientation is another principle whereby services of cyber-physical systems, humans and smart factories are provided by means of the Internet of Services. It provides necessary tools so that immediate amendments can be made with respect to customer preferences. The incorporation of service orientation in business can improve transparency in the value chain, which will thus increase customer focus. Service orientation provides benefits such as facilitating the flow of information within and between enterprises, the establishment of internal software to increase its values since its functionality is visible and lastly its ability to adapt to changes. Thus, service orientation facilitates the effective sharing of knowledge across supply chains [17,18].

1.2.6 Real-Time Capability

Real-time capability can collect and analyse data, thus providing better knowledge about those data. Similarly, like the other principles, real-time capability can easily adapt to changes. Collecting data and real-time processes through constant checking of products can significantly minimise reaction time and increase productivity [17,18].

1.3 ENABLING TECHNOLOGIES OF INDUSTRY 4.0

A brief review of the enabling technologies for Industry 4.0 is given in the following subsections.

1.3.1 Internet of Things

IoT was introduced in the 1980s and gained momentum in 1999 with radio-frequency identification in companies like Olivetti, Xerox and IBM and colleges. Development was boosted by open-source dynamics, influencing chip producers, sensors, hardware

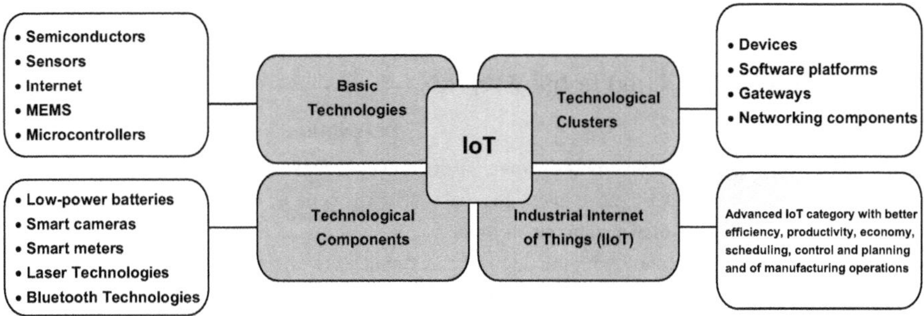

FIGURE 1.5 IoT components and technologies.

manufacturers and developers. The foundation stones of the IoT are computer science, communication, electronics and information technology. The basic technologies required to develop an IoT system are semiconductor technologies, sensor technologies, Internet and microelectromechanical systems, as shown in Figure 1.5. IoT encompasses low-powered batteries, smart camera systems and smart meters. The list also includes laser technologies and Bluetooth technologies. Three technological clusters are present in this heterogeneous assembly of different devices and solutions, namely devices, software platforms and gateways and other networking components. Because IoT technologies are still in their development stage, they function in an unstable competitive and uncertain technological landscape where they are challenged by the data exchange among large-scale heterogeneous network parts and service adaptation in a dynamic system environment [20].

Moreover, in modern industries, the Industrial Internet of Things (IIoT) is an advanced category of IoT which improves efficiency and monitoring of intelligent manufacturing operations. IIoT services offer effective scheduling, planning and control of manufacturing procedures. Unparalleled economic and productivity boosts are some results of its interconnected system which fuels analytics, fosters decision-making and reduces product time-to-market [21].

1.3.2 Robotics

Robotics has significantly progressed since George Devol's first digitally invented robot in 1954. In today's era, robotics finds its applications in the mass production of industrial goods, assembly and packing, medical surgery, weaponry, scientific study, transportation and space exploration amongst others. Modern flexible robots are being developed as a result of advancements in computer hardware and data management software. Core components of robotics include sensors, actuators, power converters, manipulators and software. According to Martinelli et al. [20], industrial robots can be grouped into six categories, namely, selective compliance assembly robot arm (SCARA), articulated, delta, cartesian, dual arm and cobots and are explained in Table 1.1.

TABLE 1.1
Categories of Robots Used in Industry 4.0

Type of Robot	Description
Delta robots	Parallel robots with parallelogram used for selection, lifting and packaging.
Cobots	Designed to interact in real time with operators in a common workplace, unlike autonomous robots.
Articulated robots	Robots with rotating joints as simple as two-jointed mechanisms or complicated elaborate structures having several interacting joints.
Dual arm robots	Have six joints and anthropomorphic (human-like) elbows.
Cartesian robots	Used for pick-and-place tasks, machine tool handling, assembling and sealant functions.
SCARA robots	A high-speed repeatable robot designed for serial assembly.

1.3.3 5G

5G, the fifth generation of mobile telecommunications networks, replaces existing networks. This technology is intended to alleviate the problem of the massive increase in daily devices in all aspects of modern life. 5G enables faster transmission of data and considerably fosters the Things IoT, which comprises an enormous number of devices [22,23]. 5G offers services for a wide range of vertical applications via network slicing and edge computing. Several features of existing industrial requirements are linked to time sensitivity. 5G and further networks are expected to effectively address such demands in terms of time sensitivity by allowing ultrareliable and low-latency communications. QoS requirements are investigated for the integration of time-sensitive networking with 5G and beyond. The 5G core network architecture is service based. This means that the core network can accept any type of radio access, and it can also provide a wide range of services including voice, video, autopilot, industrial manufacturing and telemedicine [22].

5G is developed to enhance capacity, flexibility and performance. Massive Multiple Input Multiple Output (MIMO), edge computing, network slicing, millimetre wave frequencies, beamforming, low latency, high reliability and duplex transmission are some foundational technologies that it offers. Higher bandwidth and faster data rates are the assets of these technologies; however, line-of-sight data transmission between the transmitter and receiver remains its liability. Moreover, beamforming is achieved by directing the radio signal toward a particular device, which boosts signal strength and lessens interference. While at the network edge, edge computing supports real-time analysis and processing, lowering latency and boosting performance, network slicing builds dedicated virtual networks for various applications. Furthermore, full duplex communication allows simultaneous data transmission and reception. Moreover, for real-time communication and mission-critical applications, especially for IoT services and high-bandwidth virtual and augmented reality, low latency and high reliability remain critical [23].

The main requirements of 5G are as follows [24]:

1. A reliable connection with the least amount of downtime is 99.999% available.
2. Real-time communication is feasible due to its latency of 1 ms.
3. Autonomous operation will become long term if we use lower power consumption for IoT devices that can last up to ten years on batteries.
4. With its lower power requirements and longer-range capability over previous technologies like Bluetooth or Zigbee wireless protocols, IoT devices may have an up to ten-year battery life when used with low power wide area networks supported by fifth-generation cellular networks (5G).
5. Universal access, even in remote areas, is 100% covered.
6. Making infrastructure more sustainable means reducing network energy usage by 90%.
7. More data traffic capacity will be possible when implementing 5G since it allows for up to a thousand times more bandwidth per unit area than previous generations did.
8. Enhanced device connectivity can be achieved through coupling many devices in proximity; this would require allowing for about 100 times as many coupled devices per unit area as were allowed under prior generations such as 4G or WiFi6.

1.3.4 Blockchain

Blockchain technology comprises a sophisticated database which enables transparent data communication within a network. Data are stored in blocks that are connected in a chain within the blockchain database. Blockchain, an immutable ledger, allows decentralised transactions in numerous Industry 4.0 fields like financial systems, reputation systems and IoT to improve security, privacy and data transparency for all sizes of enterprises. However, its drawbacks include scalability and security concerns [25].

Blockchain has the following basic characteristics:

- **Decentralisation:** Blockchain employs consensus algorithms to foster data consistency in distributed networks, thereby reducing the need for a central trusted agency for the validation of transactions.
- **Persistency:** Blockchain transactions can be promptly verified by trustworthy miners, and invalid transaction blocks can be immediately tracked.
- **Anonymity:** Blockchain does not disclose the identity of users; instead, users are provided with generated addresses for interaction. However, due to intrinsic limitations, absolute privacy preservation cannot be guaranteed.

Blockchain delivers an almost real-time record that is copied across a network of business partners and never changes. Blockchain offers several benefits that help organisations better understand their clients, especially on the demand side [26]. Figure 1.6 shows the basic operations of blockchain.

Moreover, blockchain has several applications in Industry 4.0 such as logistics and transportation, data privacy, fraud prevention, record keeping and tracking transactions.

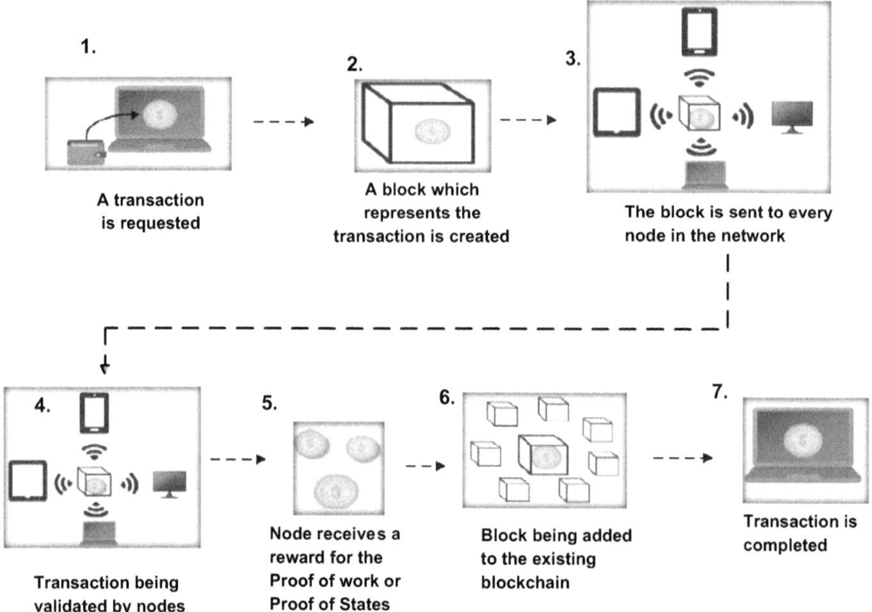

FIGURE 1.6 Steps in a blockchain transaction.

1.3.5 Cloud Computing

Cloud computing gives users Internet-based, on-demand access to computational resources, enabling scalable and cost-effective solutions for data storage and processing. IT is an outsourcing that uses many servers, often with a lot of computing power and resources brought together to provide computer programs, high-level services and on-demand or pay-per-cycle-basis real-time resources; thus, it has the capability to enhance and transform the current industry. Its three tiers are Infrastructure as a Service (IaaS), Platform as a Service (PaaS) and Software as a Service (SaaS). Cloud computing accommodates modern apps and services that grow with users' numbers in a dynamic manner. What this implies therefore is that users can access applications, programs and services quickly without much management overhead. For their function systems like human resource management and customer relationship management, industrial enterprises employ cloud-based apps. In addition to this cloud computing leads to lower development costs, continuous accessibility, high availability, maintainability and product time-to-market. This is why all types of businesses all over the world have adopted this state-of-the-art technology very quickly to maximise their potentiality while minimising costs incurred during operations [21]. In Industry 4.0, cloud-based platforms support the storage, analysis and sharing of large volumes of data generated by IoT devices and other connected systems. Advanced analytics, machine learning (ML) and AI algorithms deployed on cloud infrastructure empower manufacturers with actionable insights and decision support. The three main service classes of cloud computing are as follows:

i. **SaaS:** It essentially represents the applications which a cloud service provider offers to its clients via a thin interface, making it possible for clients to use applications such as Google Apps, DropBox, YouTube and Office 365.
ii. **PaaS:** It is basically an application hosting environment which a cloud service provider offers to clients who wish to deploy their applications on the cloud. Examples include Microsoft Azure and Google App Engine.
iii. **IaaS:** In this case the cloud service provider provisions processing, storage, networks and other fundamental computing resources for clients to deploy and run arbitrary software, which can include operating systems and applications.

1.3.6 Other Technologies

In addition to the aforementioned technologies, the following enabling technologies are very important for Industry 4.0:

i. **Cybersecurity:** Cybersecurity is paramount in Industry 4.0 to protect against cyber threats and ensure the integrity and confidentiality of data [27]. Security measures include encryption, authentication, intrusion detection and regular security audits. Collaboration between industry stakeholders and cybersecurity experts is essential to address evolving threats.
ii. **Digital Twins:** A digital twin can provide a near real-time connection between both the physical and virtual worlds by creating a virtual model to simulate a physical entity or operation during its life cycle. With a digital twin, the industry can more quickly identify hardware problems, make better forecasts and generate enhanced goods and services. Digital twin integration is driven by the IoT across several industries, providing them with diverse benefits such as mass customisation along with mass personalisation, simultaneously ensuring mass production efficiency [28].
iii. **AI and Big Data Analytics:** These two technologies constitute the "brain" of Industry 4.0 and have a very major impact on all its processes. A detailed review of these technologies and their applications in Industry 4.0 will be given in the next section.

1.4 AI, ML AND BIG DATA FOR INDUSTRY 4.0

As previously mentioned, AI and Big Data have a huge impact on Industry 4.0 and are crucial to several Industry 4.0 applications. The following subsections define these technologies and describe their applications in the context of Industry 4.0.

1.4.1 Artificial Intelligence

AI is defined as the ability of machines to demonstrate cognitive functions. As AI is now applied in every other field, it has become a remarkably familiar concept to the world. The emergence and influence of AI have been pivotal in shaping the development of the Internet, and its prominence has grown over the past ten years, impacting numerous facets of everyday life with its widespread applications in society. AI is an amalgam of multiple fields such as computer science, logic, biology, psychology and philosophy. It has made significant strides in domains like the development of intelligent robotics, speech recognition, photo processing, language processing and theorem proving. Currently, the main sources through which people mainly acquire knowledge about AI include news reports, movies and applications in daily life [29].

With the integration of AI, the probability of obtaining the desired outcomes increases as machines are capable of sensing and understanding their surroundings. They make use of these data to determine future actions concerning products' demand in terms of temporal, spatial and socioeconomic factors. Several issues, such as uncertain decision-making and lack of skills among employees within an industry, can be resolved with the application of AI. Making use of a combination of AI and robotics in the production process has transformed and helped evolve mass production for industries. With the results being provided by AI analysis, overall productivity and efficiency can be enhanced together with the quality of the products [26,30].

Industry 4.0 exploits the use of AI, which is an advanced computing technology. AI plays a crucial role in analysing, sensing and providing accurate responses to real-time data and high-level tasks within industrial processes without the necessity of human intervention. The integration of AI has truly ensured efficient productivity since the data provided by these machine algorithms not only are accurate but are also generated at a much higher rate. AI, combined with digital technology, has undoubtedly made Industry 4.0 more smart and productive.

AI has proved to be useful in many ways. One of the major uses of AI would be that it can effortlessly bind with IoT devices and provide profound analysis based on the data collected from IoT devices and sensors. The data measured and registered by the sensors are obtained from different parts of the manufacturing process. Through frequent monitoring, AI can easily identify and alert for potential issues that can arise within manufacturing processes, thus allowing operators to take the right and necessary measures accordingly.

Another key application of AI in Industry 4.0 is that it facilitates human tasks. It does not need human intervention since it can increase productivity more accurately and at a faster rate than humans. AI, along with digital technologies, can make Industry 4.0 more efficient and smart. Moreover, AI is deemed to be advantageous when it comes to the improvement of energy efficacy. By analysing the trends in the use of energy, it has the capability to come up with solutions to minimise the waste of energy. AI-powered robots and cobots partnered with human workers have immensely helped to achieve higher productivity in the industry. In the same line of thought, AI empowers the development of sophisticated robotic systems that can specialise in complex tasks such as assembly, packaging and quality assurance.

In Industry 4.0, the yield of products is greatly enhanced using predictions from AI frameworks. Analysis is made based on goods that are already available, further helping in improving their quality. Manufactured products' quality and yield are efficiently predicted by sophisticated analytics, using data collected from different smart machines. Continuous monitoring and quality prediction, which involve a repeated cycle of predicting a product's yield throughout its life cycle, ensures an increase in manufacturing yield, customer satisfaction and quality management [31].

The combination of AI and robotics has brought major changes to Industry 4.0. Robots perform multiple tasks such as quality assurance, strategic planning and reducing errors and provide more efficient and reliable outcomes [30]. AI applications have enhanced many processes through automation. One application of AI in robotics is assembling and manufacturing. AI, combined with other technologies, such as ML, cognitive computing systems and edge computing, has effectively improved the performance of robots in assembling and manufacturing processes. With the integration of AI, robots' ability to adapt to changes in their environment has increased as they can understand and optimise complex tasks more efficiently. For example, for the process of chip assembly, AI is used to create an advanced production system with robotics technology using cognitive manufacturing [32].

Predictive maintenance is another key area in Industry 4.0 where AI is playing a major role. The aims of predictive maintenance are reducing expenses and averting failures. AI is achieving this goal by locating the specific component in a product that is prone to break down, and, hence, maintenance is done at the right time. In Industry 4.0, predictive maintenance is being implemented through a combination of technologies such as AI, sensors and IoT. Real-time monitoring of manufacturing processes, prediction of remaining useful life (RUL) and optimisation of maintenance cycle and scope are some of the key components of predictive maintenance. Despite its advantageous use of forecasting defaults in products beforehand, a drawback of predictive maintenance is that a huge number of resources in terms of finance and human knowledge are required [33].

1.4.2 Machine Learning

A revolutionary element of AI is ML, which allows computers to learn from data and enhance their capabilities without the need for explicit programming. Unlike traditional programming methods where humans provide explicit instructions for tasks, ML enables computers to discern patterns in data and make predictions autonomously.

ML originated in the mid-20th century, notably with Alan Turing's exploration of intelligent machines. Yet, the formal recognition of ML as a distinct field took shape during the 1950s and 1960s. Arthur Samuel coined the term "machine learning" in 1959, referring to computers' ability to gain knowledge through experience. Over subsequent decades, the field of ML experienced alternating phases of progress and stagnation. Advancements were hindered by constraints such as limited computational capabilities and data accessibility until the end of the 20th century. The development of ML algorithms has driven this shift, moving from traditional statistical methods to complex artificial neural networks [34].

Some of the main capabilities of ML are summarised in Figure 1.7.

- **Data-Driven Insights:** They enable organisations to extract large volumes of data, which could be impractical to analyse manually. Businesses can make data-driven decisions, increasing productivity and obtaining a competitive edge, by seeing patterns and trends.

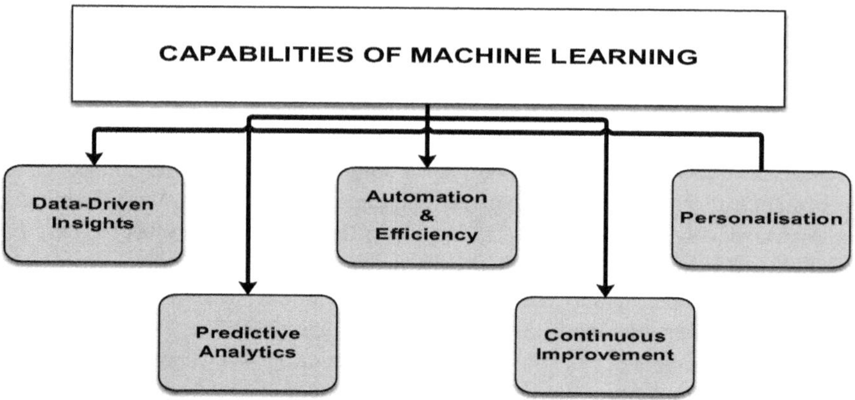

FIGURE 1.7 Capabilities of machine learning.

- **Automation and Efficiency:** These algorithms are excellent at learning from experience and automating repeated processes. As a result, they may boost capacity and efficiency across a range of industries, reduce waste and improve product quality.
- **Personalisation:** Through the analysis of user behaviour and preferences, these systems can customise recommendations, thereby improving user satisfaction.
- **Continuous Improvement:** Models are able to modify and enhance their predictions in response to fresh data, thereby maintaining the system's efficacy and relevance over time.

It is important to note that ML systems typically don't entirely replace entire jobs, processes or business models. Instead, they often complement human activities, enhancing the value of their work [35].

ML techniques can have a leading role in revolutionising smart manufacturing processes. They can identify and analyse complex patterns of data and help in decision-making procedures in several areas of manufacturing and production [36]. ML algorithms analyse user behaviour and use the results to learn, adapt and continuously improve themselves over time. These algorithms are used in many applications such as fault detection, health monitoring systems and prediction of RUL. To develop alternatives that can address the challenges faced with traditional ML methods, several deep learning algorithms have been analysed. The algorithms are as follows: convolutional neural network (CNN), deep belief network and stacked auto encoder have been analysed for fault detections and recurrent neural network and the support vector regression for predictive processes [30].

Since Industry 4.0 highly promotes the implementation of smart machines, smart sensors and devices, ML plays a crucial role in seamlessly allowing these devices to communicate and collect data continuously related to the production of the industry. The various techniques of ML allow the development of insights based on the data obtained, thus enhancing the effectiveness of manufacturing processes without the need to make amendments to the available resources. Also, due to its ability to make predictions, the breakdown of complex tasks has become much easier. Therefore, ML provides a way to make more strategic and effective decisions.

One of the crucial features of ML that demarcates it from AI is that it has different algorithm techniques used in manufacturing, namely, supervised learning, unsupervised learning, deep learning and reinforcement learning. Supervised learning enables the detection of defects in equipment and prediction based on labelled data. Supervised learning is used in the automobile industry, whereby the model is trained based on images and can be used to classify defects in car parts. Unsupervised learning groups products and determines defects based on patterns and structure of unlabelled data. This machine language technique is used to identify groups of the same products based on features such as weight, size and material. Another technique is deep learning, whereby it uses neural networks to configure relationships for complex data, which facilitate quality prediction and process optimisation. For instance, it is significantly used in the food industry to give a better quality of baked food through the monitoring of data received from sensors such as humidity temperature and pressure sensors to determine the optimal conditions for baking. Lastly, we have reinforcement learning, which allows models to make decisions depending on the environmental feedback. For example, this technique can be applied to enhance production based on a scheduled demand, thus minimising energy consumption and reducing equipment downtime. Therefore, the techniques used by ML have scope for innovation and efficiency in the manufacturing process, contributing to the growth of Industry 4.0.

The manufacturing industry is adopting the use of cloud storage for handling and storing Big Data. ML techniques, together with the cloud, not only provide an expanded storage capacity and a decrease in cost but also lead to an increase in profits within production processes. A profound understanding of the significance of edge and cloud computing in ML-based solutions is deemed beneficial for the future. An experiment was conducted using both cloud and edge computing scheduling learning models. The two-level hybrid model allows for predicting outcomes at a faster rate. The model consists of two levels: at the first level, scheduling procedures are decomposed and divided according to the first-in first-out principle; and, at the second level, further decomposition occurs, and the tasks are divided into atomic units, which are then scheduled into different devices. For scheduling problems at the second level, a combination of bat algorithm and long- and short-term memory networks is used. The results of the analysis demonstrate that the two-level model enhances performance in real-world applications for cloud manufacturing [36].

Monitoring failure in machines effectively is a significant task in the industrial world as any malfunctions can cause serious safety risks to workers and the environment, halt the manufacturing process and increase expenditure. ML techniques are the most common type of ML applications used for scalable automation of fault detection and diagnosis (FDD) systems. An aspect of FDD systems is visual inspection as various faults such as imperfections on the surface of products and signs of deterioration in products' condition are visible to the naked eye. CNNs are usually employed to automate analysis within the system. However, there are challenges associated with the use of CNNs in FDD systems. Due to the diverse range of machines available in the industry, creating large image datasets to train CNNs is exceedingly difficult [37].

1.4.3 Big Data

Since the invention of computers, there has been a massive rise in the generation of the amount of data at a rapid pace. The primary drive for these phenomena is the source of both present and future study projects. Gathering information has been made easier by breakthroughs in digital sensors, mobile devices, communications as well as computation and storage of data.

Big Data refers to extensive datasets characterised by their large, diverse and complex structures, posing challenges in storage, analysis and visualisation for subsequent procedures or outcomes [38]. The Industrial Development Corporation projects that the amount of data in the world will double every two years, having expanded nine times in the past five years. Doug Laney categorised Big Data using three main attributes, known as the three Vs: volume, velocity and variety. Volume indicates the sheer size of the data, velocity refers to the speed at which data are generated and processed, while variety encompasses the diversity of data sources and types. IBM and Microsoft introduced veracity, or variability, as the fourth V to further define Big Data. Veracity pertains to the reliability and accuracy of data, as well as its degree of inconsistency or "messiness" [38].

Big Data is a prerequisite for AI and serves as a foundational element by playing a pivotal role in enhancing recognition rates and accuracy. The widespread adoption and evolution of the IoT have led to an exponential increase in data generation, with a remarkable annual growth rate. This proliferation of large volumes of high-dimensional data enriches its comprehensiveness, providing ample support for various [38]. An essential factor of Industry 4.0 is Big Data, which offers significant benefits by generating, gathering and analysing quantitative data internally within the system. Real-time monitoring of the data is very useful for handling any issue faced during production and, hence, allowing automated processes and manufacturing to occur without the need to halt any procedure [26].

Big Data is used in Industry 4.0 for its predictive maintenance, production planning and material requirements planning. From the smart factories' perspective, Big Data plays a crucial role since it helps to better forecast and analyse those datasets that it receives from a plethora of sensors related to production machines. Big Data becomes advantageous in situations where maintenance and repair activities will be mandatory. This will therefore allow manufacturers to have better efficiency in terms of production, have a better knowledge about their real-time data and automate production with the aid of automated platforms and many more.

Big Data is known to be reliable when it comes to the management of huge amounts of data since it can handle terabytes/petabytes of data. The integration of Big Data allows the industry to constantly evaluate and enhance their supply chain management and to make large networks safer to use. Therefore, Big Data gives the industry the opportunity to make more reliable and accurate decisions. For instance, the telecom industry has adopted the use of Big Data to make predictions on customer churn. By making use of the data pertaining to the customer such as payment methods, survey data and customer site visit data, the latter's churn can be anticipated.

Risk management allows industries to use their large datasets to obtain reliable and detailed information. This can further enhance risk management by implementing real-time strategies. The application of Big Data is significant for risk assessing, capturing and identifying hidden values through sophisticated data quality monitoring. These data measures are important in managing risks that cannot be quantified. Data collected by sensors in smart industries is increasing enormously; this represents an excellent opportunity to leverage AI-based methods for risk evaluation and effectively extracting values from Big Data. Despite the significant advantage provided by Big Data in risk evaluation for real-time data monitoring in Industry 4.0, together with new mechanisms and tools, real-time data processing requires an advanced level of mathematical and statistical techniques and knowledge to analyse data quantitatively and qualitatively [31].

1.5 INDUSTRY 4.0 AND THE UN SDGs

The UN SDGs can be positively impacted by developments in Industry 4.0 and particularly in the enabling technologies of Industry 4.0 discussed in Section 1.3. In this section we first provide an overview of the UN SDGs followed by the impact which Industry 4.0 and its technologies can have on the SDGs.

1.5.1 The UN SDGs

The 2015 United Nations General Assembly created the UN Sustainable Development Goals (UNSDGs) – a series of 17 global goals that are meant to be achieved by 2030. They aim to improve various parts of the world, such as healthcare services, education standards, ecological balance maintenance and social justice promotion. This gathering led to a plan named "Transforming our world: the 2030 Agenda for Sustainable Development", which established 17 goals for addressing worldwide socioeconomic and environmental concerns. These goals are meant to tackle an array of issues such as poverty, hunger, health, education, gender equality, clean water and sanitation, affordable and clean energy, decent work and economic growth, industry innovation and infrastructure, reduced inequalities, sustainable cities and communities, responsible consumption and production, climate action, marine lives, life on land, peace, justice, strong institutions and partnerships for the goals. Table 1.2 shows the different goals that the UN aims to accomplish by 2030. Each goal is tracked with different measurements and indicators as summarised in Table 1.2 [39,40].

1.5.2 Impact of Industry 4.0 on the UN SDGs

Industry 4.0 has greatly revolutionised the manufacturing industry with the application of digital technology. The process of implementing more digital technologies has helped to maximise productivity, better economic growth, develop innovative products at a rapid rate and manage customer preferences and many more. To have a better insight into what extent this evolution leads to sustainability, it is primordial to gain knowledge about the interconnection between Industry 4.0 and the Sustainable Development Goals (SDGs).

There has been a remarkable use of Industry 4.0 technologies accounting for up to 72%. This surge is due to the ease of accessibility to these technologies,

TABLE 1.2
The UN SDGs [39,40]

SDG	Description
SDG 1: No poverty	By 2030, SDG 1 aims to eliminate extreme poverty for all people, presently defined as those living on less than $1.25/day and to implement social protection systems that are suitable for the nation with the goal of achieving considerable coverage for the poor and vulnerable.
SDG 2: Zero hunger	In addition to ending hunger and all types of malnutrition, SDG 2 strives to increase food security, agricultural production and just and sustainable farming practices.
SDG 3: Good health and well-being	Significant progress has been made recently towards accomplishing the overall goals of SDG 3 (good health and well-being), especially regarding lowering maternal and newborn mortality rates. Even now, it is estimated that less than half of the world's population has access to basic medical care.
SDG 4: Quality education	Equal and inclusive access to high-quality education is the focus of SDG 4. To prepare children for suitable jobs, early childhood care, pre-primary education, primary and secondary school, higher education and vocational training are all covered.
SDG 5: Gender equality	To achieve gender equality, SDG 5 seeks to end all forms of violence, discrimination and harmful behaviours towards women and girls.
SDG 6: Clean water and sanitation	Goal 6 of the SDGs aims to provide equal access to water and sanitation, which is acknowledged as a fundamental human right and an essential need for existence. Along with the provision of drinking water, sanitation and hygiene services, SDG 6 places a strong emphasis on the long-term management of water resources.
SDG 7: Affordable & clean energy	The aim of SDG 7 is to ensure reliable and cost-effective access to modern energy services.
SDG 8: Decent work and economic growth	The objectives of SDG 8 are to promote full and effective employment, equitable and sustainable growth in the economy and jobs for everyone.
SDG 9: Industry, innovation & infrastructure	SDG 9 talks about three things mainly constructing robust infrastructure, fostering sustainable industrialisation and encouraging innovation.
SDG 10: Reduced inequalities	The objective of SDG 10 is to decrease inequalities within and among countries. This includes reducing financial inequality, encouraging social inclusion and ensuring equitable opportunities for all people, regardless of age, gender, handicap, race, ethnicity, origin, religion or socioeconomic status.
SDG 11: Sustainable cities & communities	Providing affordable housing, promoting sustainable transportation and enhancing urban planning and administration are all part of SDG 11's efforts to make towns and communities equitable, secure, resilient and environmentally friendly.
SDG 12: Responsible consumption & production	The objective of SDG 12 is to maintain sustainable consumption and production practices, which involves improving resource efficiency, decreasing waste output and supporting environmentally friendly production and consumption.

(*Continued*)

TABLE 1.2 (*Continued*)
The UN SDGs [39,40]

SDG	Description
SDG 13: Climate action	The objective of SDG 13 is to respond promptly to address climate change and its consequences. This includes strategies to lower greenhouse gas emissions and attempts to strengthen resilience to climate-related hazards.
SDG 14: Life below water	SDG 14 aims at protecting and in a sustainable manner using the oceans, seas and marine resources to promote sustainability, which involves decreasing marine pollution and regulating overfishing.
SDG 15: Life on land	SDG 15 aims to combat desertification, prevent land degradation and biodiversity loss as well as safeguarding, restoring and encouraging the long-term preservation of agricultural environments and forests.
SDG 16: Peace, justice & strong institutions	SDG 16 seeks to establish fair access to justice for all, support inclusive and peaceful societies for future generations and establish productive organisations at all levels.
SDG 17: Partnership for the goals	SDG 17 calls for mobilisation of resources, enhanced international cooperation and the development of multi-stakeholder partnerships to strengthen implementation mechanisms and revive global partnerships for sustainable development.

environmental awareness, global sustainability goals and many more. In fact, after 2015, numerous companies had already started aligning their strategies with SDGs, which entailed a highly competitive environment for businesses. As a result, more cost-effective products are being manufactured. This transition reflects the goals of fostering sustainability and maximising efficiency in today's business world [41].

Moreover, Industry 4.0 is not only about the integration and application of digital technologies but also about sustainable practices in industry which aligns with the SDGs. UN SDGs are empowered through the application of resource-efficient and sustainable manufacturing processes. There are a total of 17 goals, and some of them can be achieved through Industry 4.0. For instance, technologies such as IoT sensors, AI algorithms and Big Data Analytics are the core of Industry 4.0, and these new advances can help reduce energy consumption, decrease the level of waste and minimise environmental impacts [42].

These pro-actions of technology fall in SDG 9: Industry, Innovation and Infrastructure, which promotes sustainable industrialisation. Also access to proper information and infrastructure in underlying areas has been possible due to the rise of Industry 4.0. Through digital platforms and connectivity solutions, businesses have enabled proper access to essential services such as healthcare, education and financial support in different communities, thus contributing more to achieve the ninth goal.

In the same vein, SDG 8, Decent Work and Economic Growth, is promoted by Industry 4.0. The use of automation to produce new goods and digital technologies for the creation of new job prospects greatly helps meet the eighth goal. The utilisation of IoT devices provides blockchain technology, tracking systems and data analytics, allowing companies to monitor their productivity more effectively, ensuring sustainable patterns of consumption. This leads to the contribution to SDG 12 pertaining to Responsible Consumption and Production.

Similarly, other UN goals are still being achieved. For example, Industry 4.0 encourages clean energy approaches and development, contributing to the goals of SDG 7 and SDG 13, Affordable and Clean Energy and Climate Action, respectively. Also, the use of Big Data Analytics gives better insights and assurance about decision-making process, facilitating the process to achieve SDG 4 (Quality Education). One of the many examples of how Industry 4.0 has made great progress in contributing to achieve the UN SDGs would be smart agriculture. The integration of IoT sensors, drones and software to better control irrigation, plant growth and less usage of pesticides can help achieve SDG 2 (zero hunger) since crop yield will increase [43–45].

Smart cities have brought a leap in the lifestyle of people. The implementation of more eco-friendly transportation as well as better waste management and energy-efficient buildings SDG 11 can be attained, which is about Sustainable Cities and Communities. Nevertheless, the integration of AI, Big Data Analytics and IoT in the healthcare sector has proved to be more effective and accurate in terms of data management, better resource management and better medicines development for disease prevention adds up majorly to SDG 3 (Good Health and Well-being). These examples demonstrate how some SDGs can be achieved by harnessing Industry 4.0 technologies in businesses and companies in different sectors.

More research has been conducted to get an in-depth analysis of how Industry 4.0 contributes to achieving the most out of the 17 UN goals. A strategy is used to identify the various technologies being used in Industry 4.0, and these elements are mapped into SDGs. This is done to get insight into which extent that particular element can contribute to achieving a specific SDG. As a result of the mapping, it was noticed that Big Data has a prominent effect of a score value of 1.49 on the SDGs compared to Additive Manufacturing (AM) with a score of 0.54. SDG 9 at a score value of 4.48 occupies the highest position, followed by SDGs 11, 12 and 3. The following elements related to the revolution in digital technologies were identified, and their capability to meet UN SDGs goals were analysed [43–45].

1.5.2.1 IIoT and the SDGs

Based on the facts, IoT has an average score between 0.82 and 1.29, proving to bring a positive impact in achieving the SDG goals. From irrigation techniques to enhanced farming activities, AI has helped to achieve SDG 2 (Zero Hunger). Effective measures which increase productivity have been undertaken ever since animals were closely monitored and tracked via AI. Service-oriented architecture has the highest score of 1.29, which allows multiple collaborations between components in the industry, health sector and smart cities, and has shown to contribute significantly to enhancing human health and well-being (SDG 3), industry (SDG 9) and sustainable and smart cities (SDG 1). Therefore, IoT empowers the industrial sector to contribute to sustainable and smart cities; however, it does not contribute much to alleviating gender equality (SDG 5) and income equality (SDG 10) [46–48].

1.5.2.2 Big Data and Analytics and the SDGs

According to average scores of the enabler's effects on UN SDGS, Big Data has been shown to have a much greater impact on SDGS at a score value of 1.49. For instance, the collection of data to analyse the energy consumption of machines and other aspects of the working industry indubitably contributes to sustainable industry, innovation and infrastructure goals (SDG 9). Big Data gives the possibility to identify water quality, hazards and monitoring of crops through the use of sensors, which in turn contributes to achieving zero hunger and clean water and sanitation (SDG 2 and SDG 6, respectively). Through the collection of these data, predictions can be made for future predicaments. Sensors also give enough information, for example, the pulse of a person and many more, which can be used to gather updates on a person's health, which contributes to SDG 3. In such cases, some health issues can be diagnosed at an early stage by monitoring the data collected. Big Data contributes efficiently to sustainability goals while lacking to meet the UN SDG goals such as gender equality.

1.5.2.3 Cloud Computing

Cloud systems are highly known for their seamless connection and interaction among components and technology elements, which promote interoperability and accessibility through interfaces and protocols. This cloud concept, which can be extended to the manufacturing domain, promulgates the use of affordable and clean energy, thus contributing to SDG 7. Through global scheduling and optimisation, higher energy efficacy can be achieved with cloud computing. Nevertheless, in the context of sustainable cities and communities (SDG 11), the deployment of smart devices and monitoring systems is supported by cloud computing, which gives rise to resilient, reliable, inclusive and sustainable urban development [46–48].

1.5.2.4 Simulation

As a result of mapping, simulation has shown positive impacts towards achieving the SDGs. Simulation has helped to achieve SDG 9 to quite some extent through its ability to manufacture robust infrastructure and promote the generation of new ideas and inclusive industrialisation. Additionally, according to experts' rates, it was highlighted that simulation has significantly contributed to achieving SDG 8 (economic growth and decent work) and SDG 12 (sustainable consumption and production) with scores of 1.81 and 1.9, respectively. However, simulation has shown minimal contribution in achieving goals such as SDG 16 (inclusive societies and effective institutions) since it mainly focuses on legal frameworks compared to industry. Similarly, referring to its score of 0.14, the simulation does not contribute to SDG 5 (gender equality) and SDG 17 (global partnership). Therefore, simulation has an overall moderate positive effect towards achieving the SDG concerning Industry 4.0 and environmental impact.

1.5.2.5 Augmented Reality

Based on the facts of the seven groups, augmented reality brings an average impact towards promoting the SDGs. It mostly has a much more positive effect in the industrial, innovation and infrastructure arena, which is related to SDG 9, and it also contributes to the enhancement of the education system related to SDG 4. The use of augmented reality helps to achieve better worker abilities, productivity and minimum pressure on mental health through expressing their concerns individually. Augmented reality also promulgates communication skills, which indirectly contributes to achieving the SDGs. For instance, awareness can be spread through documentaries of aquatic life related to SDG 14 and the empowerment of renewable energy (SDG 7). Therefore, augmented reality is efficient when it comes to spreading awareness through communication which contributes to achieving some SDGs.

1.5.2.6 Additive Manufacturing

AM has shown a slight positive impact on SDGs according to the mapping process. It has significantly made progress in achieving SDG 9. It has an average impact score of 2.92 on SDG 9, which shows the prospect of transforming manufacturing processes and giving small businesses to come up with innovative products even though they have access to limited resources. On the contrary, AM has also been problematic in the sense that it is a disruptive technology. It requires constant designing and effort in bid to design large, customised products.

1.5.2.7 Autonomous Robots

Autonomous robots have shown indirect positive effects on SDGs. They have an average score that varies from −0.1 for reduced inequality (SDG 10) to 2.38 for industry for its influence on SDG 9. Autonomous robots are well known for their ability to develop good quality, reliable and sustainable infrastructure. Nevertheless, it has also greatly helped to achieve SDGs such as good health and well-being (SDG 3), whereby it monitors the health of elderly people and other clinical processes. However, it was noted that autonomous robots have a negative impact on SDG 10, which is linked to reduced inequality. It has a low effect on SDG 16 (Peace and Justice) and SDG 17 (Partnership for Goals). This is due to the use of autonomous weapons, which does not bring peace, and human rights can be potentially violated. Autonomous robots do not contribute significantly to SDGs 1 and 6, which are elated to Poverty and Clean Water, respectively. Thus, autonomy has brought a positive impact on achieving SDG goals to some extent.

1.5.2.8 Cybersecurity

From an in-depth analysis, it was noted that cybersecurity has brought great advances in achieving the SDGs. Cybersecurity has a strong influence on SDG 9 for the construction of strong and resilient infrastructure and ease of access to various technologies. Not only that but it also contributes to SDG 11, related to smart cities with a view of promulgating sustainable urbanisation and a reduced impact on the environment. However, it was noticed that cybersecurity holds lesser importance in the aspect of achieving SDGs 5 and 10 since there already exist gender inequality and limited access to necessary tools. Therefore, cybersecurity has an indirect influence on SDGs [46–48].

1.6 OPEN ISSUES AND CONCLUDING REMARKS

This chapter presented an in-depth overview of Industry 4.0 by elaborating on its evolution, main pillars and enabling technologies. The major impacts of AI, ML and Big Data Analytics on Industry 4.0 have also been described. Moreover, how Industry 4.0 can contribute to several SDGs has been analysed. A mapping process for each element related to digital technologies with respect to a corresponding SDG has been discussed. It can be concluded that these technologies being highly integrated and employed in Industry 4.0 have indubitably contributed to achieving most SDGs directly or even indirectly. While some of them lack to contribute or meet certain goals, there are other elements, for instance, Big Data, which help to achieve most of the SDGs. In general, Industry 4.0 is going to play a crucial role in the attainment of the UN SDGs in the coming yours. It is therefore vital that development in the enabling technologies of Industry 4.0 and beyond, particularly AI, is sustained to make it more impactful for achieving the UN SDGs.

REFERENCES

1. Arnold, T. (1884) *Lectures on the Industrial Revolution in England: Excerpts.* Available at: https://www.sjsu.edu/people/cynthia.rostankowski/courses/HUM2AF13/s3/Reader-Lecture-27-Industrial-Revolution.pdf
2. Miah, M.T., Szilvia, E.-G., Anita, D., and Mária, F.-F. (2024) A systematic review of Industry 4.0 technology on workforce employability and skills: Driving success factors and challenges in South Asia. *Economies*, 12(2), p. 35. https://doi.org/10.3390/economies12020035
3. Yavari, F., and Pilevari, N. (2020) Industry revolutions development from Industry 1.0 to Industry 5.0 in manufacturing. *Journal of Industrial Strategic Management*, 5(2), pp. 44–63. Available at: https://www.noormags.ir/view/en/articlepage/1757315/industry-revolutions-development-from-industry-1-0-to-industry-5-0-in-manufacturing
4. Thangaraj, J., and Lakshmi Narayanan, R. (2018) *Industry 1.0 to 4.0: The Evolution of Smart Factories, Researchgate.* Available at: https://www.researchgate.net/publication/330336790_INDUSTRY_10_TO_40_THE_EVOLUTION_OF_SMART_FACTORIES
5. Sharma, A., and Singh, B.J. (2020) Evolution of industrial revolutions: A review. *International Journal of Innovative Technology and Exploring Engineering*, 9(11), pp. 66–73. https://doi.org/10.35940/ijitee.i7144.0991120
6. Brynjolfsson, E., and McAfee, A. (2016) *The Second Machine Age: Work, Progress, and Prosperity in a Time of Brilliant Technologies.* W. W. Norton & Company; Reprint edition (January 25, 2016), ISBN-10:0393350649.
7. Shaji, G.A., and Hovan, G.A.S. (2020) Industrial revolution 5.0: The transformation of the modern manufacturing process to enable man and machine to work hand in hand. *Seybold Report*, 15(9), pp. 217–221. https://doi.org/10.5281/zenodo.6548092
8. Gailey, R. (2007) As history repeats itself, unexpected developments move US forward. *The Journal of Rehabilitation Research and Development*, 44(4), p. vii. https://doi.org/10.1682/jrrd.2006.11.0148
9. Frieden, J. (2008) *The Future of Globalization.* Available at: https://scholar.harvard.edu/jfrieden/publications/globalization-and-exchange-rate-policy
10. Zakoldaev, D.A., Korobeynikov, A.G., Shukalov, A.V., Zharinov, I.O., and Zharinov, O.O. (2020) Industry 4.0 vs industry 3.0: The role of personnel in production. *IOP Conference Series: Materials Science and Engineering*, 734(1), p. 012048. https://doi.org/10.1088/1757–899x/734/1/012048
11. Folgado, F., Calderón, D., González, I., and Calderón, A.J. (2024) Review of Industry 4.0 from the perspective of automation and supervision systems: Definitions, architectures and recent trends. *Electronics*, 13(4), p. 782. https://doi.org/10.3390/electronics13040782

12. Vogel-Heuser, B., and Hess, D. (2016) Guest editorial: Industry 4.0–prerequisites and visions. *IEEE Transactions on Automation Science and Engineering*, 13(April 2), pp. 411–413.
13. Bauernhans, T., Vogel-Heuser, B., and ten Hompel, M. (2017) In: Grundlagen, A., (ed.), *Handbuch Industrie 4.0*, Bd.4. Springer; ISBN: 978-3-662-53254-53256.
14. B. Vogel-Heuser, G. Bayrak, U. Frank, Research questions in "Production automation of the future"Discussion paper for the acatech project group "ProCPS – Production CPS". acatech materials (2012)
15. Dalenogare, L.S., Benitez, G.B., Ayala, N.F., and Frank, A.G. (2018) The expected contribution of Industry 4.0 technologies for industrial performance. *International Journal of Production Economics*, 204, pp. 383–394. https://doi.org/10.1016/j.ijpe.2018.08.019
16. Wan, Z., Gao, Z., Renzo, M.D., and Hanzo, L. (2022) The road to Industry 4.0 and beyond: A communications-, information-, and operation technology collaboration perspective. *IEEE Network*, 36(6), pp. 157–164. https://doi.org/10.1109/mnet.008.2100484
17. Hall, R., Schumacher, S., and Bildstein, A. (2022) Systematic analysis of Industrie 4.0 design principles. *Procedia CIRP*, 107, pp. 440–445. https://doi.org/10.1016/j.procir.2022.05.005
18 Hermann, M., Pentek, T., and Otto, B. (2015) Design principles for Industrie 4.0 scenarios. *2016 49th Hawaii International Conference on System Sciences (HICSS)*, Koloa, HI, USA [Preprint]. https://doi.org/10.1109/hicss.2016.488
19. Rosin, F., Forget, P., Lamouri, S., and Pellerin, R. (2021) Industry 4.0 and decision making. In: Lionel, R., Paredes, M., Eynard, B., Morer, P., and Rizzi, C., (eds.), *Advances on Mechanics, Design Engineering and Manufacturing III, Proceedings of the International Joint Conference on Mechanics, Design Engineering & Advanced Manufacturing, JCM 2020*, June 2–4, 2020. Lecture Notes in Mechanical Engineering, pp. 400–405. https://doi.org/10.1007/978-3-030-70566-4_63
20. Martinelli, A., Mina, A., and Moggi, M. (2020) The enabling technologies of Industry 4.0: Examining the seeds of the fourth industrial revolution. *SSRN Electronic Journal* [Preprint]. https://doi.org/10.2139/ssrn.3726922
21. Lampropoulos, G., Siakas, K., and Anastasiadis, T. (2019) Internet of Things in the context of Industry 4.0: An overview. *International Journal of Entrepreneurial Knowledge*, 7(1), pp. 4–19. https://doi.org/10.2478/ijek-2019-0001
22. Chowdhury, A.T. (2023) *The Impact of 5G Network on Industry* 4.0 [Preprint]. https://doi.org/10.13140/RG.2.2.25308.16001
23. Pons, M., Valenzuela, E., Rodríguez, B., Nolazco-Flores, J.A., and Del-Valle-Soto, C. (2023) Utilization of 5G technologies in IOT applications: Current limitations by interference and network optimization difficulties—A review. *Sensors*, 23(8), pp. 3–14. https://doi.org/10.3390/s23083876
24. Fowdur, T.P., Indoonundon, M., Milovanovic, D.A., and Bojkovic, Z.S. (2024) *5G NR Modelling in MATLAB: Network Architecture, Protocols, and Physical Layer.* Taylor & Francis Ltd, ISBN 9781032720753. Available at: https://www.routledge.com/5G-NR-Modelling-in-MATLAB-Network-Architecture-Protocols-and-Physical/Fowdur-Indoonundon-Milovanovic-Bojkovic/p/book/9781032720753
25. Zheng, Z., Xie, S., Dai, H., Chen, X., and Wang, H. (2017) An overview of blockchain technology: Architecture, consensus, and future trends. *2017 IEEE International Congress on Big Data (BigData Congress)*, Honolulu, HI, USA [Preprint]. https://doi.org/10.1109/bigdatacongress.2017.85
26. Rosenzweig, S., Crum, J., Ezer, J., and Nwankpa, J. (2020) Cybersecurity in the fourth industrial revolution: Challenges and opportunities. *IEEE Transactions on Industry Applications*, 56(6), pp. 6971–6981.

27. Stavropoulos, P., and Mourtzis, D. (2022) Digital twins in Industry 4.0. In: Mourtzis, D., (ed.), *Design and Operation of Production Networks for Mass Personalization in the Era of Cloud Technology*. Elsevier, pp. 277–316, ISBN 9780128236574, https://doi.org/10.1016/B978-0-12-823657-4.00010-5
28. Zhang, C., and Lu, Y. (2021) Study on artificial intelligence: The state of the art and future prospects. *Journal of Industrial Information Integration*, 23, p. 100224. https://doi.org/10.1016/j.jii.2021.100224
29. Nguyen, H.D., Tran, K.P., Zeng, X., Koehl, L., and Castagliola, P. (2019) Industrial Internet of Things, big data, and artificial intelligence in the smart factory: A survey and perspective. *ISSAT International Conference on Data Science in Business, Finance and Industry*, Da Nang, Vietnam. pp. 72–76. ⟨hal-02268119⟩.
30. Javaid, M., Haleem, A., Singh, R.P., and Suman, R. (2021a) Artificial intelligence applications for Industry 4.0: A literature-based study. *Journal of Industrial Integration and Management*, 07(01), pp. 83–111. https://doi.org/10.1142/s2424862221300040
31. Jagatheesaperumal, S.K., Rahouti, M., Ahmad, K., Al-Fuqaha, A., and Guizani, M. (2022) The duo of artificial intelligence and big data for Industry 4.0: Applications, techniques, challenges, and future research directions. *IEEE Internet of Things Journal*, 9(15), pp. 12861–12885. https://doi.org/10.1109/jiot.2021.3139827
32. Hu, L., Miao, Y., Wu, G., Hassan, M.M., and Humar, I. (2019) IRobot-factory: An intelligent robot factory based on cognitive manufacturing and edge computing. *Future Generation Computer Systems*, 90, pp. 569–577. https://doi.org/10.1016/j.future.2018.08.006
33. Lee, S.M., Lee, D., and Kim, Y.S. (2019) The quality management ecosystem for predictive maintenance in the Industry 4.0 era. *International Journal of Quality Innovation*, 5(1). https://doi.org/10.1186/s40887-019-0029-5
34. Goar, V., and Yadav, N.S. (2024) Foundations of machine learning. In: Tavana, M., (ed.), *Advances in Business Information Systems and Analytics*. pp. 25–48. https://doi.org/10.4018/979-8-3693-1598-9.ch002
35. Bi, Q., Goodman, K.E., Kaminsky, J., and Lessler, J. (2019) What is machine learning? A primer for the epidemiologist. *American Journal of Epidemiology*, 188(12), pp. 2222–2239. Available at: https://academic.oup.com/aje/article/188/12/2222/5567515?login=false
36. Rai, R., Tiwari, M.K., Ivanov, D., and Dolgui, A. (2021) Machine learning in manufacturing and Industry 4.0 applications. *International Journal of Production Research*, 59(16), pp. 4773–4778. https://doi.org/10.1080/00207543.2021.1956675
37. Mazzei, D., and Ramjattan, R. (2022) Machine learning for Industry 4.0: A systematic review using deep learning-based topic modelling. *Sensors*, 22(22), p. 8641. https://doi.org/10.3390/s22228641
38. Yaqoob, I., Hashem, I.A.T., Gani, A., Mokhtar, S., Ahmed, E., Anuar, N.B., and Vasilakos, A.V. (2016) Big data: From beginning to future. *International Journal of Information Management*, 36(6), pp. 1231–1247. https://doi.org/10.1016/j.ijinfomgt.2016.07.009
39. *The 2030 Agenda for Sustainable Development*. (2022) https://sustainabledevelopment.un.org/content/documents/21252030AgendaforSustainableDevelopmentweb.pdf
40. *Sustainable Development Goals*. (2022). https://www.undp.org/content/undp/en/home/sustainabledevelopment-goals.html
41. Caiado, R.G.G., Scavarda, L.F., Azevedo, B.D., de Mattos Nascimento, D.L., and Quelhas, O.L.G. (2022) Challenges and benefits of sustainable Industry 4.0 for operations and supply chain management—A framework headed toward the 2030 agenda. *Sustainability*, 14(2), p. 830. https://doi.org/10.3390/su14020830
42. Konstantinidis, F.K., Myrillas, N., Mouroutsos, S.G., Koulouriotis, D., and Gasteratos, A. (2022) Assessment of Industry 4.0 for modern manufacturing ecosystem: A systematic survey of surveys. *Machines*, 10(9), p. 746. https://www.mdpi.com/2075-1702/10/9/746

43. Alhammadi, A., Alsyouf, I., Semeraro, C., and Obaideen, K. (2024) The role of Industry 4.0 in advancing sustainability development: A focus review in the United Arab Emirates. *Cleaner Engineering and Technology*, 18, p. 100708. ISSN 2666-7908, https://doi.org/10.1016/j.clet.2023.100708
44. Hung, H.-C., and Chen, Y.-W. (2023) Striving to achieve United Nations sustainable development goals of Taiwanese SMEs by adopting Industry 4.0. *Sustainability*, 15, p. 2111. https://doi.org/10.3390/su15032111
45. Mabkhot, M., Ferreira, P., Maffei, A., Podržaj, P., Mądziel, M., Antonelli, D., Lanzetta, M., Barata, J., Boffa, E., Finžgar, M., Paśko, Ł., Minetola, P., Chelli, R., Nikghadam Hojjati, S., Wang, X., Priarone, P., Lupi, F., Litwin, P., Stadnicka, D., and Lohse, N. (2021) Mapping Industry 4.0 enabling technologies into United Nations sustainability development goals. *Sustainability*, 13, pp. 1–35. https://doi.org/10.3390/su13052560
46. Fowdur, T.P., Indoonundon, M., Hosany, M.A., Milovanovic, D., and Bojkovic, Z. (2022) Achieving sustainable development goals through digital infrastructure for intelligent connectivity. In: Boulouard, Z., Ouaissa, M., Ouaissa, M., and El Himer, S., (eds.), *AI and IoT for Sustainable Development in Emerging Countries. Lecture Notes on Data Engineering and Communications Technologies*, vol 105. Springer, Cham. https://doi.org/10.1007/978-3-030-90618-4_1
47. Fowdur, T.P., Hurbungs, V., and Babooram, L. (2024) Leveraging the power of blockchain in Industry 4.0 and intelligent real-time systems for achieving the SDGs. In: Fowdur, T.P., Rosunee, S., Ah King, R.T.F., Jeetah, P., and Gooroochurn, M. (eds.), *Artificial Intelligence, Engineering Systems and Sustainable Development*. Emerald Publishing Limited, Leeds, pp. 109–121. https://doi.org/10.1108/978-1-83753-540-820241009
48. Fowdur, T.P., Rosunee, S., Ah King, R.T.F., Jeetah, P., and Gooroochurn, M., (eds.). (2024) *Artificial Intelligence, Engineering Systems and Sustainable Development: Driving the UN SDGs*. Emerald Publishing Limited, ISBN: 9781837535415.

2 5G/6G-Based Sustainable Systems for Industry 4.0

Bhuvaneshwar Doorgakant, Tulsi Pawan Fowdur, Dragorad A. Milovanovic, and Zoran S. Bojkovic

2.1 INTRODUCTION

Industry 4.0 is regarded as a revolution that harnesses the capabilities of advanced technologies to pave the way for more efficient, reliable, productive, and customer-oriented manufacturing operations. The Industry 4.0 ecosystem links data, processes, people, and IoT-based systems in a digital manner across the physical and cyber domains [1]. In other words, Industry 4.0 relates to the principle whereby production systems, in the form of cyber-physical production systems, can reach intelligent decisions for maximising the efficiency in the mass production of high-quality and personalised goods via real-time communication between manufacturing entities [2].

The technological framework of Industry 4.0 was defined as being composed of several key technologies, namely, big data analytics, autonomous robotics, simulations, artificial intelligence (AI) computing, Industrial Internet of Things (IIoT), cybersecurity, additive manufacturing, cloud technologies, and augmented reality (AR) [3]. This is illustrated in Figure 2.1.

Connectivity and interoperability constitute two major pillars for sustaining Industry 4.0. In fact, the fourth industrial revolution has often been dubbed the era of connectivity. In the industrial context, connectivity lays the foundation for linking warehousing systems, human operators, machines, and operations to allow institutions to optimise the various steps in the supply chain. Moreover, the setting up of networks allows IIoT devices to connect to servers, routing devices, and gateways that render machine-to-machine communication possible for further autonomous, smart, and efficient manufacturing processes. The main industrial network connectivity features include network coverage, bandwidth/data throughput, power consumption, cost, mobility, latency, and security. However, the interoperability of network components represents another sine qua non pre-requisite for achieving the connectivity requirements of Industry 4.0. Interoperability is defined as the ability of network equipment from multiple vendors to interwork across multi-carrier interconnections as per the established standards and specifications. Therefore, companies need to ensure that the technologies being deployed to set up their networks apply open, transparent, and standardised protocols for efficient end-to-end connectivity.

The connectivity requirements of Industry 4.0 over interoperable network systems can be rendered possible via a wide range of industrial network protocols and cellular technologies.

DOI: 10.1201/9781003511298-2

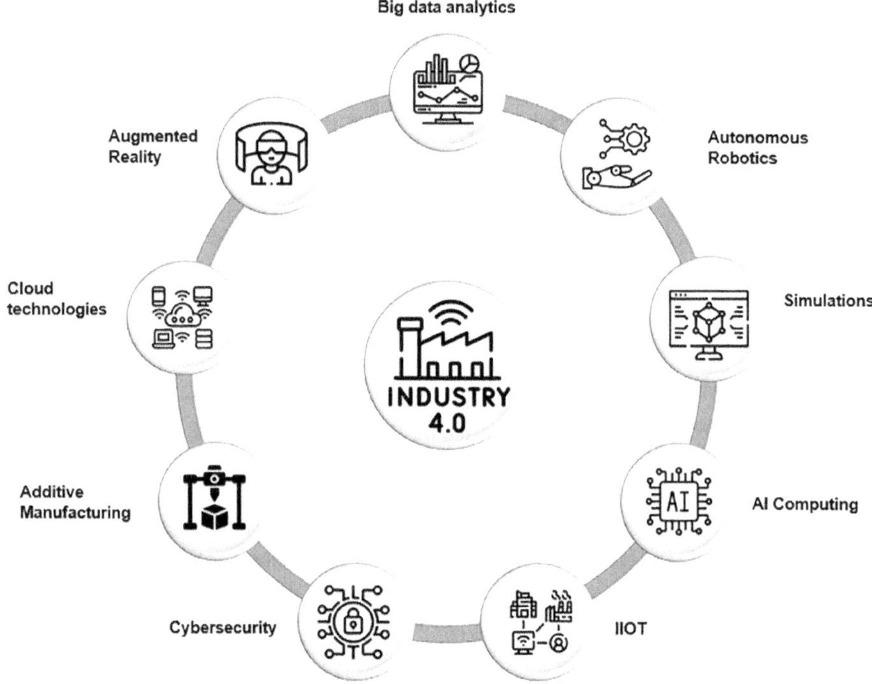

FIGURE 2.1 Technological framework of Industry 4.0 [3].

Several supply chain organisations and other businesses are favouring connectivity through mobile networks over Wi-Fi networks. According to the report by Doyle [4], mobile networks offer the optimum blend of bandwidth, security, and security for enterprise use cases. In fact, the mobile standards of the Third Generation Partnership Project (3GPP) for 5G and other cellular technologies guarantee a higher degree of reliability and interoperability compared to Wi-Fi technology. Moreover, the 3GPP Release 16 5G service sets in terms of enhanced mobile broadband (eMBB), ultra-reliable, low-latency communications (URLLC), and machine-type communications are considerably better than the corresponding features of the latest Wi-Fi standard, namely Wi-Fi 6E.

Table 2.1 compares the performances of 5G and the Wi-Fi 6E based on the most relevant network key performance indicators (KPIs) for Industry 4.0. The Wi-Fi 6E specifications are derived from the report [5].

"*5G is a standard of superlatives*" [6]. There are various factors that propel 5G as an ideal candidate for harnessing the immense potential of Industry 4.0. In fact, 5G private networks offer the possibility of dedicated data highways and allow businesses to manage the capacity, speed, and data flow according to specific organisational requirements [7]. The main attributes of 5G technology that make it attractive for the requirements of Industry 4.0 are as follows:

TABLE 2.1
Comparison of 5G and Wi-Fi 6E Capabilities [5]

Key Performance Indicators	5G	Wi-Fi 6E (802.11 ax)
Data rate	Up to 20 Gbps	Up to 10 Gbps
Range	The coverage area can extend over hundreds of square kilometres	Up to around 100 m from the broadcast point
Latency	Less than 1 ms	3–5 ms
Security	Robust end-to-end encryption with SIM card authentication	WPA3 certification

1. **Ultra-High Speed and Low Latency**
   Connectivity is at the heart of Industry 4.0. Boasting peak data rates of up to 20 Gbps, 5G technology is 20 times faster than 4G. On top of that, 5G also allows data transmission with an almost non-existent delay of the order of 1 ms. These features make 5G data transmission almost as reliable as over wired networks [6]. Moreover, the high data rates and low latency permit real-time communication between different components of the production chain through seamless data exchange. This has paved the way for significant improvements in operational efficiency from instantaneous decision making, whereby manufacturers are now able to synchronise the end-to-end production processes in a very precise manner [8].
2. **Density of Connected Devices**
   With the advent of revolutionary concepts such as the Internet of Things (IoT), the number of connected devices is expected to rise exponentially. For instance, the number of connected devices is expected to reach 70 billion by 2025 [8]. Wi-Fi technology can only support up to a few hundred devices over traditional access points. On the contrary, 5G networks can support up to 1 million devices over a single wide-area cell whilst guaranteeing high bandwidth and low latency by leveraging enabling technologies such as millimetre waves [4].
3. **Stability and Reliability**
   Network availability is a key requirement to ensure uninterrupted connectivity between devices. This significantly reduces the risk of downtime and ensures the proper execution of processes to boost productivity. The challenge facing organisations is that any wireless network is susceptible to some extent of interference, which can impede the achievable throughput speeds, latency, and reliable communication. Traditional Wi-Fi networks are usually operated in unlicenced spectrum making them more prone to interference from adjacent networks and even non-Wi-Fi equipment such as microwave ovens. This situation is particularly prevalent in the 2.4 GHz band due to the limited number of non-overlapping channels. In contrast, 5G networks operate in a more controlled environment whereby the licenced spectrum is managed by regulatory authorities, minimising any risk of interference and downtime [4]. This ensures stable and reliable connectivity with a guaranteed quality of service in terms of throughput and latency for critical operations.

4. **Security and Privacy**
   According to a survey published in British Telecom [7], data security is the most critical requirement when assessing the viability of any network solution. It is primordial for organisations to ensure the security and privacy of information being communicated over their networks, especially when dealing with confidential and proprietary content. Whilst Wi-Fi networks provide security features such as Wi-Fi Protected Access 2 (WPA2) or WPA3 and Advanced Encryption Standard (AES), 5G networks incorporate more robust protection mechanisms. The 5G Universal Subscriber Identity Module is elaborated in 3GPP Release 16 along with the Extensible Authentication Protocol authentication framework [4]. Moreover, network slicing and encryption features at the 5G core network level offer further protection against the risks of cyberattacks.

2.1.1 6G – Brief Introduction

As the worldwide deployment of 5G around the world reaches cruising speed, the next major advancement in wireless communication looms on the horizon in the form of 6G technology. Although it is still at an embryonic stage, with its standardisation and commercialisation not expected earlier than 2028 and 2030, respectively, 6G is already being regarded as the natural evolution of 5G for industrial use cases to such an extent that it is expected to become one of the major infrastructural pillars of the intelligent industry of tomorrow [9].

The essence of the 6G era is to switch to a more human-centric approach. The fusion of the digital, physical, and human worlds will create immersive experiences, and the combination of intelligent knowledge systems with powerful computational resources will propel efficiency and influence the life and work of individuals as well as the way industries operate, as shown in Figure 2.2 [10].

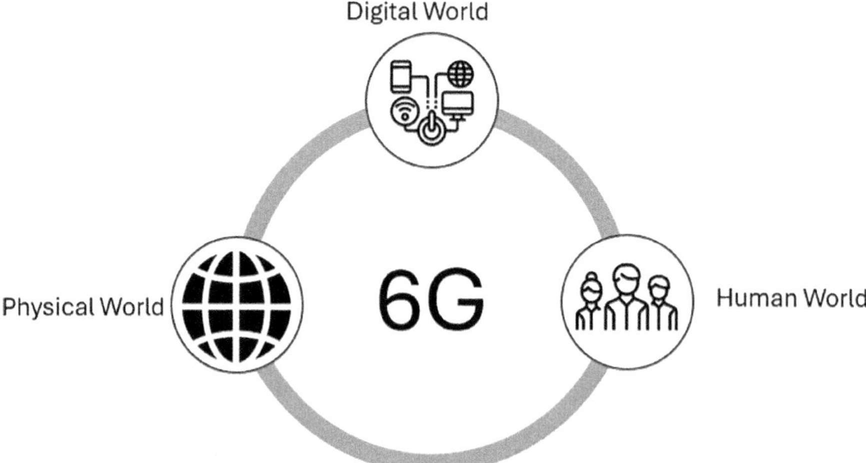

FIGURE 2.2 Unified experience with 6G across the human, digital, and physical worlds [10].

The massive deployment of sensors, coupled with AI, will enable 6G to interconnect the physical, digital, and human worlds to provide enhanced and more cost-effective connectivity compared to what 5G is offering today.

2.1.2 Related Works

According to the research of Chowdhury [11], 5G will emerge as a major enabler of Industry 4.0. The study explained the connectivity requirements that would stem from the Internet of Everything concept to enable communication between humans and devices for allowing new methods of production and value creation. The various attributes of 5G technology are enumerated in terms of eMBB, with 10 Gbps throughput, uRLLC, with a latency of the order of 1 ms and massive machine-type communication (mMTC), with over 1 million connections per square kilometre. The research also highlights the suitability of 5G technology for Industry 4.0 in terms of power efficiency compared to 4G LTE for reducing the assembly time of goods, improving flexibility, and boosting productivity. Several use cases to illustrate the impact of 5G technology on Industry 4.0 are explained, namely, the Internet of Vehicles (IoV) for supporting self-driving cars, Smart Manufacturing applications for revamping industry business models, as well as Smart Grid Applications for multidirectional supply of electricity.

The impact of 5G technology on industry is further echoed in the study [12]. Their research reiterates the commitment of businesses to continually strive for higher productivity and efficiency. In fact, organisations aim at a yearly increase of at least 3% in productivity. Moreover, latency was identified as a pivotal pre-requisite for improving the connectivity standards in Industry 4.0 through tactile Internet to enhance human-to-machine interactions. The ultimate goal of Industry 4.0 was described as the ability to improve efficiency by leveraging the features of technological advances into processes and assets and to provide a holistic understanding of manufacturing processes across different sites almost in real time. 5G enables unprecedented download speeds, ultra-low latency, and higher capacity. Industries are looking at combining the latter service sets with AI to perform decision making in a more precise manner and to undertake the automation of physical tasks according to historical data. According to Mourtzis et al. [13], it was estimated that 5G would increase the global manufacturing gross domestic product by 4% by 2030 by improving the performance of a wide range industries including agriculture, healthcare, energy harvesting, and supply chain logistics.

The research by Alhayani et al. [14] evaluates the application of 5G technology in Industry 4.0 applications by designing and implementing a novel Smart Healthcare System. In this work, the various advantages of 5G technology in terms of high bandwidth and low latency were explained through their enabling technologies such as millimetre waves, massive MIMO, software-defined networking, network function virtualisation, and device-to-device communications. Furthermore, the research elaborated on the suitability of 5G technology for IoT applications by offering stable and rapid interconnections among various sensors and other gadgets. The role of AI in improving the proactiveness of 5G networks was also highlighted. However, the research also analysed the design challenges with regard to deploying 5G technology for Industry 4.0 applications in terms of criteria such as optimising resource utilisation, cost-effectiveness, interference management, and spectral efficiency. For this purpose, Alhayani et al. [14] presented a design for a wireless sensor network for a Smart HEalthcare SYstem according to 5G and IoT standards. Whilst the simulation results demonstrated the superior results achieved by 5G for throughput and latency, there was still room for improvement regarding the optimisation of radio resources and interference management.

The study by Ghildiyal et al. [15] is centred on the United Nations Sustainable Development Goal (SDG) 9, which focuses on innovation for building resilient infrastructure and promoting sustainable industrialisation. 6G technology is touted as the major forthcoming advancement in wireless communication that will pave the way for the next evolution of Industry 4.0 systems. Taking the relay from 5G, 6G is widely expected to be at the core of the future smart industry. 6G is considered an improvement in many ways, namely through eMBB with user bit rates of up to 1 Tb/s. Moreover, the ability to use higher frequencies compared to 5G will allow 6G to offer much higher capacity as well as reduced latency of the order of microseconds, resulting in superior performance and quality of experience. The pairing of 6G and other emerging technologies such as AI, digital twin (DT), and blockchain will catalyse the evolution of Industry 4.0 applications and systems. In other words, 6G is expected to vastly improve the performance of Industry 4.0 solutions in terms of spectrum management and enhanced mobility AI-based mobile edge computing (MEC) by providing very high reliability, unprecedented data rates, ultra-low latency, and improved energy efficiency.

In the same breath, Han et al. [9] also studied the use cases of 6G in Industry 4.0. The study reiterated the prediction regarding 6G technology being at the heart of the next evolution of Industry 4.0, with a special focus on DT systems. In fact, it was forecast that 6G would be covering every corner of the planet by 2030 and propose reliable intelligent services with high performance. The research also discussed the European Union's project, Hexa-X, wherein 6G is predicted to enable multiple emerging use cases that can be categorised as sustainable development, collaborative robots, massive twinning, collaborative robots, hyperconnected resilient network infrastructures, and reliable embedded networks. The essence of DT is to create a virtual replica of any system that is connected to it and make the information associated with it accessible and actionable. The DT can find its application in different phases of the lifecycle of a physical system, from the prototyping phase to providing sufficient details for describing, designing, and developing the system. In other words, DT can assist organisations with the design, development, and commissioning of any production and business process.

2.2 OVERVIEW OF 5G TECHNOLOGY

5G is the fifth generation of cellular technology that has been designed to provide unprecedented data rates and enhanced coverage; networks are being designed to provide pervasive networking, high data rates, coverage, improved reliability and flexibility in wireless services, and low latency communication [16].

5G technology significantly demarcates itself from previous generations of mobile networks as it is expected to radically transform the influence of telecommunications

TABLE 2.2
Evolution of Cellular Networks [17]

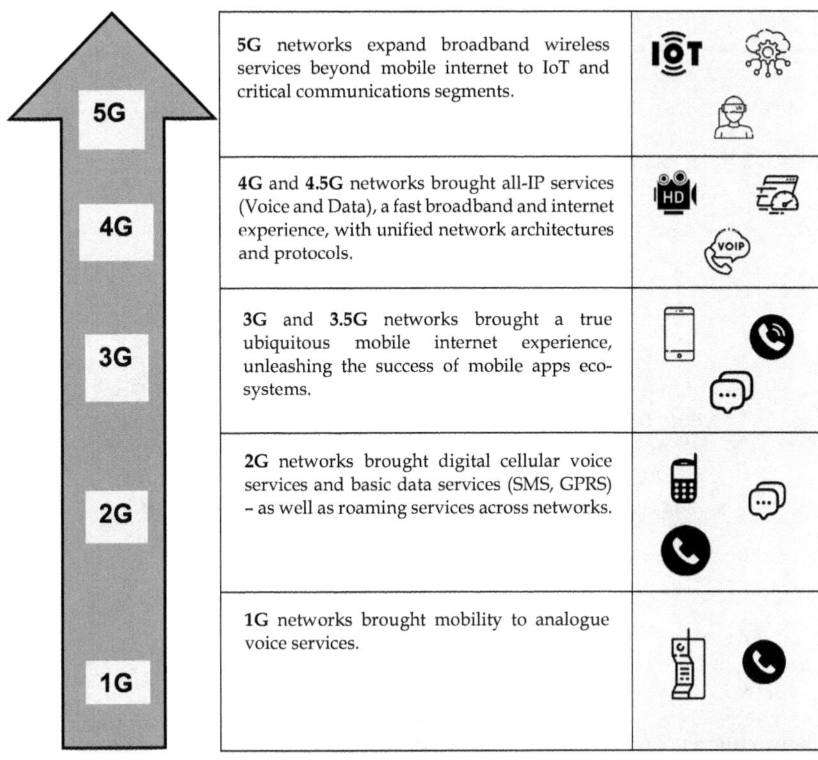

5G	**5G** networks expand broadband wireless services beyond mobile internet to IoT and critical communications segments.
4G	**4G** and **4.5G** networks brought all-IP services (Voice and Data), a fast broadband and internet experience, with unified network architectures and protocols.
3G	**3G** and **3.5G** networks brought a true ubiquitous mobile internet experience, unleashing the success of mobile apps eco-systems.
2G	**2G** networks brought digital cellular voice services and basic data services (SMS, GPRS) - as well as roaming services across networks.
1G	**1G** networks brought mobility to analogue voice services.

in industries and society at large. Through its worldwide deployment, 5G is paving the way for further economic progress and pervasive digitalisation whereby the number of people and devices connected to the network will continue to grow exponentially to fulfil the concept of the Internet of Everything. In this way, 5G will give rise to new use cases such as smart agriculture, smart manufacturing, digitalised logistics and supply chain, and smart healthcare. The main spectrum bands available for 5G implementation are sub-1 GHz, 1–6 GHz, and above 6 GHz [17]. Table 2.2 summarises the evolution of cellular networks from the first generation to the fifth generation.

The evolution of the data rates for the different mobile technologies is illustrated in Figure 2.3.

2.2.1 Requirements of 5G Technology

The third Generation Partnership Project classifies any system applying the 5G New Radio (5G NR) software as "5G" technology as specified in Release 15. Moreover, the International Telecommunications Union (ITU) has also published a standard referred to as "IMT-2020" that defines the different requirements of 5G networks.

2 Kbps
20-200 Kbps
2Mbps
200Mbps
20 Gbps
1G
1980s
2G
1990s
3G
2000s
4G
2010s
5G
2020s

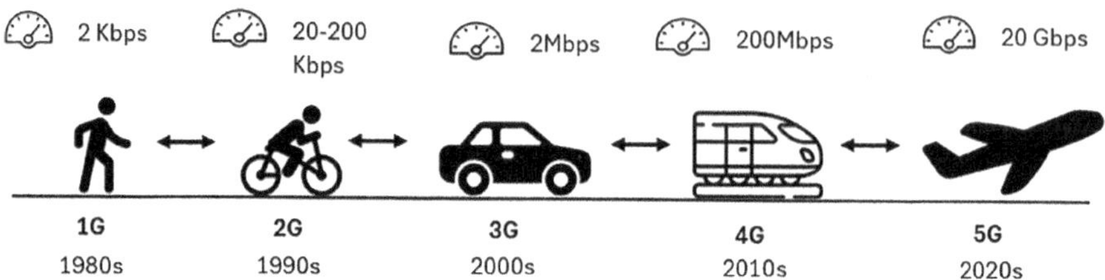

FIGURE 2.3 Evolution of mobile technologies.

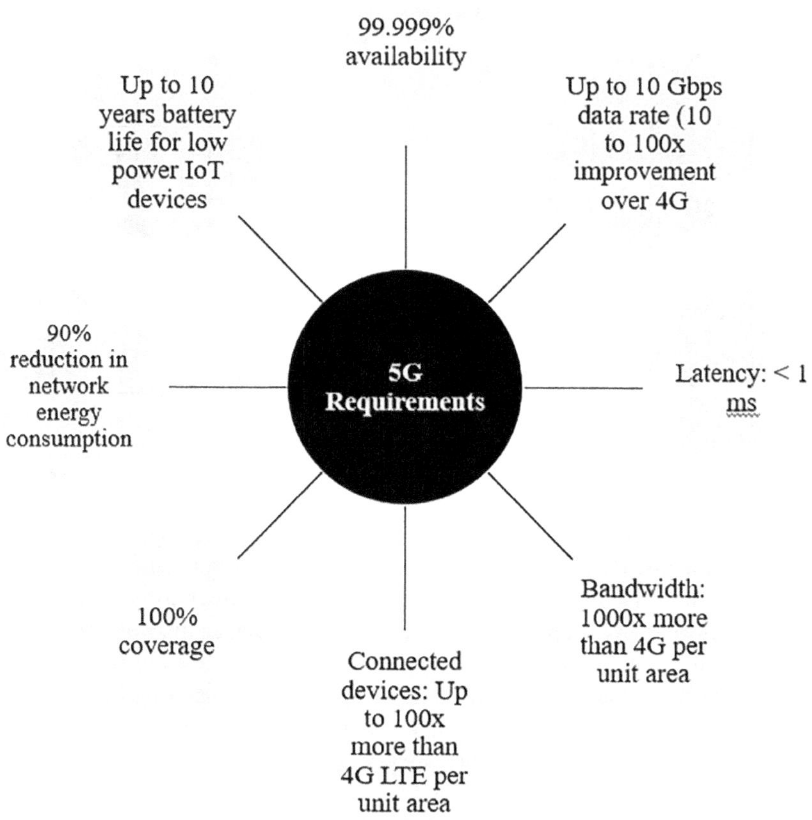

99.999% availability
Up to 10 years battery life for low power IoT devices
Up to 10 Gbps data rate (10 to 100x improvement over 4G
90% reduction in network energy consumption
5G Requirements
Latency: < 1 ms
100% coverage
Connected devices: Up to 100x more than 4G LTE per unit area
Bandwidth: 1000x more than 4G per unit area

FIGURE 2.4 Requirements of 5G technology [18].

These include enhanced network and device capabilities, faster data rates, reduced energy consumption, low latency, higher number of devices connected per unit area, and high network availability, as shown in Figure 2.4 [18].

2.2.2 5G Service Classes

The ITU-R has defined the following three main service classes for 5G:

1. **eMBB**
   The aim of this use case is to provide unprecedented data rates of the order of 10 Gbps, which constitutes an increase of 10 to 100× compared to 4G networks. eMBB will play an important role in supporting rich media applications such as AR, virtual reality (VR), and 4K video streaming.
2. **URLLC**
   This use case is applicable for services that demand extremely high reliability and low latency. High data rates may not be essential. Examples include mission-critical scenarios such as automation in smart factory environments, remote surgery by robots in healthcare environments, and self-driving vehicles.
3. **mMTC**
   mMTC is particularly applicable for IoT scenarios whereby there is a requirement to connect a huge number of low-power consumption devices in wide areas. These use cases usually need to cater to scalability and an increased battery lifetime of several years.

2.2.3 5G Enabling Technologies

The delivery of attractive and promising service sets of 5G technology requires the implementation of several enabling technologies [16]. The main ones are described as follows.

Millimetre Waves (mmWaves): With frequencies falling under 30–300 GHz, millimetre waves offer more bandwidth and, consequently, superior data rates to end users. Due to high frequencies being more vulnerable to attenuation, mmWaves are however constrained to transmission over short distances in small cells.

Massive MIMO: This technology involves connecting multiple antennae to a single base station with the objective of enhancing spectrum usage and transmission rate. Moreover, attributes such as efficient beamforming and spatial multiplexing aid in mitigating interference.

Ultra-Dense Network: Network coverage and throughput can be significantly improved by the dense deployment of small cells. Ultra-dense networks are usually built up of several small cells, are implemented using the base station or remote radio heads, and offer functionalities similar to typical macro cells. However, the improved coverage and bandwidth result in higher costs associated with the deployment of a larger number of equipment.

Device-to-Device Communication (D2D): The D2D principle in mobile networks refers to direct communication between user devices, independently of the base station. The direct interactions between 5G-enabled devices have several advantages in terms of optimising spectral efficiency, minimising latency, and reducing energy consumption [19].

Network Slicing: The segregation of the physical network into distinct logical networks, referred to as slices, is known as network slicing. In this scenario, each slice is configured in a way to provide some specific network capabilities and characteristics. In this way, network slicing allows the creation of virtual end-to-end connections to implement 5G-based services with specific quality-of-service (QoS) requirements [20].

Energy Harvesting: 5G networks consume much more energy than previous mobile generations. Radiofrequency energy harvesting techniques can be employed to produce energy from natural sources, thereby supplying the electrical power required for 5G network elements such as base stations and user devices. This could aid towards making 5G networks more self-sustaining whilst reducing carbon emissions around the world [21].

2.2.3.1 6G Requirements and Potential

The futuristic 6G technology is expected to cause a major revolution in the way communication happens globally. In essence, 6G will broaden the horizon for wireless communication through the introduction of more stringent performance metrics and new use cases. For instance, 6G is expected to provide global coverage in every corner of the globe, including areas that were previously deemed impossible to access [22].

In comparison to 5G, 6G will also significantly improve the spectral and energy efficiency whilst remaining economically viable. The foundation for this marked improvement from 5G capabilities will require enabling technologies that rest on four important pillars, namely global coverage, all spectral bands, smart applications, and security as illustrated in Figure 2.3.

To achieve global coverage, 6G networks will complement conventional terrestrial communication networks with satellite and unmanned aerial vehicle communication systems. This will allow the integration of holistic space-air-ground-sea communication. Moreover, a wide range of frequency spectra will have to be considered for improving data rates and coverage. The sub-6 GHz, millimetre wave, Terahertz, as well as optical frequency bands will be taken on board. Thirdly, 6G will leverage the huge computational capabilities of AI and big data technology to provide a portfolio of smart applications that will be able to process huge data sets generated by the different network elements. Security constitutes another key requirement, especially with the evolution of cyber threats. 6G technology is expected to incorporate vital features such as robust end-to-end encryption and authentication mechanisms to maximise protection against unauthorised access and other forms of network attacks [23]. The pillars of 6G technology are illustrated in Figure 2.5.

It is anticipated that 6G networks will outperform 5G networks in terms of performance metrics and application scenarios. Whilst 5G limits itself to the service sets comprising eMBB, mMTC, and uRLLC, 6G networks take a major leap in adopting the enhanced scenarios including further-eMBB (feMBB), ultra-mMTC (umMTC), and enhanced-uRLLC (euRLLC). These scenarios are further amplified by additional applications including AI and mobile ultra-wideband [23].

According to Rojek et al. [22], the main enabling technologies for 6G networks include:

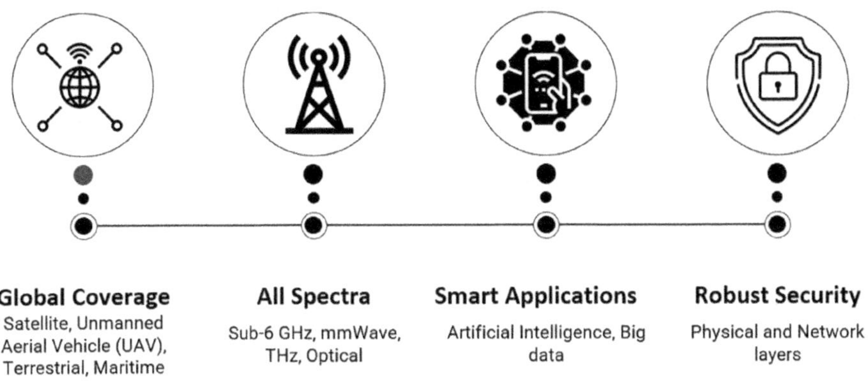

FIGURE 2.5 Pillars of 6G communication [23].

- **THz Band Communications:** To exploit the huge bandwidth available between the mmWave and infrared radiation.
- **Holographic Beamforming:** The objective is to go beyond mMIMO using hardware architectures that are of small sizes with lower power consumption to optimise and manage the wireless environment.
- AI/machine learning for a more dynamic and efficient spectrum utilisation.
- **Quantum Communications:** A strong catalyst to simplify the computing requirements of 6G networks in terms of their security and to help achieve the desired QoS levels.
- **Blockchain-Based Wireless Accessing and Networking:** A promising technology for optimising resource sharing among different components in wireless networks in terms of cooperative transmissions and D2D communications.
- **Multi-Connectivity (MCo) for URLLC:** A promising concept for supporting higher reliability and faster data rates by setting up a connection framework that caters to diverse routes to required destinations.
- **New Multiple Access Techniques:** The advent of techniques such as 6G NOMA (non-orthogonal multiple access) to support massive connectivity whilst ensuring optimum resource sharing and ultimate end-user experience.

The aforementioned service sets and enabling technologies give 6G the edge over 5G and 4G communications over several performance KPIs, as illustrated in Table 2.3.

In a nutshell, the benefits of 6G technology for Industry 4.0 can be summarised as follows:

- Enhanced-uRLLC will be pivotal for real-time M2M interactions in the industrial setup to allow machines to communicate with one another in a reliable manner with minimal delay to face any situation arising in the factory premises.

TABLE 2.3
Comparing Network Performance for 4G, 5G, and 6G [22]

Network Parameter	4G	5G	6G
Latency (ms)	<100	<10	<0.1
Density of connectivity (devices/km^2)	10^5	10^6	10^7
Maximum data rate (Tbps)	0.002	0.02	>1
Maximum user experience data rate (Gbps)	0.01	0.1	1

- Through mMTC, a huge number of devices including sensors, actuators, and other gadgets will be able to connect and communicate seamlessly for efficient coordination and automation of processes.
- The smooth and real-time transmission of huge volumes of data originating from sensors, video streams, and VR/AR applications will be greatly facilitated by eMBB.
- Enhanced edge computing will be instrumental to the M2M data processing closer to the source, thereby minimising delay for quick responses to evolving situations on the factory floor.
- Enhanced security and privacy integrated in 6G technology such as robust encryption will be a major asset to Industry 4.0 for mitigating the risks of cyberattacks and other forms of threats to ensure secured communication.

It is forecast that the commercial launch of 6G technology will happen around 2030 and will extrapolate the application of mobile technologies to several fields and industries, including IIoT for the requirements of Industry 4.0 and 5.0 with the human and environmental considerations at the core of its development.

2.3 5G AND 6G FOR SMART LOGISTICS

Recent years have witnessed a significant shift in customer expectations and behaviour regarding the purchase of goods and services. Customers favour personalised experiences by buying and selling products online in small batches whilst expecting minimum shipment times at the expense of conventional shopping in retail stores. This new trend has spurred logistics supply chains to undergo considerable evolution to cater to those demands and embrace the new era of Industry 4.0 [24].

Figure 2.6 compares conventional retail logistics and smart logistics. The challenges facing the existing logistics chain, in terms of adapting the movement of transit stock according to market trends and unforeseen events, call for novel modes of operation with more flexibility, efficiency, and financial viability. 5G technology positions itself as a strong technological enabler for smart logistics to bring in the required level of digitalisation and flexibility. The improved connectivity and faster data speeds provided by 5G are vital attributes that allow the flow of data covering operations in distribution centres, manufacturing processes, and sales of processes in parallel with the flow of products [25].

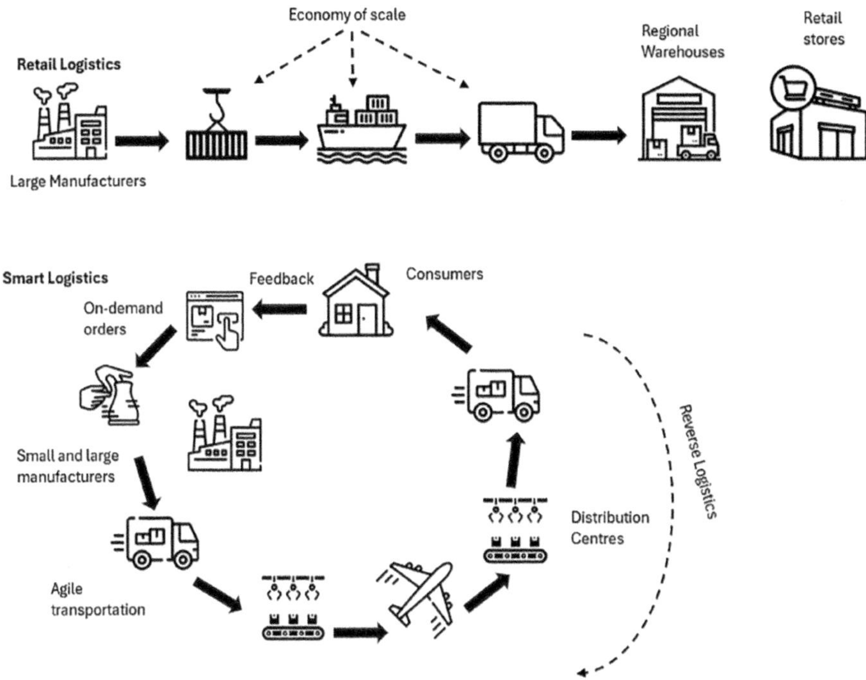

FIGURE 2.6 Comparison between conventional retail logistics and smart logistics.

According to Khatib and Barco [24], the main Industry 4.0 applications that are rendered possible via wireless connectivity over 5G networks include:

- The use of smart tags to track objects from the manufacturing stage and throughout the delivery process. This relies on the massive machine-type communication principle whereby numerous devices transmit packets with low latency.
- The movement of parcels, pallets, and other tools in distribution centres using automated guided vehicles (AGVs) and drones. This application is based on a combination of eMBB and URLLC service sets.
- AR for providing remote support to workers whereby real-time video can be complemented by images generated from computers. This principle also combines the eMBB and URLLC properties of 5G.
- Ensuring the monitoring and control of robots and other machinery in an accurate and safe manner. This requirement also relies on URLLC connectivity whilst less critical data sources can apply mMTC.
- The deployment of ambient sensors to continuously monitor the operating conditions of distribution centres.

Similarly, Kong [25] applied 5G technology to realise the connection of massive data from business organisations to end users by mapping out a DT world with structured features for validating intelligent decisions to be implemented in the physical world. This methodology paved the way for constant improvement and innovation in logistics design.

In their research, Khatib and Barco [24] proposed a solution that applied network slicing to customise 5G network components along a smart logistics distribution chain. A traffic modelling method was devised based on big data predictions network slicing was employed for the allocation of resources for different traffic profiles. The study also highlighted certain limitations of the proposed methodology over 5G networks such as big data analytics failing to fully optimise system performance.

6G technology opens up a whole new array of possibilities for advanced logistics. It is predicted that 6G networks will integrate three revolutionary scenarios, namely AI, integrated sensing and communication, as well as ubiquitous connectivity to support groundbreaking advances in transportation systems such as autonomous driving. 6G will supersede 5G for transportation methods that require higher speed or altitude that are currently unattainable using existing technologies.

In their research, Liu et al. [26] predicted an evolution from a technology-driven intelligent transportation system (ITS) to a data-driven ITS. The application of AI techniques for data processing enables ITS to proactively analyse traffic trends and efficiently detect incidents, allowing drivers to make informed decisions for improving the efficiency and safety of transportation infrastructures. Moreover, the study demonstrated the huge potential of 6G networks in the realm of heterogeneous space-air-ground transportation, with particular emphasis laid on the railway industry, airplanes, and satellite communications.

6G powered connectivity is set to lay new tracks for the railway industry, whereby trains, passengers, merchandise, and logistics infrastructure will be interconnected. Five main concepts have been identified to provide unparalleled travelling experience to passengers, namely, train-to-infrastructure, inside station, train-to-train, infrastructure-to-infrastructure, and intra-wagon. These applications will require very high bandwidth that will be fulfilled by the service sets of 6G. For instance, a combination of all available frequency bands will be used to fulfil the different communication needs. Radio waves of lower frequencies will be deployed for areas requiring coverage over long distances due to lower attenuation. On the contrary, sub-THz and THz bands will be used for applications requiring very high data rates. On top of that, fast and adaptive beamforming supported by 6G will be of particular importance for high-mobility scenarios involving the movement of high-speed trains [26].

Similarly, 6G technology will fuel the take-off of enhanced features in air-to-ground (ATG) communication whereby connectivity will be extended to passengers in addition to essential interactions between the crew and air traffic controllers. The cellular-based ATG communication will be a significant improvement over traditional non-terrestrial networks by ensuring connectivity with the required throughput and latency to guarantee an excellent customer experience in almost any corner of the world. The cellular ATG required for this use case will enhance coverage to reduce the frequency of handovers to ensure a better user experience. This will be rendered possible through technologies such as physical random access channels that will allow the 6G-based ATG networks to have cell radii extending up to 300 km [26]. Figure 2.7 illustrates the architecture of a cellular-based ATG whereby signals

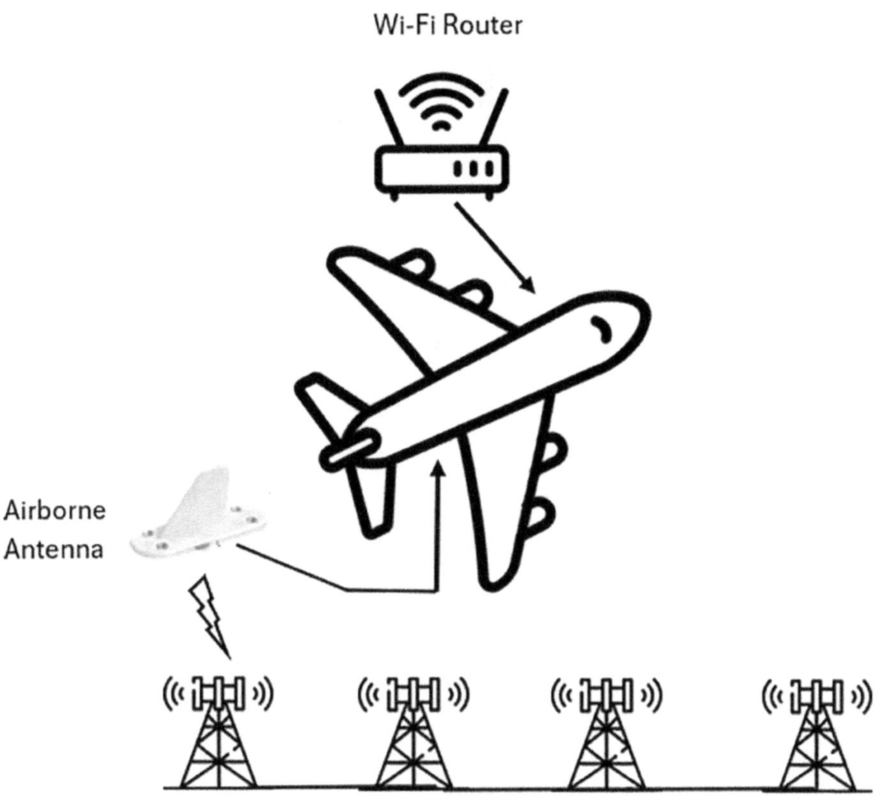

FIGURE 2.7 6G-powered ATC [26].

are transmitted from the base stations to the airborne antenna installed on the plane, which, in turn, is connected to an appropriate switch or router for propagating Wi-Fi signals to the passengers and crew.

2.4 SMART LOGISTICS USING SELF-DRIVING ROBOTS

The emerging technologies of Industry 4.0 are providing manufacturing industries with much-needed flexibility to respond to the ever-increasing needs of customers in a timely manner and to improve their productivity at reasonable costs [27]. Another such use case is the transportation of materials across the factory floor via autonomous mobile robots (AMRs) and AGVs, avoiding obstacles, coordinating with fleet mates, and identifying where pickups and drop-offs are needed in real time. An AMR needs to be able to navigate without any disruption and simultaneously avoid obstacles whilst operating in indoor or outdoor environments that are unpredictable or even unknown. Thus, the main challenges of AMR systems include navigation, path planning, obstacle avoidance, and localisation [28].

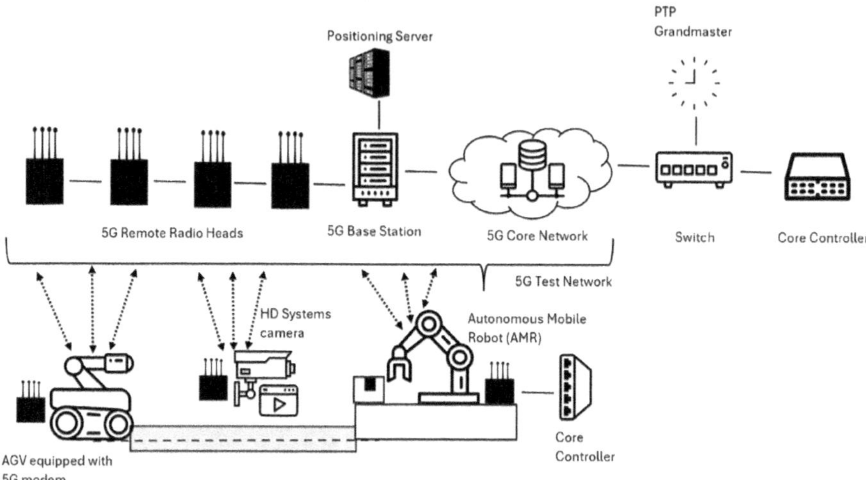

FIGURE 2.8 5G-enabled autonomous mobile robot (AMR) and automated guided vehicle (AGV) system [29].

It is essential that the AMRs receive the correct assignment at the right time, execute the task at the appropriate time, coordinate properly with one another, and adapt to site conditions to ascertain that processes and operations are undertaken efficiently. High-speed and reliable connectivity are vital pre-requisites for this. With its diverse service sets, 5G is quickly emerging as the preferred technology for implementing AMR and AGV systems. Figure 2.8 illustrates a 5G-enabled AMR and AGV installation, incorporating flexible manufacturing with time-sensitive networking, low latency, and ultra-high reliability along with 5G precise positioning by Qualcomm and Bosch Rexroth.

The system basically inspects manufactured automotive parts. An AGV is responsible for navigating to the AMR, which has to segregate defective parts from good ones in separate containers on the AGV. The time-synchronised system introduces flexibility in the manufacturing process by allowing the AGV and AMR to any appropriate location in the factory. Similarly, 5G positioning allows the AGV to navigate to the AMR in a precise manner. 5G-based systems with time-sensitive networking allows manufacturers to ensure wireless connectivity on factory floors for maximising space and time for better operational efficiency and cost savings [29].

Moreover, 5G technology significantly improves mobility support and coverage to ensure continuous connectivity between AGVs, AMRs, and supervisory systems. The high data rates made available by 5G allow the further relay of camera perspectives with the added benefits of low latency and high resolution. This allows for more informed decision making for the coordination of traffic [4].

Factories also have the possibility of re-using their existing 5G infrastructure for AMR/AGV fleet management, integration of warehouse, as well as to perform real-time analytics. For instance, 5G-based MEC (5G MEC) system can redeploy AGVs based on data collected depending on production volumes and requirements for a more cost-efficient and agile operation mode. It is also worth noting that the integrated hardware units for AGV and AMR navigation are usually limited regarding processing power. This can be a major handicap in cases where the AGVs have to be deployed in remote locations beyond the reach of factory infrastructure. 5G MEC helps to overcome this hurdle by making it possible to offload the data and processing functionalities to the MEC. The wider coverage offered by 5G allows the remote management of AGVs operating outside the factory premises [30].

6G technology is expected to extend further the range of use cases for AMRs through its diverse service sets. This is illustrated in the research conducted by Costa et al. [31] regarding the application of mobile robotic platforms (MRPs) to provision dynamic communication cells on an ad-hoc basis. The study proposed a methodology for the obstacle-aware positioning of an MRP to set up line-of-sight links for wireless communications. The solution comprised two main components, namely a vision module to determine the precise location of obstacles and devices in the vicinity of the MRP and a control module for positioning the MRP in the appropriate location as per the data gathered from the vision module.

The main challenge identified by Costa et al. [31] consisted of the efficient positioning of the MRPs to provide the optimum wireless connectivity. The inherent properties of 6G technology in terms of millimitre waves, terahertz, and visible light networks greatly helped fulfil the connectivity requirements to perform the obstacle identification and location processes as accurately and timely as possible. Simulation results demonstrated that the proposed model was successful in controlling the movement of the MRP towards the desired position by circumventing obstacles successfully. The extremely low latency values of 0.1 ms also ensured the efficient exchange of data between the vision and control modules for quick decision making. Moreover, additional 6G KPIs such as high reliability and extremely massive instant access anytime and anywhere were identified as avenues to explore to further improve on the classification of obstacles detected, making the solution more scalable. Thus, it can be safely said that 6G will act as a strong catalyst and enabler for AGV/AMRs mission-critical services and will exceed the barriers of conventional communications [32].

2.5 COBOTS (COLLABORATIVE ROBOTS)

Collaborative robots, also referred to as cobots, are considered one of the major breakthroughs in Industry 4.0 [33]. A collaborative robot is basically an industrial robot that can perform operations safely in conjunction with humans in a common work environment. Contrary to autonomous robots, which are configured to perform a single task repeatedly in a static and independent manner, cobots are designed with more flexibility to assist a wide range of human interventions [34]. Recent developments in mobile technology, AI, machine vision, and cognitive computing empower cobots to be aware of their surroundings and perform multiple tasks such as packaging automation, material handling, machine tending, and even welding safely in close proximity to human workers. In addition to being programmed to protect the safety of their human coworkers, cobots can quickly learn tasks through demonstration and reinforcement learning. Cobots are currently being deployed in various sectors such as healthcare, automobile, aerospace, pharmaceutical, science research, furniture and equipment, food, and agriculture due to their simplicity, reliability, safety, and accuracy [35].

In the past decades, robots were primarily deployed as stand-alone machines in confined production workspaces. Industry 4.0 has reversed this trend in favour of collaborative robots through concepts such as Industrial IoT and cooperating objects. These changes have been fuelled by the advent of new technologies such as smart sensors for better visualisation of a robot's environment, enhanced communication at the industrial level for better real-time interaction, and synchronisation among robots and their environment. The wide range of cloud and edge computing has facilitated the acquisition of information and processing from stand-alone devices to enterprise level. Hence, wireless communications have a pivotal role with regard to the mobility of robots and the integration of a wide array of sensors for supporting the actual robot [36].

5G technology is proving to be a sound enabler for the adoption of cobots in Industry 4.0. The rapid adoption of robot–human collaboration will result in micro-factories with dozens of production cells. Each production cell will need specific latency from the 5G network slice based on the requirement of each cobot machine cell. This calls for a solution that enables the robotics application to request QoS from the network for particular actions in a dynamic way. The two main pre-requisites for such applications are network exposure function and network slicing.

2.5.1 5G Network Exposure Function for Cobots

Network Exposure Function can be defined as a network entity that exposes the 5G core network capabilities to third parties or non-3GPP environments such as collaborative robots. Network Exposure Function (NEF) is vital to devise programmable networks for establishing communication with all IoT devices and fulfil the requirements of 5G use cases such as smart manufacturing. The programmability of the 5G-based networks is enabled via the application programming interface (API), which makes it possible for developers to exploit the different functionalities and services of the network. In other words, novel applications and services can be developed by harnessing the capabilities of 5G technology. APIs allow developers to customise several aspects of the 5G infrastructure such as network functions, devices, and network slicing to develop applications with use cases in various industries ranging from healthcare to automobile and logistics management. In fact, NEF also provides security when services or application functions (AFs) access 5G core nodes and can be thought of as a proxy or API aggregation point for the 5G network. Figure 2.9 illustrates the network exposure function [33].

2.5.2 Network Slicing

Network slicing refers to a fundamental capability of 5G, which allows the division of the network into multiple virtual connections, which can be customised according to the traffic requirements of different use cases. Different communication services can be provided by network slices such as access network and core network functions that are supported by transport networks [37].

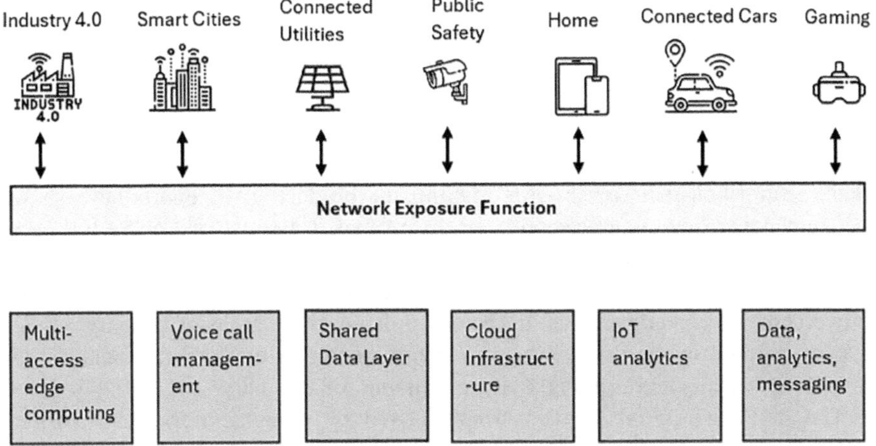

FIGURE 2.9 Network exposure function for 5G [33].

Whilst 5G technology remains the most popular enabler for the deployment of cobot use cases currently, continuous technological developments are opening up a new array of possibilities for further innovations and improvements in the robotics industry. In their study, Calandra et al. [38] described the important role that 6G could play in new solutions applying DT and extended reality (XR) due to its human-centric approach and reliability. As mentioned earlier, the roles of cobots are not limited to routine procedures but have to be adapted continuously according to the actions of operators sharing the workspace. XR offers new opportunities to improve the current programming of collaborative robots for such scenarios via a combination of real and virtual elements. Moreover, Calandra et al. [38] claimed that a collaborative approach to cobots programming consisting of users teaming up on-site or remotely via VR technologies would be of paramount importance to the successful deployment of next-generation robotics solutions. However, the limitations in terms of the connectivity provided by current technologies in terms of peak data rates and end-to-end latencies could jeopardise the feasibility of such a collaborative approach. With vastly superior data rates and latency, 6G lays the foundation for users to have immersive experiences in their working environments. The simulations carried out also demonstrated that advanced features such as AI-powered air interfaces in 6G led to a 50% reduction in transmit power over the current mobile networks for the same bandwidth and data rates, thereby confirming the sustainability of the proposed collaborative approach.

Moreover, the case for 6G technology in the robotics industry is further strengthened by the research carried out by Kerboeuf et al. [39]. With reference to the European-level 6G flagship project Hexa-X-II, the major performance and value indicators for 6G end-to-end systems, as applicable to cobots in Industry 4.0, are discussed. For instance, the higher frequencies at mmWave and sub-THz communication proposed by 6G are reckoned as valuable features to provide connectivity and flexibility to cobots that are expected to operate in confined areas. Furthermore,

Joint Communicating and Sensing in 6G would be very helpful for cobots to sense and navigate their surroundings to move around successfully.

2.6 PREDICTIVE MAINTENANCE

Unforeseen and unplanned downtimes are one of the major hurdles encountered by manufacturers on the way towards maximising productivity. Reports have shown that unplanned downtimes account for an estimated amount of US$50 billion in losses each year globally, and hardware failures have been identified as the main root cause of such breakdowns. Several avenues are being explored at the industry level to overcome such predicaments. In the era of data-rich domains of Industry 4.0 and IIoT, traditional maintenance schemes seem outdated with predictive analysis via connected systems representing a very promising solution [40].

The maintenance mechanisms implemented by manufacturers have a direct impact on their productivity. It is estimated that appropriate predictive maintenance schemes can increase machine uptime by 10–20% whilst simultaneously reducing the overall maintenance cost by 5–10% [12]. The adoption of big data analytics and predictive analytics techniques is becoming an increasingly popular trend in Industry 4.0. In essence, these schemes monitor deployment equipment and machinery on a permanent basis to determine the current working status of the devices. Furthermore, the collected data is analysed using specific algorithms to predict the future operating status of the equipment based on projections made on current trends. However, the sensors deployed at factory floor level for monitoring equipment generate huge volumes of data. Therefore, predictive analysis frameworks necessitate very fast and stable connections between the data acquisition devices and servers responsible for processing the data. Moreover, there is also a requirement to ensure that equipment and machinery installed in remote locations are able to report the collected data to central processing locations in a reliable manner. This will allow accurate decisions to be made by responsible parties to ensure that timely directives are given regarding maintenance and repairs with the least disruption to operational activities. Enhanced data handling technologies such as DTs and emerging concepts such as edge computing that rely on the data processing capabilities of mobile devices have been identified for the efficient use of inputs from the data collection devices to predict any future equipment malfunction.

An edge computing-based predictive maintenance solution has been proposed by Mourtzis et al. [13] to determine KPIs such as the remaining useful life of critical equipment. The proposed network topology comprised the deployment of microcontroller units for integrating the data from sensors for equipment monitoring. The data was relayed to edge nodes for preliminary processing before being relayed to support vector machines for classification. The latter process determines the information transferred to the cloud server, which hosts the DT of the equipment under monitoring. In case any issue is detected in the operating status of the equipment, appropriate instructions are initiated from the cloud server to the concerned edge node. Figure 2.10 illustrates the proposed monitoring framework.

The above framework relies a lot on high data rate transmission and low latency communication, which again highlights the starring role of 5G technology in this use

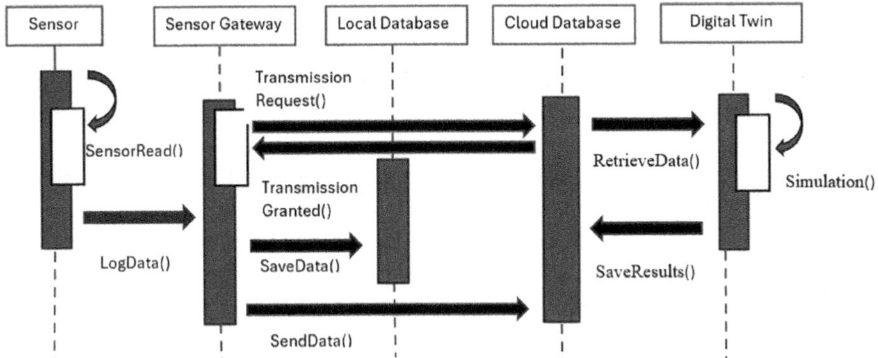

FIGURE 2.10 Data sequence of the proposed architecture [12].

case of Industry 4.0. The various requirements of predictive maintenance solutions in terms of flexibility, reliability, and environment predictability can be met by 5G technology in verticals such as agriculture, manufacturing, mining, and healthcare. 5G networks can cater to the connectivity requirements for the large-scale deployment of sensors that collect data to be shared with analytic systems. These in turn proactively detect any potential malfunction and inefficiencies in machinery. Technologies like 5G are particularly helpful in remote environments where conventional means of connectivity like Wi-Fi are not available. The collection of data makes it possible to implement predictive maintenance during production and further down the product life cycle using a single technology [41].

The advantages of 5G technology for predictive maintenance applications are listed below.

- **Higher Data Rates and Bandwidth:** The superior data rates offered by 5G networks support higher frequency sampling and precision sampling for important data such as low/high voltage, current, temperature and noise. The larger bandwidth offered by 5G also permits higher sensor data volume transmission.
- **More Robust Security:** Confidentiality is of paramount importance for controlling remote management access to machinery in high-value production lines. Business organisations benefit from the inherent security features of 5G as detailed by the 3GPP.
- **Massive Sensor Deployment:** 5G can support lower data rate sensors, which can therefore be deployed in larger quantities, for monitoring more nodes across the production line.

Another flavour of predictive maintenance was presented by Giannakidou et al. [42]. A network application (NetApp) was proposed for implementing predictive maintenance in power plants by applying the containerisation principle, 5G and AI. Resting on the underlying principle of forecasting failure in advance and mitigating impacts of equipment damage, AI-based predictive maintenance applies intelligent data to make predictions accurately for instantaneous interventions on major assets. The aim of predictive maintenance applications is to flag hardware issues sufficiently in advance to give decision enforcers sufficient time to undertake required preventive or remedial actions for minimising risks of actual damage to equipment. The most accurate predictions however require the collection and analysis of as much data as possible in a timely manner. This reiterates the importance of big data analytics for deploying predictive maintenance solutions.

The proposed NetApp development employs an autoencoder that is responsible for detecting outliers in data sets collected to report anomalies as early as possible. The NetApp consists of six main components, namely, an on-premise Kafka engine, on-premise data collector, edge data collector, Edge engine, cloud-based data collector, and cloud visualisation engine. Each of the latter components is provided in a container format under the control of an application orchestrator.

Figure 2.11 illustrates the proposed 5G-based NetApp.

The proposed NetApp was deployed for predicting the outliers with regard to electricity generators. The autoencoder was integrated into the scenario of the Edge Analytics Engine. The simulation results demonstrated that the accuracy of the autoencoder was superior to that of other AI-based models with a peak value of 0.866.

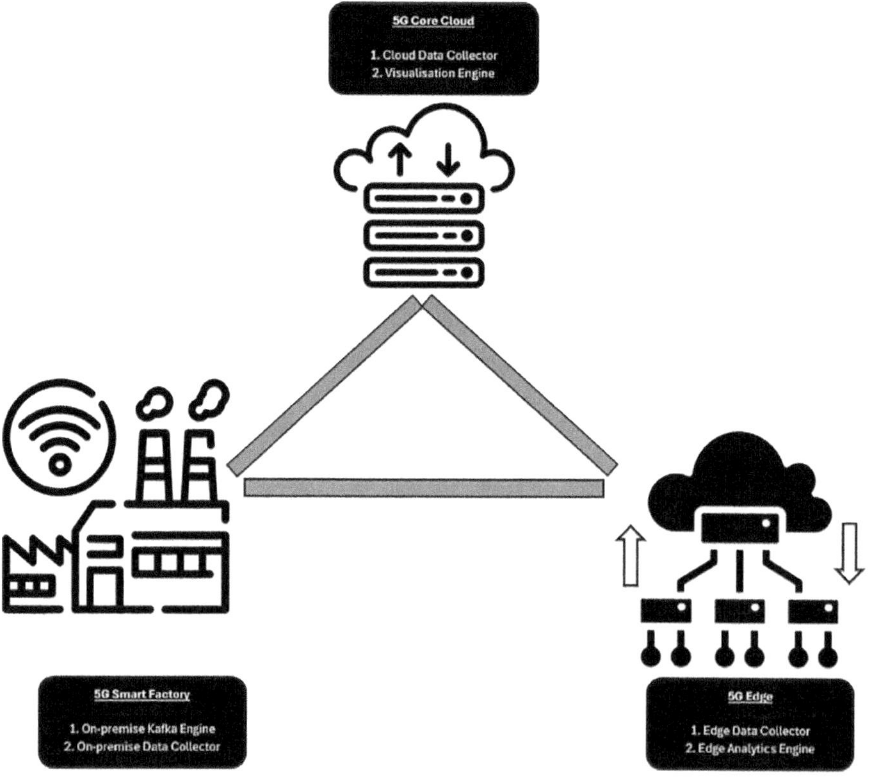

FIGURE 2.11 5G-based NetApp for predictive maintenance [42].

6G technology offers a level up in terms of possibilities for predictive maintenance through its enhanced service sets and AI. More precisely, machine learning algorithms are very promising with regard to predictive maintenance by gathering data over time, despite privacy concerns over the exchange of a considerable amount of data. Li et al. [43] proposed a lightweight privacy-preserving predictive maintenance solution that was founded on machine learning in 6G–IIoT scenarios to report the risk of breakdowns for industrial components. Binary neural networks were applied for training the predictive models to safeguard the privacy of all participants. Basically, 6G-based communication is expected to offer IIoT with unprecedented data rates, massive connections, and real-time intelligence. It is projected that 6G technology will offer much more considerable AI capabilities compared to existing 5G technology through the deep integration of novel techniques. The 6G-powered IIoT systems are forecast to connect a larger number of advanced devices with faster speeds, sensing the human and physical worlds, whereby the connectivity will be up to 1,000 times superior to 5G. Nevertheless, the exchange of vast amounts of data is expected to make 6G IIoT more vulnerable to security threats, which have to be mitigated accordingly. IIoT sensors operating in wireless mode can sense data continuously to collect key machine parameters such as temperature and operation status. The data can then be transferred to the cloud to be processed by the privacy-preserving predictive maintenance algorithm to output the status of equipment health. New techniques such as deep learning and big data analytics can detect anomalies more rapidly and provide additional predictive insights [43].

In their study, Mezair et al. [44] proposed an advanced deep learning framework for beyond 5G/6G-based preventive fault diagnosis in Industry 4.0 to overcome some of the challenges facing the adoption of such schemes in terms of handling heterogeneous data captured in different formats by sensing devices. The proposed framework, referred to as ADL-FDI4, involves the combination of mainly long short-term memory and convolutional neural networks to deal with heterogeneous data. The proposed principle is illustrated in Figure 2.12.

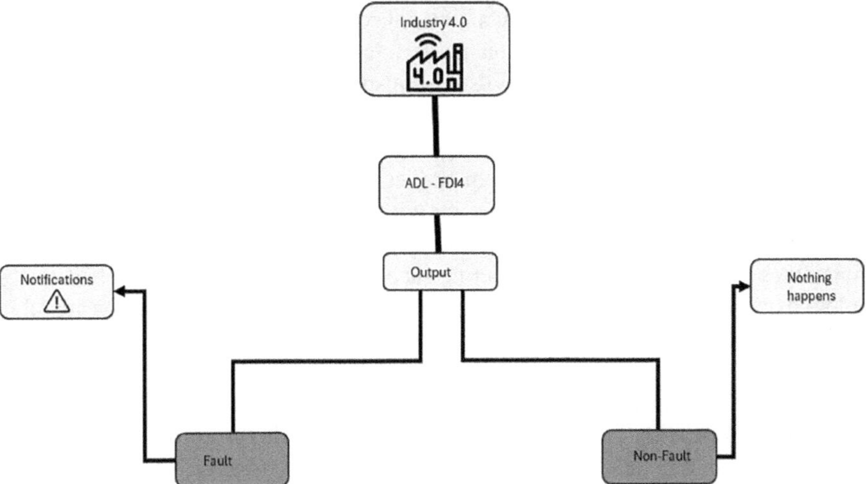

FIGURE 2.12 ADL-FDI4 framework [44].

The data is initially extracted from various sensors. The next steps consist of executing deep learning algorithms to identify faults in the system and proceed with the triggering of relevant alarms, if applicable, to the Industry 4.0 monitoring system. This architecture involves the combination of deep learning mechanisms to successfully handle heterogeneous data.

2.7 SUPPORTING VIRTUAL AND AR IN MANUFACTURING

Extended reality, often referred to as XR, is a revolutionary concept of immersive technologies that fuse the physical and virtual worlds, comprising technologies such as VR, AR, and mixed reality (MR). The innovative frameworks have the ability to transform the way we interact with the digital world encompassing various fields such as business, entertainment, gaming, and educational applications. Recent advances in extended reality have laid the foundation for various possibilities across different fields for allowing users to experience computer-generated environments like never before. VR technology creates a scenario whereby users are completely immersed in a computer-generated environment such that they are completely disconnected from the physical world. Conversely, AR integrates the simulated world with the real one to enhance the user's perception of reality.

In most applications the user relies on a smartphone or tablet screen to accomplish this, aiming the phone's camera at a point of interest and generating a live-streaming video of that scene on the screen. MR is basically a hybrid version of AR and VR, allowing users to interact with both the digital and physical worlds simultaneously [45].

One main area of interest is the way XR is expected to revolutionise industries and radically change the interaction of people with the digital world to improve productivity and overall user experiences. The application of VR in Industry 4.0 would significantly aid organisations to bring down design and production costs, improve product quality and suitability, and decrease the lead time for moving from product conception to production. The four main areas of application of VR and AR in Industry 4.0 include design optimisation, factory maintenance and control, operations instruction and staff training, incident prevention, and resolution. In other words, manufacturing systems applying VR/AR could be used to inspect products at preliminary design stages, assessing the interaction of clients with the finished products, optimising the manufacturing processes, and remote supervision of operation [46].

The promises held by VR/AR technologies in delivering immersive experiences to businesses, and end customers rely on unprecedented requirements for connectivity ensuring ultra-low latency and ultra-high throughput. 5G mobile technology constitutes a viable candidate for meeting such stringent requirements [47]. 5G technology is considered to be a game-changer for XR use cases. Offering considerably faster data rates than previous mobile generations, 5G allows seamless and real-time interactions in XR environments. Consequently, the end-user experience will be improved by the superior quality of AR/VR content. Moreover, the inherent ability of 5G to efficiently handle large volumes of data will permit the integration of advanced XR components for the development of innovative XR applications [45]. Moreover, the ability to transfer large files like 8K video streams without literally no lag ensures high-quality displays on VR sets. Improved mobility is another important consideration. 5G contributes to more reliability by allowing users to move around more freely in the VR environment whilst staying connected.

According to Carter [48], the VR industry is growing so fast that it is expected to reach a value of $184.66 billion by 2026. The main applications of VR in Industry 4.0 are described as follows:

1. **Design and Prototyping:**
   VR allows us to see how a product would appear before we even build a prototype. Using its own Ford Immersive Vehicle Environment (FIVE) VR technology, Ford Motors has been meticulously simulating future automobiles down to the last detail in the automotive sector. Engineers can experience what the future vehicle will be behind the wheel by transmitting designs into virtual automotive surroundings. This enhances the manufactured product's quality while decreasing the time spent on product design and review [49].
2. **Planning of Plant Layout:**
   Engineers can try out various plant layouts in a virtual setting using augmented and VR technology to find issues before making practical adjustments.

   Manufacturers can boost productivity, decrease expenses, and enhance efficiency by maximising the structure of their plants. Also, AR overlays can show plant performance in real time, which helps management spot inefficiencies and bottlenecks so they can fix them to boost production [49].
3. **Visualisation of Equipment:**
   Engineers and technicians can have a better understanding of the equipment's operation and potential for improvement with the use of augmented and VR visualisations of manufacturing equipment. For instance, engineers can utilise VR to replicate a production process and find possible issues with equipment location or workflow.

   Furthermore, technicians may swiftly detect issues and fix them with the help of real-time information regarding equipment performance provided by AR overlays.

   This has the potential to enhance equipment performance and decrease downtime [49].

The emergence of 6G technology heralds a new era of immersive experiences and transformative potential in the realm of VR and AR. Boasting unmatched data rates, very high reliability, and almost negligible latency, 6G will profoundly impact the VR and AR applications [50].

1. **Ultra-High Speed and Minimal Latency**
   With user data rates superior to 10 Gbps and end-to-end latency of 0.1 ms, 6G will catalyse the creation of high-quality, real-time VR content. VR gaming with imperceptible delay will improve seamless experiences and make VR applications more engaging and user-friendly.

2. **Enhanced Connectivity and Real-Time Interactivity**
   Ubiquitous connectivity via 6G will act as a gateway for interactions between the digital and physical worlds, thereby causing a revolution in several industries such as enabling remote surgeries in healthcare and real-time distance learning for students.
3. **Personalisation via AI Integration**
   The onboarding and incorporation of AI is another vital aspect of 6G technology. AI can be a powerful enabler for personalising user experience by analysing customer behaviour and preferences and adapting the VR/AR provided to them accordingly. Personalisation will ultimately boost user satisfaction and loyalty.
4. **Enhanced Security and Privacy**
   Ensuring the security and privacy of user data in AR and VR applications is a primordial requirement. The enhanced security features of 6G in terms of robust encryption protocols and reliable authentication methods will significantly help in the protection of sensitive user data for use cases such as medical diagnostics and financial transactions.

2.8 CONCLUSION

Over the past decade, the manufacturing industry has been undergoing a major digital revolution branded as "Industry 4.0". This transformation has been fuelled by the advent of revolutionary technologies such as cloud computing, AI/machine learning, DTs, novel connectivity technologies such as 5G/6G/Wi-Fi 6e, and advanced robotics. The primary goals of business organisations include increasing efficiency and productivity year after year amidst operating in challenging worldwide conditions. As the latest commercialised mobile technology, 5G is currently being massively deployed in different regions of the world. 5G has emerged as a key enabler for the transition to Industry 4.0 through its enhanced service sets of eMBB, URLLC, and massive mMTC. 5G has thus set the stage for innovative industrial use cases such as smart logistics, for bringing in digitalisation and flexibility in supply chains for the efficient delivery of goods from distribution centres to customer premises, self-driving robots in the form of AMRs, and AGVs for transporting materials across factory floors, collaborative robots to perform operations in conjunction with humans in common work environments, predictive maintenance for proactively detecting degradation in machinery to prevent unwanted downtimes and maintain productivity, and XR to offer groundbreaking immersive experiences to customers through the fusion of the physical and digital worlds [13].

There is no limit to advancements in technology. The futuristic 6G technology, expected to be commercialised by 2030, is already positioning itself to supersede 5G. Through a more human-centric approach, 6G will combine intelligent knowledge systems with powerful computational resources to influence the lives of individuals and the way industries function. Moreover, the incorporation of AI and mass deployment of sensors will provide enhanced and more cost-effective connectivity compared in terms of higher data rates and spectral efficiency, lower latency, and sub-centimetre geo-location accuracy [23].

REFERENCES

1. M. Javaid, A. Haleem, S. Rab, R. P. Singh, R. Suman, and S. Mohan, "Progressive schema of 5G for Industry 4.0: features, enablers, and services," *Industrial Robot: The International Journal of Robotics Research and Application*, vol. 49, no. 3, pp. 527–543, Jan. 2022, https://doi.org/10.1108/ir-10-2021-0226
2. X. Xu, Y. Lu, B. Vogel-Heuser, and L. Wang, "Industry 4.0 and Industry 5.0—inception, conception and perception," *Journal of Manufacturing Systems*, vol. 61, no. 1, pp. 530–535, Oct. 2021, https://doi.org/10.1016/j.jmsy.2021.10.006
3. M. Rüßmann et al., "*Industry 4.0: The Future of Productivity and Growth in Manufacturing Industries*," Boston Consultation Group, USA, Apr. 2015. Accessed: Feb. 15, 2024. [Online]. Available: https://www.bcg.com/publications/2015/engineered_products_project_business_industry_4_future_productivity_growth_manufacturing_industries
4. S. Doyle, "*5G for Industry 4.0 Operational Technology Networks*," GSMA, London, Mar. 2021. Accessed: May 05, 2024. [Online]. Available: https://www.gsma.com/solutions-and-impact/technologies/internet-of-things/wp-content/uploads/2021/03/2021-03-GSMA-5G-Industry-4.0-Op-Tech-Networks.pdf
5. C. Lukaszewski, "*Wi-Fi 6E: The Details You Need to Know*," Wi-Fi Alliance, Jul. 2022. Accessed: May 16, 2024. [Online]. Available: https://www.wi-fi.org/beacon/chuck-lukaszewski/wi-fi-6e-the-details-you-need-to-know
6. A. Müller, "*5 Reasons for 5G*," Bosch, May 2023. Accessed: May 18, 2024. [Online]. Available: https://www.bosch.com/stories/5g-industry-4-0/
7. British Telecom, "*5G Private Networks: Building the Bridge between Secure Connectivity and Competitive Advantage*," British Telecom, England, Jun. 2023. Accessed: May 18, 2024. [Online]. Available: https://business.bt.com/content/dam/bt-business/pdfs/insights/unlocking-the-benefits-of-industry-4-0-with-5G-june2023.pdf
8. S. Russel, "*The Role of 5G in Industry 4.0*," InstaNet, Dec. 07, 2023. Accessed: May 19, 2024. [Online]. Available: https://www.linkedin.com/pulse/role-5g-industry-40-stanley-russel-cdvrc
9. B. Han et al., "Digital twins for Industry 4.0 in the 6G Era," *IEEE Open Journal of Vehicular Technology*, vol. 4, pp. 820–835, Jan. 2023, https://doi.org/10.1109/ojvt.2023.3325382
10. Nokia, "*6G Explained*," Nokia, Apr. 2023. Accessed: May 22, 2024. [Online]. Available: https://www.nokia.com/about-us/newsroom/articles/6g-explained/#:~:text=In%20the%206G%20era%2C%20the,take%20care%20of%20the%20planet
11. A. T. Chowdhury, "The impact of 5G network on Industry 4.0," In *STI*, Bangladesh: Institute of Electrical and Electronics Engineers (IEEE), Sep. 2023. Accessed: Jan. 23, 2024. [Online]. Available: https://www.researchgate.net/publication/373897746_The_Impact_Of_5G_Network_On_Industry_40
12. D. Mourtzis, J. Angelopoulos, and N. Panopoulos, "Design and development of an edge-computing platform towards 5G technology adoption for improving equipment predictive maintenance," *Procedia Computer Science*, vol. 200, pp. 611–619, 2022, https://doi.org/10.1016/j.procs.2022.01.259
13. D. Mourtzis, J. Angelopoulos, and N. Panopoulos, "Smart manufacturing and tactile internet based on 5G in Industry 4.0: challenges, applications and new trends," *Electronics*, vol. 10, no. 24, p. 3175, Dec. 2021, https://doi.org/10.3390/electronics10243175.
14. B. Alhayani et al., "5G standards for the Industry 4.0 enabled communication systems using artificial intelligence: perspective of smart healthcare system," *Applied Nanoscience*, Jan. 2022, https://doi.org/10.1007/s13204-021-02152-4

15. Y. Ghildiyal et al., "An imperative role of 6G communication with perspective of industry 4.0: challenges and research directions," *Sustainable Energy Technologies and Assessments*, vol. 56, p. 103047, Mar. 2023, https://doi.org/10.1016/j.seta.2023.103047
16. A. Mughees, M. Tahir, M. A. Sheikh, and A. Ahad, "Towards energy efficient 5G networks using machine learning: taxonomy, research challenges, and future research directions," *IEEE Access*, vol. 8, pp. 187498–187522, 2020, https://doi.org/10.1109/access.2020.3029903
17. GSMA, "*Road to 5G: Introduction and Migration*," GSMA, London, Apr. 2018. Accessed: Apr. 23, 2024. [Online]. Available: https://www.gsma.com/solutions-and-impact/technologies/networks/wp-content/uploads/2018/04/Road-to-5G-Introduction-and-Migration_FINAL.pdf
18. V. Singh and G. Sharma, "*What is 5G Technology – Requirements, Initiatives, and Enabling Technologies*," *5G*, Mar. 17, 2023. Accessed: Feb. 16, 2024. [Online]. Available: https://www.greyb.com/blog/5g-technology/
19. S. Hashima, B. M. ElHalawany, K. Hatano, K. Wu, and E. M. Mohamed, "Leveraging machine-learning for D2D communications in 5G/beyond 5G networks," *Electronics*, vol. 10, no. 2, p. 169, Jan. 2021, https://doi.org/10.3390/electronics10020169
20. S. Wijethilaka and M. Liyanage, "Survey on network slicing for Internet of Things realization in 5G networks," *IEEE Communications Surveys & Tutorials*, pp. 1–1, 2021, https://doi.org/10.1109/comst.2021.3067807
21. A. O. Ercan, M. O. Sunay, and I. F. Akyildiz, "RF energy harvesting and transfer for spectrum sharing cellular IoT communications in 5G systems," *IEEE Transactions on Mobile Computing*, vol. 17, no. 7, pp. 1680–1694, Jul. 2018, https://doi.org/10.1109/tmc.2017.2740378
22. I. Rojek, P. Kotlarz, J. Dorozynski, and D. Mikołajewski, "Sixth-generation (6G) networks for improved machine-to-machine (M2M) communication in Industry 4.0," *Electronics*, vol. 13, no. 1832, pp. 1–17, May 2024, https://doi.org/10.3390/%20electronics13101832
23. X. You et al., "Towards 6G wireless communication networks: vision, enabling technologies, and new paradigm shifts," *Science China Information Sciences*, vol. 64, no. 1, Nov. 2020, https://doi.org/10.1007/s11432-020-2955-6
24. E. J. Khatib and R. Barco, "Optimization of 5G networks for smart logistics," *Energies*, vol. 14, no. 6, p. 1758, Mar. 2021, https://doi.org/10.3390/en14061758
25. Z. Kong, "Impact of 5G technology on digital transformation of Jingdong Smart Logistics," *Journal of Computing and Electronic Information Management*, vol. 12, no. 1, pp. 12–16, Feb. 2024, https://doi.org/10.54097/jbtgj70prx
26. R. Liu et al., "6G enabled advanced transportation systems," *IEEE Transactions on Intelligent Transportation Systems*, pp. 1–17, Jan. 2024, https://doi.org/10.1109/tits.2024.3362515
27. G. Fragapane, D. Ivanov, M. Peron, F. Sgarbossa, and J. O. Strandhagen, "Increasing flexibility and productivity in Industry 4.0 production networks with autonomous mobile robots and smart intralogistics," *Annals of Operations Research*, vol. 308, Feb. 2020, https://doi.org/10.1007/s10479-020-03526-7
28. M. B. Alatise and G. P. Hancke, "A review on challenges of autonomous mobile robot and sensor fusion methods," *IEEE Access*, vol. 8, pp. 39830–39846, 2020, https://doi.org/10.1109/access.2020.2975643
29. Everything RF, "*Qualcomm and Bosch Rexroth Accelerate Industry 4.0 with 5G and TSN Technologies*," EverythingRF, Mar. 07, 2022. Accessed: Apr. 25, 2024. [Online]. Available: https://www.everythingrf.com/news/details/14048-qualcomm-and-bosch-rexroth-accelerate-industry-4-0-with-5g-and-tsn-technologies

30. E. A. Oyekanlu et al., "A review of recent advances in automated guided vehicle technologies: integration challenges and research areas for 5G-based smart manufacturing applications," *IEEE Access*, vol. 8, pp. 202312–202353, 2020, https://doi.org/10.1109/access.2020.3035729
31. A. Costa, P. Duarte, A. Coelho, and R. Campos, "*Obstacle-Aware Positioning of a Mobile Robotic Platform for 6G Networks*," Jan. 2024. Accessed: Mar. 17, 2024. [Online]. Available: https://arxiv.org/pdf/2401.06717
32. J. He, K. Yang, and H.-H. Chen, "6G cellular networks and connected autonomous vehicles," *IEEE Network*, pp. 1–7, 2020, https://doi.org/10.1109/mnet.011.2000541
33. S. Nafees, "*Using 5G Network Exposure to Mitigate Risks in Industrial Cobot Environments*," Jun. 02, 2021. Accessed: Apr. 13, 2024. [Online]. Available: https://www.ericsson.com/en/blog/2021/6/using-5g-network-exposure-to-mitigate-risks-in-industrial-cobot-environments
34. K. Yasar, "*Collaborative Robot (Cobot)*," TechTarget, Mar. 2023. Accessed: May 12, 2024. [Online]. Available: https://www.techtarget.com/whatis/definition/collaborative-robot-cobot
35. M. Javaid, A. Haleem, R. P. Singh, and R. Suman, "Substantial capabilities of robotics in enhancing Industry 4.0 implementation," *Cognitive Robotics*, vol. 1, no. 1, pp. 58–75, 2021, https://doi.org/10.1016/j.cogr.2021.06.001
36. A. Grau, M. Indri, L. Lo Bello, and T. Sauter, "Robots in Industry: the past, present, and future of a growing collaboration with humans," *IEEE Industrial Electronics Magazine*, vol. 15, no. 1, pp. 50–61, Mar. 2021, https://doi.org/10.1109/mie.2020.3008136
37. M. C. Soveri et al., "*5G Network Slice Management*," 3GPP, Jul. 2023. Accessed: May 19, 2024. [Online]. Available: https://www.3gpp.org/technologies/slice-management
38. D. Calandra, F. G. Pratticò, A. Cannavò, C. Casetti, and F. Lamberti, "Digital twin- and extended reality-based telepresence for collaborative robot programming in the 6G perspective," *Digital Communications and Networks*, Oct. 2022, https://doi.org/10.1016/j.dcan.2022.10.007
39. S. Kerboeuf et al., "Design methodology for 6G end-to-end system: hexa-X-II perspective," *IEEE Open Journal of the Communications Society*, pp. 1–1, Jan. 2024, https://doi.org/10.1109/ojcoms.2024.3398504
40. Singtel, "*7 Use Cases for 5G in Manufacturing*," Singtel, Singapore, Aug. 2021. Accessed: May 20, 2024. [Online]. Available: https://www.singtel.com/business/articles/7-use-cases-for-5g-in-manufacturing#:~:text=Studies%20have%20shown%20that%20unplanned,predict%20and%20prevent%20unwanted%20downtimes
41. J. Diekmann, "*What is the Impact of 5G on Predictive Maintenance?*," Ericsson, Aug. 2022. Accessed: May 27, 20241. [Online]. Available: https://www.ericsson.com/en/enterprise/cellular-curious/predictive-maintenance
42. S. Giannakidou et al., "5G-Enabled NetApp for Predictive Maintenance in Critical Infrastructures," *Zenodo (CERN European Organization for Nuclear Research)*, Sep. 2022, Nagoya, Japan. https://doi.org/10.1109/wsce56210.2022.9916037
43. H. Li, S. Li, and G. Min, "Lightweight privacy-preserving predictive maintenance in 6G enabled IIoT," *Journal of Industrial Information Integration*, vol. 39, pp. 100548–100548, May 2024, https://doi.org/10.1016/j.jii.2023.100548
44. T. Mezair, Y. Djenouri, A. Belhadi, G. Srivastava, and J. C.-W. Lin, "A sustainable deep learning framework for fault detection in 6G Industry 4.0 heterogeneous data environments," *Computer Communications*, vol. 187, pp. 164–171, Apr. 2022, https://doi.org/10.1016/j.comcom.2022.02.010
45. P. Martínez, "*Extended Reality: Exploring the Future of Immersive Technologies*," Augmented Reality, Nov. 13, 2023. Accessed: May 28, 2024. [Online]. Available: https://www.onirix.com/extended-reality/

46. V. Liagkou, D. Salmas, and C. Stylios, "Realizing virtual reality learning environment for Industry 4.0," *Procedia CIRP*, vol. 79, pp. 712–717, 2019, https://doi.org/10.1016/j.procir.2019.02.025
47. A. Paszkiewicz, M. Salach, P. Dymora, M. Bolanowski, G. Budzik, and P. Kubiak, "Methodology of implementing virtual reality in education for Industry 4.0," *Sustainability*, vol. 13, no. 9, p. 5049, Apr. 2021, https://doi.org/10.3390/su13095049
48. R. Carter, "*How 5G Will Accelerate the VR Revolution*," Virtual Reality, Jul. 08, 2021. Accessed: May 16, 2024. [Online]. Available: https://www.xrtoday.com/virtual-reality/how-5g-will-accelerate-the-vr-revolution/
49. D. Pavlov, "*AR & VR Applications in Manufacturing: Redefining Productivity and Quality*," AR & VR Development, Apr. 07, 2023. Accessed: Apr. 26, 2024. [Online]. Available: https://smarttek.solutions/blog/ar-and-vr-in-manufacturing-use-cases-and-benefits/
50. J. Musk, "*The Impact of 6G on Virtual and Augmented Reality*," Oct. 27, 2023. Accessed: May 12, 2024. [Online]. Available: https://medium.com/@johnymuski990/the-impact-of-6g-on-virtual-and-augmented-reality-c85572124907

3 Industrial Internet of Things for Industry 4.0

Mussawir Ahmad Hosany

3.1 INTRODUCTION

The Industrial Internet of Things (IIoT) provides an increase in productivity and efficiency by combining the physical world handled by humans with the cyber world. The integration of the physical and cyber domains is changing the industrial automation and control process systems into cyber-physical manufacturing systems.

Presently, there are millions of industrial devices that are embedded in the physical world, and all these are connected to the Internet through industrial networks. These devices, together with human inputs, enable the integration of data generated by them into information systems and business processes and services. The fourth industrial revolution, known as Industry 4.0, has created the digitization of industrial manufacturing systems. IIoT includes highly efficient technology arrangements in numerous industrial systems such as manufacturing, transportation, healthcare, and offices. IIoT ensures that the digitization and connectivity among all productive units as well as human resource systems are maintained, simultaneously retaining all the advantages of the traditional manufacturing processes.

IIoT can be broadly defined as intelligent or smart operations that are handled by machines, computers, and humans, and these are made possible using advanced data analytics for transformational business outcomes [1]. To enable IIoT in intelligent industrial processes, there are wireless sensor networks, wireless sensor and actuator networks, communication protocols, and Internet infrastructure for monitoring, analysis, and management, increasing productivity, efficiency, and safety of industrial operations [2].

In an IIoT setting, various machines, smart devices, and physical sensors such as RFIDs send data to a central network, which decreases downtime, thereby increasing efficiency. In today's world complicated marketing conditions require that the manufacturing industry is effective and adaptable as well as sustainable, and the IIoT technology has made these possible. As a result, Internet of Things (IoT) devices have been used to monitor and control many aspects of the quality and production processes. For example, sensors can be used to measure the temperature, humidity, and pressure of a production unit in the manufacturing system. The data obtained are then analyzed using machine learning algorithms to generate patterns and optimize various production processes [3].

DOI: 10.1201/9781003511298-3

3.2 IIoT IN MANUFACTURING INDUSTRY

IIoT enables a novel age of advanced technology for chains in supply and value production and models of businesses in manufacturing industries. The fundamental difference between a customized IoT platform and an IIoT one is that for an IIoT operational framework more emphasis is laid on the use of smart devices as well as data analytics for optimization using machine learning. The IIoT manufacturing sector is presently undergoing enormous changes in the assembly and procedural systems with the objective of creating new business interests and markets all around the globe.

In October 2017, there was agreement among all stakeholders in the Amazon Web Services (AWS) public sector SUMMIT in Singapore that the IoT infrastructure will consist of three pillars, namely, things, cloud, and intelligence [1,4]. Without the cloud and intelligent systems, IoT technology will be just a transmitter and receiver of information across the globe. The present manufacturing scene comprises advanced automation robot technologies implemented in an IoT architecture. This framework has the crucial requirement of combining smart data sensors and state-of-the-art machine-to-machine (M2M) technologies with the existing automation ones. Such an intelligent architecture is required to create an IIoT technology as elaborated in Caesarendra et al. [4].

Essential environments for IIoT in manufacturing systems include wireless sensor and actuator networks and middleware technologies that monitor and customize them. Moreover, money-saving choices are facilitated by real-time data using big data analytics such as recording, managing, and monitoring procedures. In Caesarendra et al. [4] a potential IIoT architecture consisting of four layers is proposed for the manufacturing industry. This IIoT architecture is composed of various manufacturing processes. In the setting of a conventional IoT architecture for the manufacturing industry, a rigid option and a more specific adaptation of the architecture are imperative. The various industry processes associated with each layer of the IIoT are described next.

3.2.1 LAYER 1: DATA LAYER

As illustrated in Caesarendra et al. [4], the data layer consists of data generated from physical sensors, radio frequency identifications (RFIDs), and real-time robot data with other shop floor equipment. Also, asset/resource information and the location of items in the manufacturing industry provide crucial data for sensing and localization procedures. The data layer has the primary goal of identifying data sources and acquiring relevant data from these sources. Data acquisition devices are used to retrieve data from physical sensors. The RFID tags of IoT-enabled assets can support information acquisition with respect to the localization and sensing of assets.

3.2.2 LAYER 2: APPLICATION LAYER

Since the information emerging from the data layer comes from different sources and sizes, its classifications and preliminary assessments have to be performed, and in this regard a human–machine interface and an application program interface are essential at the application layer. These interfaces can be installed by off-the-shelf systems such as distributed control systems or custom data interfaces. In recent days, most of the manufacturing processes have the software ability to monitor themselves. These interfaces have been developed on a scalable domain and therefore can cater to the IIoT setting as approved by the user.

3.2.3 Layer 3: Security Layer

The digitization of an IIoT network in the manufacturing industry inevitably generates big data and can be used to improve business productivity and efficiency. It is vital to secure all these big data at the security layer using data management systems. The safety of data starts by recording the total number of connected devices with their properties, which can be accessed from a centralized location. As a basic security procedure, all the data provided by the application layer have to be password protected. A third-party information security solution that can be adopted by the manufacturing industry to address the various security aspects of the application data is also described in Caesarendra et al. [4].

3.2.4 Layer 4: Service Layer

The proposed IIoT architecture of Caesarendra et al. [4] consists of a topmost layer called the service layer, which keeps the traditional functions of manufacturing operations. This layer acts as a platform to establish connections to remote devices and services. With advancements in computing technologies, the cloud, fog, and edge servers are delivering cost-effective data analysis with intelligent algorithms. The service layer is used to interface with such smart platforms to deliver various manufacturing services.

3.3 BLOCKCHAIN APPLICATIONS FOR IIoT

Blockchain and the Internet of Things in an industry setting environment are regarded as highly potential technologies that can provide the utmost security to the manufacturing organization. It is well known that blockchain is a technology that decentralizes transaction purposes by providing novel and smart models to store and manage data [5,6]. Moreover, IIoT is related to an environment consisting of connected machines, devices, humans, and the Internet. A combination of both blockchain and IIoT is therefore crucial to the manufacturing business, which, by adopting both systems, can experience a digital revolution.

A blockchain-based IIoT system enables the various parties of the manufacturing industry to communicate with each other without the need for a trusted intermediary in a decentralized, unreliable, peer-to-peer network. A disseminated ledger that expands over an entire distributed system in an industry database is called a blockchain. It has been shown that blockchains are capable of conducting decentralized confirmation of contracts, thereby significantly reducing the cost and operational traffic at the core industry center [7–10].

A blockchain consists of well-organized components termed blocks that are composed of headers with transactions. There are various types of blockchains, namely, public, private, and consortium. If no permissions are required to access a database, then a public blockchain is used, but if credentials are required to access a private database, then a private blockchain is implemented [5]. Also, consortium networks span various manufacturing industries and support in maintaining clarity among all the parties. Hence, a consortium blockchain is used as an auditable and dependably coordinated dispersed database that maintains a smart path placed between the consortium entities. There are indeed potential benefits of applying blockchain to IIoT networks in an industrial setting [11].

3.3.1 Blockchain for Healthcare Internet of Things

Blockchain can be employed in an operative way to share crucial patient information in the healthcare industry and among all stakeholders. The delivery of healthcare services can be fundamentally improved using blockchain, for instance, by eliminating mismatched patient data [5]. Regarding the various stakeholders such as medical doctors and healthcare providers, it has been shown that patients' health records can be perfectly authenticated and secured by them [12–14]. Moreover, several infectious and severe diseases can be monitored and managed by the implementation of blockchain-enabled IoT technology. Some of the implementations include wearable devices for tracking vital signs and providing feedback, smart pills, and improved quality control [13].

In the healthcare industry, the management of patient data is vital for ensuring safety and effective healthcare delivery. To maintain patient records while they leave their data spread across a range of stakeholders, the design of electronic health record (EHR) systems has been already implemented. However, shortcomings of the EHR such as interoperability and resource coordination between hospitals and healthcare service providers create barriers to successful data sharing. In this context the technology of blockchain is critical to be implemented to provide a structure for data sharing along with the security of these data [5]. Hence, patients' information such as date of birth, name, medical prescriptions, and diagnosis can be securely shared among healthcare providers. These data can be stored, processed, and saved on the cloud/fog/edge computing systems and in the existing databases of the healthcare organization.

Blockchains in the management of patient data make use of a cryptographic hash from every data source that is calculated and then transmitted along with the patient's ID and stored in the blockchain database. Also, healthcare providers in hospitals or medical clinics can access and manage patient data through an application program interface by simply querying the blockchain and locating the information of this patient without even revealing the identity of the patient. The medical history of a patient can be shared with or without recognizable data to any stakeholder if required.

In today's world of the pharmaceutical industry, there is a major issue of drug counterfeiting where around 30% of the drugs are sold in the developing world as fake drugs. The counterfeit drug market has a budget of $200 billion annually, and the online market is expanding exponentially [5]. It has been reported that these fake drugs give rise to undesirable effects and can result in sudden death for those patients who have complications such as diabetes and heart disease [5]. There are two primary issues of drug traceability that can be handled using blockchain technologies. First, fake drugs can be detected and located in the supply chain of the pharmaceutical industry using specific IDs stored in the blockchain. Second, healthcare providers such as medical laboratories are given access by the blockchain database to the exact drug location, and hence traceability can be guaranteed.

In Alladi et al. [5] a general pharmaceutical supply chain infrastructure is proposed that can reinforce drug traceability and hence reduce counterfeiting. In Alangot and Achuthan [13], a proposed infrastructure where blockchain is embedded in the supply chain operations is proposed. In this work effective communication of data has been guaranteed using a GDP IoT framework, with the entire supply chain infrastructure consisting of physical sensors such as RFIDs, barcode scanners, and smartphones. In the architecture, the pharmaceutical infrastructure consists of three planes, namely the data, control, and application layers. The control plane is equipped with a smart controller that can detect drug counterfeiting with the help of the blockchain database.

Blockchain has been successfully implemented and deployed in the healthcare industry for storing EHRs safely and very effectively. The commercial deployment is termed as MedRec [15] and has an integrated blockchain patient health record capability to enable the patient and healthcare providers to establish a longitudinal healthcare record system. MedRec is a novel application that employs blockchain technology for medical access, permission management, and trend analysis [15].

3.4 APPLICATIONS OF IIoT TO INDUSTRY 4.0

There have been numerous recent developments in the application of IIoT to smart factories with the primary objective of driving the digital transformation of manufacturing and enabling companies to operate effectively and sustainably. The applications of IIoT technologies to Industry 4.0 requires a large investment in designing and installing large infrastructures and electronic and computer resources [1]. Also, the workforce requirement is a highly qualified one to have the proper knowledge to manipulate various devices and machines in the manufacturing system.

There are several benefits of linking IIoT to Industry 4.0 such as boosting the energy efficiency of the smart factory. By employing IoT sensors many industrial companies have increased their efficiency targets and monitored the energy consumption in real time as well as measured the carbon dioxide released by each industrial facility [2]. Moreover, IIoT technology in Industry 4.0 enables a reduction in cost and enhanced service as well as decreasing the requirement to carry out item inspections by humans. The installation of IoT in manufacturing systems is capable of detecting and diagnosing any type of failure precisely and informing the technical service quickly.

In an industrial setting, the IIoT system provides smart monitoring of machines by enabling constant workflow control of each machine, detecting operational issues and reporting, and alleviating critical workplace accidents. With the integration of

IIoT at the production stage of a manufacturing industry, the enterprise can monitor issues associated with quality of production and customer service. IoT sensors can be used to uphold quality standards and help control supply tasks and detect risks linked to thefts.

3.4.1 Asset Tracking and Inventory Management

With the introduction of IIoT in smart factories, it has been shown that optimization of production workflows and asset lifespan is possible by implementing asset tracking of the entire organization with smart IoT sensors [16]. Moreover, asset tracking can be achieved for the location and condition of assets in an IIoT setting. The supply chain and manufacturing operations can be optimized as well with the applications of IoT sensors to factory units. IoT devices and smart devices can be leveraged so that decision makers in manufacturing industries can track assets in real time. In this way better visibility, profitability, and productivity are achieved [17].

An IoT-assisted asset tracking system generally makes use of wireless sensors attached to assets and sends information concerning the exact location and status as well as condition to the central industry repository. Moreover, RFID tags are attached to the assets to gather data on their movements where subsequently this information is analyzed and evaluated using machine learning algorithms [18]. The trained data can then be used to provide better perceptions of the enactment and deployment of the assets of the industry. Asset tracking is also employed to guarantee that equipment is readily available as and when needed to reduce latencies and increase productivity.

Furthermore, an IoT-enabled asset tracking system can give precise procedures to manufacturers in complying with international regulations and standards in the process of part production. For a manufacturing industry with raw materials and designed parts, these can be easily monitored by tracking areas where pollution in the production department can be minimized. As reported in Soori et al. [1] an IoT-based asset tracking system is a critical component to optimize the supply chain processes in a manufacturing industry. This system can provide manufacturers with real-time information on the location and status of assets to optimize their processes and reduce environmental aspects of their production processes.

To analyze the impact of asset management in the electrical industry, application fields and innovations have been matched to the international standard ISO 55001 regulation for asset management. The asset tracking of the electrical industry sector for Industry 4.0 was proposed by Soori et al. [1], in which the main component that drives the tracking is big data coupled with machine learning computation algorithms.

Manufacturers in the industry sector can monitor real-time data about various inventories on the stock levels by employing IoT-enabled sensors. The monitoring procedure can eventually help them to optimize their production processes and reduce environmental pollution in part production [19]. The following methods, based on the use of IoT technologies, are employed in Industry 4.0 to optimize inventory management.

1. Tracking real-time inventory and stock levels by employing smart sensors and RFID tags gives accurate and present information on stock levels and locations.
2. In the event a certain stock level is falling below a prescribed threshold then IoT devices can be used to reorder inventory and automate replenishment.
3. Combining data from IoT sensors and big data analytics with machine learning algorithms historical trends can be analyzed and future demands can be predicted.

It can be stated that IIoT can support smart factories in Industry 4.0 to fully optimize their processes related to inventory management and therefore increase accuracy, cost-effectiveness, and productivity [20].

3.4.2 Quality Control and Production Process Monitoring

To control the quality of manufactured units in Industry 4.0, a powerful tool can be used. The tool is made up of IoT sensors that can detect defects in real time and give a precise schedule to rapidly track and change any problem to decrease the risk of defective units in the manufacturing industry [21,22]. The quality control process can be designed to meet international standards using IoT technology, humans, machines, and smear devices in Industry 4.0. IIoT can be used to gather real-time data, and together with data analytics as well as artificial intelligence systems, the quality control process can be fully optimized and the possibility of having defective products can be reduced significantly. IIoT can be employed in the following various ways in Industry 4.0.

1. To maintain the lifetime of machines and tools in real time, IoT sensors and smart devices can be used. Data collected from the maintenance procedure can be fed to the data analytics part, and with the help of machine learning, predictive maintenance can be carried out. This will provide cost-effectiveness to the manufacturers, and downtime will be reduced, with an improvement in the quality of products.
2. IoT sensors can measure various types of information such as temperature, pressure, vibrations, and humidity in the department of manufacturing, and these IoT devices can report anomalies or deviations from the normal ratings to the technical department. In this way the quality of products is guaranteed.
3. Real-time testing of products in Industry 4.0 is possible by employing IIoT technology. For instance, to measure the dimensions and weight of manufactured products, IoT sensors can be used and any defects in the packaging can be flagged to the respective quality control department.
4. During the entire production process, IIoT is used to track the products by providing a detailed history of each item. The information obtained from this analysis can be used to identify any issues related to quality.

In the work of Zaidin et al. [22], a theoretical framework that combines Industry 4.0 with total quality management principles is introduced with a view of achieving business excellence in the quality control process. To enhance the quality of manufactured units in Industry 4.0 production process monitoring is a crucial component, which can be enabled by IIoT technology. Using various connected devices, machines, and IoT sensors in the production department, it is possible to fully optimize production process by reducing downtime and enhancing efficiency as well as the profitability of part manufacturing [23].

The working principle of production process monitoring is to gather data from various IoT sensors and smart devices in the production unit that are installed, and these data are used to monitor closely key performance indicators (KPIs) [24]. Equipment utilization, output of production, and quality of products are a few examples of KPIs, and by monitoring them in real time, managers of the industry can easily and timely detect bottlenecks, defective units, and many other issues in the production process.

IIoT operations can be performed in the production unit by monitoring the online conditions of machines and equipment, thereby providing a more restructured production process [24]. Also, during the production process, it is possible to optimize the supply chain operations by employing IoT sensors installed on various machines. The application of IIoT technology to Industry 4.0 for an online production processing system is elaborated in Soori et al. [1]. In this work, it has been shown that by employing big data analytics and data processing gathered from various sensors, a smooth monitoring system of the production process can be achieved.

3.4.3 Energy Efficiency

Energy efficiency is a key factor in a manufacturing industry to minimize energy consumption and hence achieve cost-effectiveness. With the help of IoT devices and smart sensors, the energy consumption of the entire factory can be smoothly monitored and areas for improvement can be identified [25]. This process can be implemented in the production department as well to reduce the costs of energy and improve sustainability.

IIoT sensors can be used from production lines to lighting and heating systems to observe energy usage in real time. The data gathered from these sensors can be used to analyze various patterns in energy consumption, which provide inefficient utilizations, and by employing machine learning it is possible to fully optimize energy usage in the entire smart factory. An automated control system with the aim of providing autonomous energy usage can be implemented in various sections of the industry through the use of IIoT technology [26]. For instance, an automatic adjustment of lighting and heating as well as cooling systems can be made based on the level of occupancies in a specific factory unit.

Renewable energy sources such as solar or wind power can be integrated into smart industries by employing IoT technology to provide a sustainable production process. A significant reduction in reliance on fossil fuels and lower energy costs can be achieved in terms of the management of energy consumption in the manufacturing process. To monitor the lighting in smart factories IIoT can be used by switching off automatically with unoccupied spaces and reducing light intensity based on the quantity of energy in a room [1,26].

IIoT devices can provide smart factories with highly efficient heating, ventilation, and air conditioning systems by automatically adjusting room temperature based on weather conditions and reducing air temperature for cooling in the manufacturing department consisting of equipment and machines. Also, energy storage systems can be monitored using IIoT devices in manufacturing industries by charging batteries when the energy is cheaper during specific business hours and using the charged batteries to power up factory devices when energy is expensive during peak hours.

The work of Al-Obaidi et al. [27] carried out an in-depth study on the application of IoT devices in a building to optimize energy efficiency. This study was conducted in a hospital in Singapore to maximize efficiency by monitoring solar levels, operational schedule, water flow, and electricity consumption. It has been observed that overall energy saving in electricity was reduced by around 33%.

3.5 RECONFIGURABLE INTELLIGENT SURFACES FOR IIoT

Reconfigurable intelligent surfaces (RISs) have recently gained significant attention in various business models and have been targeted as an attractive approach to mitigate wireless channel interferences in future 6G networks [28]. Arrays of scattering elements which are passive by nature and have planar structures altogether constitute the RIS unit. These elements are enabled by electronic circuits that reflect and deflect unwanted electromagnetic waves using software-defined radio technology.

The RIS structure makes it possible to be reconfigured and can be manually aligned to the base station so that a line-of-sight (LOS) propagation of the transmitted signal can be obtained. Indeed, RISs provide a variety of applications in the domain of smart manufacturing in Industry 4.0. For example, RISs can be used in the manufacturing department to convey high-energy signals, where non-LOS propagation is not possible, to machines and equipment which have wireless connectivity [28].

Wireless communications in the industrial sector are presently being challenged because of the various metallic structures, electric machines, electromagnetic interferences, and movement of robots and equipment. It has been shown that RISs can mitigate interferences and provide high reliability and error performance of the wireless links. RISs have the ability to shape the signal wave that propagates in industrial wireless networks and yields large coding gains and achievable data rates [28]. The principal enabler of RIS technology is a software-defined domain that can be reconfigured to adapt itself to changes in the wireless environment.

To overcome issues such as mitigating blockages in the manufacturing room where various machines and walls are installed, severe path loss and inadequate coverage RISs have been proposed in Industry 4.0 for IIoT applications. The work of Noor-A-Rahim et al. [28] proposes a RIS deployment, also known as intelligent reflecting surfaces installation, in a manufacturing room with machines and blockages. It can be seen that there is an automatic coverage that can be adapted according to production demand, and also there is no need to redeploy network infrastructure, thereby reducing significantly the installation cost of the production department. Moreover, by employing RISs in the manufacturing room, the use of IIoT devices can be wirelessly connected with reduced path loss. The applications of RIS deployment in Industry 4.0 enable wireless power transfer, blockage mitigation, data offloading for mobile edge computing, as well as sensing and localization.

It has been shown that it is possible to install the RISs correctly to provide an LOS with transmitters and receivers [28]. Also when the beamforming abilities of such devices are fully exploited, the received power of IIoT devices can be increased, thereby providing better performance on the wireless data transfer mechanism. The wireless charging efficiency of IIoT sensors can be substantially increased by properly installing the RIS in the production department.

3.6 OPEN ISSUES AND CONCLUDING REMARKS

There are still open issues for the integration of IIoT in Industry 4.0 that require further research and development. Some of the potential research directions are briefly explained below.

- **Industrial Autonomous Systems:** To have less human interventions and hence more accurate machine operations a future research area is to focus on the possibility of employing IIoT sensors and devices to create industrial autonomous systems in smart factories.
- **Cloud/Fog/Edge Manufacturing:** Due to the remarkable advancement in computing technologies at all levels of the cloud, fog, and edge, the latter can be used to support manufacturing processes, which will result in highly effective and productive operations. Also, future research can focus on how to use the data generated from the cloud to customize the needs of both the organization and the client.
- **Machine Learning and Artificial Intelligence:** Integrating IIoT technology with machine learning and AI can provide smart factories with better intelligent and autonomous decisions in operations and management of the manufacturing industry. Hence future research can focus on developing machine learning algorithms that have the ability to learn from the data generated from IIoT sensors to optimize factory processes.
- **Security and Privacy:** With the introduction of more IIoT devices in the manufacturing system, the level of security and privacy of data become crucial concerns. Therefore, novel methods need to be explored to secure data generated by IIoT sensors and industrial networks as well. Future research can also focus on creating secure and privacy-preserving IIoT architectures for smart factories in Industry 4.0.

The development of smart manufacturing systems in Industry 4.0 requires the fundamental component of IIoT devices, which yields optimal production processes, efficiency in energy consumption, and minimal environmental pollution. Revolutionizing the industrial sector in Industry 4.0 is possible by employing smart IoT sensors and machine learning techniques to reduce the cost of production and automatically adjust items when inventory levels fall below a specific threshold.

Moreover, IIoT technology can be used to monitor environmental factors such as temperature, humidity, and air quality so that a safe environment is present. Overall, it can be concluded that IIoT deployment can modernize Industry 4.0 and provide ways for factories to operate. With IIoT-enabled devices, the entire manufacturing system can collect, analyze, and optimize real-time data using machine learning algorithms to improve business efficiency and performance.

REFERENCES

1. M. Soori, B. Arezoo, and R. Dastres. Internet of Things for smart factories in Industry 4.0, a review. *Internet Things Cyber-Phys. Syst.*, 3, 192–204 (2023). KeAi Publishing, ISSN 2667-3452, https://doi.org/10.1016/j.iotcps.2023.04.006
2. K. C. Rath, A. Khang, and D. Roy. The role of Internet of Things (IoT) technology in Industry 4.0 economy. In: A. Khang, V. Abdullayev, V. Hahanov, and V. Shah, (eds.), *Advanced IoT Technologies and Applications in the Industry 4.0 Digital Economy.* CRC Press (2024). pp. 1–28.
3. D. Serpanos. Industrial Internet of Things: Trends and challenges. *Computer*, 57(1), 124–128 (2024).
4. W. Caesarendra, B. K. Pappachan, T. Wijaya, D. Lee, T. Tjahjowidodo, D. Then, and O. M. Manyar. An AWS machine learning-based indirect monitoring method for deburring in aerospace industries towards Industry 4.0. *Appl. Sci.*, 8, 2165 (2018). https://doi.org/10.3390/app8112165
5. T. Alladi, V. Chamola, R. M. Parizi, and K.-K. R. Choo. Blockchain applications for Industry 4.0 and industrial IoT: A review. *IEEE Access*, 7, 176935–176951 (2019). https://doi.org/10.1109/ACCESS.2019.2956748
6. C. Esposito, A. De Santis, G. Tortora, H. Chang, and K.-K. R. Choo. Blockchain: A panacea for healthcare cloud-based data security and privacy? *IEEE Cloud Comput.*, 5(1), 31–37 (2018).
7. H.-T. Wu, and C.-W. Tsai. Toward blockchains for health-care systems: Applying the bilinear pairing technology to ensure privacy protection and accuracy in data sharing. *IEEE Consum. Electron. Mag.*, 7(4), 65–71 (2018).
8. M. A. Salahuddin, A. Al-Fuqaha, M. Guizani, K. Shuaib, and F. Sallabi. Softwarization of Internet of Things infrastructure for secure and smart healthcare. *Computer*, 50(7), 74–79 (2017).
9. T. Nugent, D. Upton, and M. Cimpoesu, Improving data transparency in clinical trials using blockchain smart contracts. *F1000Res*, 5, 2541 (2016). https://doi.org/10.12688/f1000research.9756.1
10. Q. I. Xia, E. B. Sifah, K. O. Asamoah, J. Gao, X. Du, and M. Guizani. MeDShare: Trust-less medical data sharing among cloud service providers via blockchain. *IEEE Access*, 5, 14757–14767 (2017).
11. F. Ghovanlooy Ghajar, A. Sikora, and D. Welte. Schloss: Blockchain-based system architecture for secure industrial IoT. *MDPI J. Electron.*, 11, 1629 (2022). https://doi.org/10.3390/electronics11101629
12. J. Zhang, N. Xue, and X. Huang. A secure system for pervasive social network-based healthcare. *IEEE Access*, 4, 9239–9250 (2016).
13. B. Alangot, and K. Achuthan. Trace and track: Enhanced pharma supply chain infrastructure to prevent fraud. In *Proceedings of the International Conference Ubiquitous Communications and Network Computing*, Bangalore, India. Cham, Switzerland: Springer (Aug. 2017). pp. 189–195.

14. R. Lakshmana Kumar, F. Khan, S. Kadry, and S. Rho. A survey on blockchain for industrial Internet of Things. *Alexandria Eng. J.*, 61(8), 6001–6022 (2022). ISSN 1110-0168, https://doi.org/10.1016/j.aej.2021.11.023
15. Azaria, Asaph, Ariel Ekblaw, Thiago Vieira, and Andrew Lippman. "MedRec: Using Blockchain for Medical Data Access and Permission Management." *IEEE Second International Conference on Open and Big Data (OBD)*, pp. 25–30, 2016, Vienna, Austria
16. I. Bisio, C. Garibotto, A. Grattarola, F. Lavagetto, and A. Sciarrone. Exploiting context-aware capabilities over the Internet of Things for Industry 4.0 applications. *IEEE Netw.*, 32, 101–107 (2018).
17. B. Elahi, and S. A. Tokaldany. Application of Internet of Things-aided simulation and digital twin technology in smart manufacturing. In: M. Ram, (ed.), *Advances in Mathematics for Industry 4.0.* Elsevier (2021). pp. 335–359.
18. E. Nica, and V. Stehel. Internet of Things sensing networks, artificial intelligence-based decision-making algorithms, and real-time process monitoring in sustainable Industry 4.0. *J. Self Govern. Manag. Econ.*, 9, 35–47 (2021).
19. Y. Mashayekhy, A. Babaei, X.-M. Yuan, and A. Xue. Impact of Internet of Things (IoT) on inventory management: A literature survey. *Logistics*, 6, 33(2022).
20. P. K. Malik, R. Sharma, R. Singh, A. Gehlot, S. C. Satapathy, W. S. Alnumay, D. Pelusi, U. Ghosh, and J. Nayak. Industrial Internet of Things and its applications in Industry 4.0: State of the art. *Comput. Commun.*, 166, 125–139 (2021).
21. N. Mostafa, W. Hamdy, and H. Elawady. An intelligent warehouse management system using the Internet of Things. *Egypt Int. J. Eng. Sci. Technol.*, 32, 59–65 (2020).
22. N. H. M. Zaidin, M. N. M. Diah, and S. Sorooshian. Quality management in Industry 4.0 era. *J. Manag. Sci.*, 8, 182–191 (2018).
23. J. L. Chong, K. W. Chew, A. P. Peter, H. Y. Ting, and P. L. Show. Internet of Things (IoT)-based environmental monitoring and control system for home-based mushroom cultivation. *Biosensors*, 13, 98 (2023).
24. K. Mohammed, M. Abdelhafid, K. Kamal, N. Ismail, and A. Ilias. Intelligent driver monitoring system: An Internet of Things-based system for tracking and identifying the driving behavior. *Comput. Stand. Interfac.*, 84, 103704 (2023).
25. A. Kumar, and A. Nayyar. (eds.). 3-Industry: A sustainable, intelligent, innovative, Internet-of-Things industry. In: *A Roadmap to Industry 4.0: Smart Production. Sharp Business and Sustainable Development.* Springer (2020). pp. 1–21.
26. M. Elsisi, M.-Q. Tran, K. Mahmoud, M. Lehtonen, and M. M. Darwish. Deep learning-based Industry 4.0 and Internet of Things towards effective energy management for smart buildings. *Sensors*, 21, 1038 (2021).
27. K.M. Al-Obaidi, M. Hossain, N. A. M. Alduais, H. S. Al-Duais, H. Omrany, and A. Ghaffarianhoseini. A review of using IoT for energy efficient buildings and cities: A built environment perspective. *Energies*, 15, 5991 (2022). https://doi.org/10.3390/en15165991
28. M. Noor-A-Rahim, F. Firyaguna, J. John, M. Omar Khyam, D. Pesch, E. Armstrong, H. Claussen, and H. Vincent Poor. Toward Industry 5.0: Intelligent reflecting surface in smart manufacturing. *IEEE Commun. Mag.*, 60(10), 72–78 (2022). https://doi.org/10.1109/MCOM.001.2200016

4 Robotics and Automation in Industry 4.0

Emerging Trends and Research Directions

Mahendra Gooroochurn

4.1 INTRODUCTION

Industry 4.0 has been a landmark framework in the industrial revolution (IR) pathway, which saw the convergence of several emerging technologies such as Internet of Things (IoT), AI, 5G, and cloud computing for enhancing the ability to optimize processes through better productivity, which has been a key driver of each version of IR. Although the term Industry 5.0 has been coined to rightly incorporate the human dimension into the equation, I4.0 remains a key objective for developing as well as developed economies, where the use of data-driven decision-making has allowed closing the loop in understanding the dynamics of processes without the need for complex system modelling. This has been made possible through the use of AI and machine learning algorithms where the control rules are actually the outcome of the system design process instead of a needed input.

In aiming to achieve better productivity, a prominent hallmark of the various phases of the IR has been the increasing use of automation, through technological advancements, recurrent in I4.0, for example, through breakthroughs in wireless sensor networks and cloud computing. With machine learning, it has now been possible to automate the decision-making process itself through the fourth IR. A key technological solution has been the field of robotics as an automation solution and the closely associated area of robot vision and tactile sensing. Robotics was actually the prominent component of I3.0 but remains a key technology for I4.0 as well given its huge potential to improve productivity in several sectors. Moreover, the integration of automated robotic systems with emerging technologies such as sensor networks has led to the development of a whole range of manipulator and mobile robotic systems able to operate autonomously or semi-autonomously, including collaborative systems able to work safely with humans. The latter, typically referred to as cobots, has opened several new avenues for man-machine interoperability, which would have been considered impossible in the I3.0 framework.

Indeed Fathi et al. [1] reviewed the integration of humans and robots on assembly lines as a growing trend for I4.0 to combine robot and human operator skills. One of their findings was that this combination is multi-objective, with a reduction in cycle

DOI: 10.1201/9781003511298-4

time being the most popular objective and meta-heuristic as the dominant approach. The ability to develop exact solutions was found to be challenging due to the computational load involved. The use of parallel and distributed computing systems was reported to be a potential avenue for improving computational efficiency. The authors remarked that the growing trend for cobot assembly configuration is correlated with the push for I4.0. The human dimension of cobot integration in assembly lines from the perspective of human welfare, extending the sphere of I4.0 to I5.0, is a crucial area attracting interest in the use of robots in work environments shared by human operators [2]. Zheng et al. [3] studied this complex interaction involved in human–robot collaboration (HRC) in unstructured manufacturing environments, proposing a vision-based and deep reinforcement learning-enabled paradigm.

Robotics has supported the evolution of smart manufacturing, with a whole range of advanced lean and flexible manufacturing possibilities taking advantage of improvements in associated technology domains such as AI algorithms, vision, sensor, and connectivity, optimized further through data-driven frameworks. Another area of manufacturing which has benefitted from robotics is additive manufacturing, where wastage can be reduced to a minimum, and the ability to employ fast and accurate robotic manipulators for the layered approach brings added flexibility and reduced cycle time. Along the same line, construction robotics, making use of 3D printing, has attracted attention in the construction industry, where Liu et al. [4] reported that the availability of accurate 3D building information through Building Information Modelling (BIM) models has been instrumental in paving the way for robotics to be applied in this major global sector. The development of BIM to better suit robotic implementation is a research direction identified by the authors to streamline the production process.

The combination of innovative sensing technologies in the manufacturing sector has brought a whole range of opportunities for technological improvement and the adoption of machine learning solutions. Guo et al. [5] presented a state-of-the-art review of sensing technologies in robotic welding applications which is widespread across several automated production lines. The challenges and opportunities cropping up in this area linked to advances in computer technology, composite sensors, machine learning, and multi-robot work cell configurations are highlighted, which clearly showcases the increasing complex and multidisciplinary/transdisciplinary dimension the use of robotic technology entails, warranting the need for the development of cross-cutting analysis and simulation methods for effective optimization underpinning I4.0. Wahidi et al. [6] describe the contribution of robotic welding in the maritime industry to achieve better marine structure and optimize processes as well as adhere to standards, operate in confined and dangerous locations, and enable innovative construction methods.

The adoption of robotics and automation in general brings along the usual question and concern about the loss of jobs and redundancy for the human workforce. Therefore, this has warranted researchers to look at this social aspect. Stemmer [7] studied the effect of robotization in the context of Brazil as an emerging economy. It was concluded that the impact is sector dependent (mining and manufacturing in this case) as well as gender-related, with an increase noted for female workers together with highly skilled workers. Bachmann et al. [8] reported the cost of labour as a significant factor in understanding the effect of robotization on employment in 16 European countries, with important cross-country differences. Sun et al. [9] found a heterogeneous relationship of robots in employment across sectors of a given country and country-wise. Liang et al. [10] studied the question of ethics for the use of AI and robotics in the broad areas of architecture, engineering, and construction (AEC), looking at the issues of "job loss, data privacy, data security, data transparency, decision-making conflict, acceptance and trust, reliability and safety, fear of surveillance, and liability", which are crucial areas to look into for a holistic and sound adoption of such I4.0 solutions. The authors highlight current gaps in understanding these impacts on our society, including the need for more stakeholder awareness of these key challenges.

Wang et al. [11] investigated robot adoption at the firm level and reported a positive influence on employment with increased productivity for the highly skilled and educated segment. Furthermore, they report little evidence of the loss of employment and an interesting phenomenon of intra-industry labour re-allocation to robot-adopting firms from those resisting robotic technology. Xin and Ye [12] produced interesting results for policymaking to ensure equitable economic growth by showing that robotization has a beneficial effect on labour conditions, especially for low-skilled and agricultural workers but has a potential negative impact on rising income inequality. The authors recommend calibrated policies through bespoke research to capture the dynamics of a given country so that the benefits of improved productivity and working conditions can be targeted while ensuring shared prosperity.

The ability and readiness of robotic systems to work seamlessly with sensor networks is a key advantage of its adoption to promote the application of I4.0 techniques. Indeed Diouf et al. [13] showcase the application of control algorithms, including neural network-based, data-driven control techniques. Luo et al. [14] reviewed the evolving field of soft robots as an innovative robotic design paradigm bringing greater safety through their elastic structure. However, the authors opined that the soft mechanical structure is part of the feature of this type of robot, and the actual benefits can be reaped by integrating soft sensors with the construction of such systems, achieving a humanlike sensory organ. Indeed, this type of robot can open further avenues for human–machine interaction as well as in minimally invasive surgeries where flexibility and the ability to record response from the interaction of human tissue can enhance interventions. The design of such soft robots poses several challenges with the sensor design itself as well as large data processing and appropriate energy sources.

The evolution and development of new robotic system configurations to fit a more diverse range of applications are inspired on the one side by the emerging concept of nature-based solutions in the form of biomimicry and motivated on the other side by the need to bring forth innovative technological solutions to improve productivity. Zhou et al. [15] presented aerial robotic systems in terms of their morphology, biomimicry, locomotion, and manipulator attachment. The potential of aerial robots in the military, construction industry, and the transport sector is highlighted. The design challenges for smaller size and more endurable systems, the ability to deal with complex real-world scenarios, and the integration of sub-systems in a mechatronic paradigm are future directions for the emergence of this area.

The need to improve food security and increase productivity in the agricultural sector has logically made this sector a candidate for I4.0, where the shortage of labour and the high potential to automate tasks using intelligent systems with robotic articulation have been key considerations. The integration of technology in agriculture is also a powerful lever to attract the new generation to enter this economic sector, as traditional, labour-intensive approaches have tended to discourage our youth from choosing agriculture as a career. Zhou et al. [15] examined the engineering of farming robots for chicken production. A detailed, task-based robotic system architecture is presented, where the major challenge reported is the lack of customization to the farm environment, where application and integration of standard market solutions do not necessarily yield effective solutions.

The betterment of medical caregiving through the complementarity of human and robot skills is a living example of technological advancement for social good. The reduction in post-surgery recovery time through minimally invasive surgery and reduction in fatigue and physiological tremor by assisting doctors in a wide range of medical interventions are well documented in the literature. Indeed, the general positive outcomes reported for the use of robotics in medical interventions have seemingly steered the attention of patients towards surgical robotics as reported by Brinkman et al. [16]. Using Google Trend analysis tools to explore the affinity for the use of robotics in arthroplasty, a linear increase was reported for robotic hip arthroplasty and an exponential increase for robotic knee arthroplasty from January 2011 to December 2021. Li et al. [17] reported an increase in research work in robotic applications for Spline Medicine using a bibliometric analysis approach. Zeng and Wang [18] showed the clear benefits of robotics in liver surgery, juxtaposing the limitations of the traditional, laparoscopic-based, non-robotic approach. A similar case was made by Erozkan and Gorgun [19] for robot-assisted colorectal surgery. Wu et al. [20] reviewed the use of machine learning and flexible surgical and interventional robots in minimally invasive procedures, showing the enabling effect machine learning has provided to solve the complex non-linear control challenges caused by the interaction of the robotic end-effector with the operative site for perception, modelling, control, and navigation purposes.

The ability of robotic systems to reduce the invasiveness of surgical operations and in so doing reduce the time of recovery and level of infection as well as provide an unparalleled degree of precision when combined with imaging techniques has been key performance metrics in this shift in consumer expectation and demand. As reported by Pinci et al. [21], people's perception of robot-assisted surgery, for cases of robot-assisted total joint arthroplasty, favours the use of robot-assisted treatment for a better outcome and faster recovery. A similar result was reported by Stower et al. [22] in general for social robots where it was found that children aged 3–6 years preferred interaction with a robot over a human agent for selective trust tasks, showing the acceptability of properly designed and performing robots, especially those with AI capabilities to achieve natural cognitive functions. In the service sector, Wu et al. [23] reported fairer customer perception of AI robots compared to human employees. Viswakarma et al. [24] reported the use of social robots mainly in the travel, tourism, hotel, and hospitality industries.

The resemblance of certain types of robotic systems to the human anatomy, either partially or wholly, has been a key attribute to the acceptance or deterrence of its acceptance by our society and has crucially been the driver behind the development of rehabilitation robots to restore limb functionality. Starting with its basic form as a passive limb prosthesis to provide some level of limb functionality to the patient, the field of rehabilitation robotics has evolved to restore as closely as possible the lost limb features with the integration of advanced sensors, multi-modal sensor fusion, and machine learning algorithms for sensor data processing. The advancement of medical robotics in the area of rehabilitation to enhance motor recovery through brain–computer interface allows bypassing of impaired neurological networks, which can be coupled to robotic-assisted systems for limb functions [25]. With the increase in the incidence of limb amputation cases cropping from diabetes cases and limb disability from cases of stroke and accidents, rehabilitation robotic devices have opened opportunities for body limb recovery [26].

Environmental degradation and pollution have been at the forefront of concerns with our current modern economic growth model, where our natural capita are relegated as a sustainability pillar in the environment-equity-economics nexus. The contribution of robotic systems to support the reduction in pollution of the environment from industrial processes is attracting the interest of researchers and can be a powerful driver in the adoption of robotics towards the adoption of an I4.0 paradigm. He et al. [27] studied the impact of imported robotic systems on the pollution emission of the Chinese manufacturing sector and came up with conclusive findings that the adoption of robotic technology drastically reduced the sulphur dioxide emission intensity through technology progress, transition to clean energy, and end-of-pipe treatment improvement. The beneficial characteristics of robotics as one facilitating data capture and providing flexibility, accuracy, and precision are hence further enhanced by additional plus points for environmental stewardship.

The built environment is a serious cause of environmental degradation, with the continued urbanization of greenfields and the associated need for natural resources to convert to building stocks and, with it, the increased resources needed to operate and maintain these man-made structures leading to a "locking effect" [28]. Dodampegama et al. [29] specifically review the issue of construction and demolition waste, reporting the increasing complexity and volume of material streams requiring automation solutions. To deal with these challenges, the need for multisensory fusion, unsupervised machine learning, and the design and development of intelligent robotic systems to learn and adapt on an ongoing basis to new material waste streams are highlighted to ensure continued sustainable management of waste.

The integration of robotics from a data-driven perspective for driving circular economy adoption, e.g., from a sustainable management of waste perspective but equally in other areas for enhancing productivity and precision supports the contribution of the key attributes provided by robotic systems. Along this line, Wang [30] used an analytic hierarchy process (AHP) approach to evaluate the effectiveness of robot technology (categorized as industrial and service robots) for pedigree analysis, followed by intelligent management and control of urban design at the mesoscale level. The results showed that the use of robotics to this end can bring positive outcomes. Moreover, the emerging field of multiple robot systems (MRS) for architectural design is leading the way for innovative construction automation systems [31], to which the diverse members of a construction project including the architects, quantity surveyors, and Mechanical, Electrical and Plumbing (MEP) engineers have to become conversant with to stay abreast of technological advancement in this major sector of the world economy.

Tian et al. [32] produced an extensive review of the use of different types of robots for structural health monitoring of bridges, applicable to health monitoring for infrastructures in general. With I4.0 relying extensively on the use of data for increasing productivity, transposed to predictive maintenance and avoidance of structure failure in this case, as well as providing valuable insights on design choices for new projects, the study provides useful guidelines into the robotic system architecture to use for a given structural health monitoring application.

With the dramatic consequences of climate change impacts on our ecosystems, it has become important to monitor their health to take corrective action. Robotic systems offer an effective solution, especially in inaccessible and difficult environments. One such area is in monitoring the health of coral reefs, which unfortunately is impaired by the rising level of acidity and temperature of our oceans. Cardenas et al. [33] reviewed the application of robotic systems for coral reef monitoring, with the integration of AI and machine learning enabling coral reef conservation and restoration. Another area with challenging operating environments where robotics had a defining influence is in aerospace research where both mobile robots and manipulators have been used in various contexts. One of the critical roles in such operations is grasping, especially in low-gravity realms. Jahanshahi and Zhu [34] reviewed this crucial control function of robots for grasping control in space applications. The authors report the important contribution of artificial intelligence, namely deep learning, reinforcement learning, convolutional neural network, and recurrent neural network to achieve autonomous, efficient, and sophisticated manoeuvres as an improvement over traditional mechanical grippers. The incorporation of sensing and AI allowed using advanced systems powered by novel algorithms to achieve unprecedented precision and adaptability.

The combination of emerging technologies such as cyber-physical systems, IoT, robotics, big data, and cloud computing is described by Mazumber et al. [35] as a key driver for I4.0, and one outcome of this fusion is the growth of digital twin (DT), with DT in robotics highlighted as a recent innovation with high potential, needing a new framework to tap into the significant benefits this paradigm can bring. Vlădăreanu et al. [36] illustrate this high multidisciplinarity involved in the design of state-of-the-art robotic systems through a DT for a high-frequency hardening robot. Furthermore, the unique capabilities offered by IoT technology have been exploited to come up with the Internet of Robotic Things (IoRT). Sandhu et al. [37] review the growing development of robotic systems interfaced using IoT for a range of applications targeted at providing assistance for daily, critical, and emergency tasks for the elderly. With the increase in the world's elderly population as well as the number of persons living with a disability, there is an expected rise in IoRT systems, shifting from the current offline mode to more real-time, online functionalities.

Robot process automation has a counterpart in the automation of business and organizational processes through the development of software packages to streamline routine daily tasks but also feeding into the generation of strategic data and information for high-level management decision-making. The archiving of large volumes of data makes these underlying processes key candidates for applying AI and machine learning tools. Ribeiro et al. [38] underline the key disciplines of research robot process automation as natural language processing, text mining, and deep learning techniques. Following the main highlights of robotics integration within an I4.0 framework provided above, this chapter presents the key attributes of robotics and I4.0, which make them suitable for each other. The major areas where robotics and automation have been applied are discussed, with a focus on highlighting the emerging trends and how they support an I4.0 transformation. The associated hardware and software solutions reported are also reviewed, with the aim of synthesizing the research directions anticipated for the application of robotic systems in all its forms and modes of operations. The challenges encountered in the adoption of robotic systems by developing countries are briefly discussed, with possible strategic and policy decisions that can help to overcome them.

4.1.1 IRs in a Nutshell

IR has occurred over several years, backed by technological advancement to improve the productivity (and profitability) of industrial processes. The first IR stemmed from the use of steam as the driving force to run turbines and pistons and in so doing became part of a large number of industrial processes, with the use of other mechanical assemblies such as gears, belts, and pulley systems, cascading down to the consumer level through steam cars. The first IR was a clear indication of the high potential mechanical systems (later modernized as mechatronics through the integration of electrical/electronics, coding, and control) held in assisting workers in manual work as well as opening opportunities for sectors of the economy which were deemed impossible or inaccessible. The development of heavy and earth-moving equipment is one such example that enabled to envisage more complex infrastructural projects within less time. However, the main flaw of IR has been the lack of consideration of environmental impacts, including air and water pollution, depletion of natural resources, and an overall degradation of natural ecosystems. Natural phenomena such as climate change and biodiversity loss have been the sad consequences of the IR, as opposed to the technological innovation and modernization that could be achieved.

The degradation of natural ecosystems happened through the use of fossil fuels partly, powered by the second IR through the discovery of electricity, which proved to be a more efficient way to run processes compared to steam. The exponential increase in vehicle usage and factories has been drivers for greenhouse gas emissions, although regulations have been stipulated to curtail emissions, leading to important technical innovations to improve fuel efficiency and emissions levels, but the sheer rise of transportation in all its forms has meant that greenhouse gas share from this sector has been on the rise. The ease of distributing electricity to the population at large meant that there has been a growing demand for electrical power to feed not only industrial but a larger share of residential end uses. Electrification became an indicator of economic growth, creating an unsustainable coupling between urbanization and environmental degradation. The recognition that the mechanization and electrification of processes using fossil sources are not sustainable has led to efforts to reduce the harm, one of them being through control and automation to modulate the operation of machines. This happened around the late 1960s with the development of power electronics using feedback sensor sensors and control algorithms for position and speed control, marking the third IR, which also consisted of the widespread adoption of automated systems, including robots.

The developments emanating from the field of artificial intelligence in the pursuit to understand and mimic human behaviour have been rich in offering a plethora of options to investigate and control the operation of machines, as well as in understanding the dynamics of processes. These two important elements of any process environment, and the ability to deploy sensors to collect data for real-time and offline processing, led to research findings that have had landmark improvements in the design, implementation, and operation of processes. With sustainable management of resources in mind, even the ultimate disposal or bringing back the products and materials into the loop through the industrial or biological cycles is a design consideration, as effectively illustrated by the butterfly diagram of Figure 4.1. This ability to use data to optimize processes using the rich set of AI and machine learning tools and techniques has led to the fourth IR, opening up possibilities to slow, if not stop or even reverse, the degradation of our natural capital.

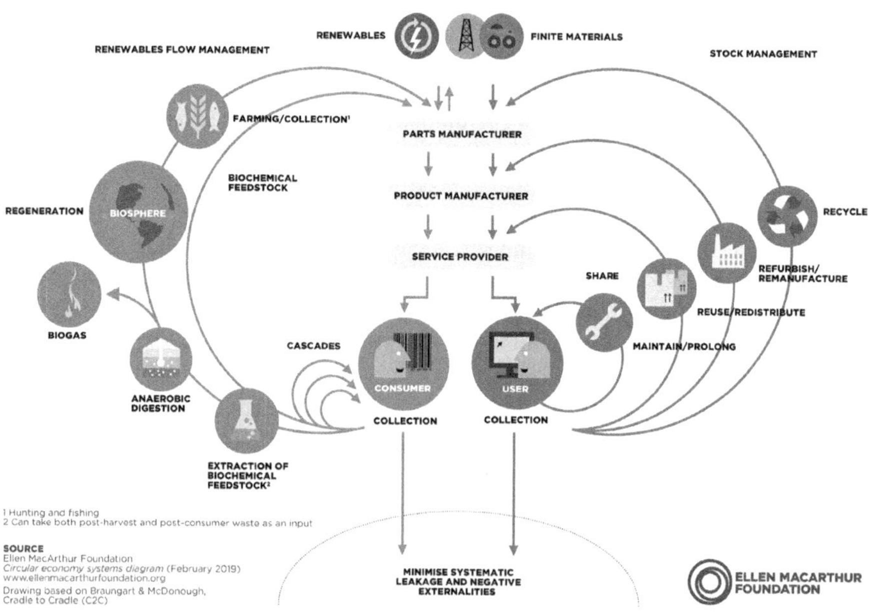

FIGURE 4.1 Technical and biological cycles of a circular economy [39].

4.1.2 Industry 4.0 and Artificial Intelligence

The area of applied research has benefitted hugely from the outcomes emanating from the field of artificial intelligence, offering tools and techniques such as artificial neural networks (multilater perception, convolution neural network, and recurrent neural networks among others), fuzzy logic, evolutionary computation, Bayesian inference, and principal component analysis. The application of these methods in supporting technical domains including computer/machine vision, big data analysis, voice recognition, and natural language processing has opened up exponentially the scope for AI applications in healthcare, finance, entertainment, agriculture, security, and finance, to name a few. The application of AI in devices and processes has been so widespread that we can even talk of its ubiquity in our modern world.

When it comes to industry processes, machine learning is used both for decision-making, for example how best to optimize production by setting the work cell configuration in a flexible manufacturing environment, and as part of the design process of systems, for example in trajectory planning and obstacle avoidance of robotic systems. With business and organizational structures becoming more and more complex, there is a need to manage the associated voluminous amount of data, and one that AI strives on, in complementarity to humans.

4.2 ROBOTICS AND INDUSTRY 4.0

Engineering research and industry practice have transitioned beyond the sphere of just the siloed traditional engineering disciplines given the unequivocal need to engage in transdisciplinary areas, and robotics is such an example. Indeed, robotics can be perceived as overwhelming given the growing number of disciplines involved. If the design, construction, and programming of robotic systems can be captured within the realm of mechatronics as an engineering practice seeking holistic design thinking, the growing influence of robotics in non-engineering applications and the need to capture social and environmental considerations as observed in the literature review section above has pushed the boundaries beyond the conventional engineering disciplines. As observed by Garg et al. [40], the task of choosing a robotic system for a given application has become a complex one given the range of advanced specifications on the market. With the adaptation of robotic systems for specific applications, the systems tend to become specialized, hence creating a niche area of practice. The following sub-sections present the main types and configurations of robotic systems and the unique advantages they offer, especially when used in an I4.0 context. The challenges to be addressed and the areas of applications are detailed in the next sections, concluding with a discussion on the emerging trends for robotics and automated systems to implement I4.0 principles.

4.2.1 Robotic System Types

The design of robotic systems is of paramount importance as the end result places constraints as well as defines the capabilities of the system, in terms of sensing modalities and achievable degrees of freedom as far as the hardware is concerned. On the contrary, the software design used determines the manoeuvres and decision-making patterns executable. As a real or virtual system, the robotic system needs to operate in a given environment, and the ability to learn and refine its decision-making pattern depends on whether an agent-oriented paradigm was followed, with the possibility to integrate expert knowledge, or whether a conventional object-oriented paradigm has been adopted to clearly define the rules the system has to follow. Clearly, an agent-oriented design is the one enabling an I4.0 workflow; hence, it is imperative that such an approach is adopted. A robotic system may have advanced hardware technologies to allow elaborate sensing of various types of signals (e.g. visual, sound, and tactile), but without the design of suitable software routines to integrate the sensor responses and properly control the system actuators, the outcome is not effective or optimal. This primordial need to integrate all the aspects of the design of a robotic system is aligned directly with the field of mechatronics, looking for synergy among the mechanical, electrical/electronic, software, and control engineering found at the base of all modern systems, including robotic and automated systems.

Robotic systems can be broadly classified as industrial robots, mobile robots, and, with recent trends for design to allow human–machine co-sharing a work environment, collaborative robots (also known as cobots). From the third IR, industrial robots are widespread and are used in the manufacturing sector to automate production lines. They are highly accurate and precise, can operate for long hours, and can perform a range of tasks from welding, painting, and assembly operations, The end-effector design and calibration are vital for achieving accurate and repetitive patterns of movement. Mobile robots are designed to move across the factory floor or a warehouse to assist in material handling, inventory control, and material movement, for which they need to be equipped with appropriate obstacle detection and avoidance functions. Mobile robots have been extensively used in different spheres apart from the industrial environment, and the taxonomies provided by Taheri and Mozayani [41] for quadruped robots and Tian et al. [32] are described briefly to present the main types of mobile robots. The quadruped robots presented in literature offer similar anatomy to animals, with better capability to navigate on real terrains compared to wheel robots. Taheri and Mozayani [41] presented different leg configurations, namely parallel and parallel-serial combined legs as well as series-joint, parallel joint, and link-distal leg designs, based on which a multitude of mobile robotic quadruped systems have been proposed over the years. Augmenting the capability to not only be restricted to the terrestrial environment, Tian et al. [32], while reviewing the use of robots for conditioning monitoring of bridges, provide a broader review of mobile robotic system types, including ground mobile robots, wall climbing robots, cable climbing robots, and flying drones. The use of underwater and overwater navigation mobile robotic systems has also been the subject of research and development for monitoring and intervention in the challenging and otherwise inaccessible realm of oceans.

Finally, collaborative robots are systems allowing the simultaneous presence of an active human operator and a robot in the same work environment, to derive the benefit of the complementarity of their skills. For this to be possible, the safe operation of the robot is vital, which departs from the way industrial robots operate where the work environment needs to be locked and ascertained to be free from any human.

Cobots are therefore equipped with sensors (visual and presence sensors) that continuously monitor the position of the human operator(s) in the workspace and actively prevent any possible collision, and the design of soft joints with compliant materials reduces the possibility and severity of injuries.

4.2.2 Advantages of Robotics for Industry 4.0

Although initially designed to assist humans in carrying out strenuous and repetitive tasks, robotic and automated systems have transcended into not just being mere assistants but have played the role of being enablers for certain applications. Given their ability to undertake accurate and precise movements in a flexible and adaptive manner, while operating for long hours even in challenging environments, robots have taken a stronghold in our modern lifestyle, and their influence can only increase as we aim for higher productivity, innovative solutions across several sectors and sustainable solutions to pressing environmental problems where both interventional and monitoring systems are needed. The ready integration of robotic systems with clean energy sources such as solar-generated electricity makes them compatible with sustainability frameworks for industrial and residential systems. Moreover, they are well poised for localized or wide connectivity given their architecture allowing integration of sensors for local use by the robot or accessible on a communication channel. Hence robots are well aligned to the goals and technical requirements of I4.0, which is about making the use of data to refine and optimize processes. Robots can operate as a monitoring system to collect data for a given environment or system, e.g., measuring seawater quality by exploring across a pre-defined set of geographical coordinates or air quality by measuring particulate matter concentration across a given floor of a building or using real-time data to optimize the process it is controlling. The advantages of robotic and automated systems in general can be categorized as follows:

4.2.2.1 Improved Quality and Productivity

Through their accurate and precise speed and position control systems, all mathematically modelled through forward and reverse kinematics, robotic manipulators can be programmed to perform repetitive tasks with very high levels of integrity and compliance. This allows high levels of quality and with the ability to run the programmed cycles at fast rates when tested and proven to be satisfactory, the productivity is greatly enhanced. The added benefit of being able to run robots tirelessly for long hours of operation means repetitive stress injuries can be reduced for workers and given that order times can be accurately pre-determined, this contributes to the valuable customer satisfaction through business process scheduling and planning.

4.2.2.2 Improved Working Conditions

The choice of suitable materials to protect the robot structure from the ambient conditions in which it has to operate means a robot can be designed to work in harsh conditions, including carrying strenuous tasks in dusty, high temperature, low temperature, underwater, terrestrial, aerial, and high radioactivity conditions. The engineering of the robot assembly therefore relies on a proper understanding of the operating environment to specify proper materials and IP protection of the electrical and electronic components. By engaging robots in this way, the working conditions of the employees are improved as they work in conjunction with the robot, either in a collaborative or separated mode, where remote control segregates the human operators from the actual work environment. A few examples of robotic systems which improved the working conditions of the staff are vehicle exhaust measurement using a robotic arm without exposure of the operator to the emissions and the use of tele-operated robotic manipulators in nuclear power plants; the extreme conditions presented by other planets in our solar system has been overcome by the use of mobile robots, equipped with sensors to collect and relay back information to earth stations.

4.2.2.3 Data Collection

A robotic system is designed with several sensors that are used to control the linkages and motoring of joints for accurate positioning in space and time. The external sensors connected to the robot in the form of tactile sensing and cameras provide data with respect to the end-effector as the robot interacts with its environment, and the robot work cell can be integrated with further sensors to gather information on the conditions of the operating environment, e.g. ultrasound and infrared sensors for collision detection. The advent of Industry 4.0, with IoT playing the crucial role of deploying wireless sensor networks, and connecting the robot itself to the Internet, has led to the coining of the term IoRT. Therefore robotics, being a system traditionally focused on achieving performance by the use of internal and external sensors, is well tuned to be a suitable candidate for automating processes aiming for I4.0. Mobile robots are also commonly used for condition monitoring of structures and environments, through which AI-based predictive maintenance can be implemented.

4.2.2.4 Flexibility

A robotic system can be likened to a base system with the necessary end-effector and tools, the requisite degrees of freedom, and sensors to complete an array of tasks. This is achieved by programming and re-programming the system through a variety of approaches, e.g., lead through and teach pendant programming modes. In a manufacturing context, this flexibility to adapt its operation is employed in reconfigurable manufacturing systems (RMS), giving the ability to cope with market and product trends dynamically, allowing bringing products to the market with minimum lead times. This has been a major trend taken by manufacturing sectors towards I4.0 in developed countries like the UK, the USA, Japan, and Germany.

4.2.2.5 Enhanced Safety

The operation of an industrial robot ensures safe operation by the use of guards and interlocks to ensure the system operates at full speed and load only when it is ascertained that there is no possibility of collisions with humans. The need for sharing the work environment between the human operator and the robot cropped up for applications which needed their skill set simultaneously to improve productivity and reduce cycle time. This was made possible through the design of soft robots, with the use of materials such as rubber and gels and sensors to detect the user position in real time and adapt the position and movement of the robot accordingly. Safety is a prime goal in the application of robotics, and the evolution of collaborative robots will further consolidate safety standards with the use of robots through an added sensor, including visual cues, and modelling and simulation of the interaction between the user and the robot.

4.2.2.6 Reduced Cost

Ironically, the use of robotics has been perceived to be cost intensive, which is true of the Capital Expenditure (CAPEX) involved, but their ability to work over long hours, the flexibility to be programmed for a range of tasks, and the achieved high and repeatable quality of products provide favourable Operational Expenditure (OPEX) values compared to alternatives. Typically, investment in robotic systems can pay back in less than a year, hence making them an attractive investment. Their compatibility with I4.0 with little or no additional electronics to extract the needed data for applying machine learning algorithms provides the added benefit of implementing data-driven decision-making, including AI-based predictive maintenance.

4.2.3 Challenges of Robotics Integration for Industry 4.0 Transformation

Although the development path of industrialized nations which have gone through the various IRs has seen the widespread use of robotic systems, continually upgraded using the latest technological advances, it remains a challenge for developing countries to follow suit, and this sub-section looks at the challenges and barriers that need to be overcome to reap the benefits of robotic systems as a step towards I4.0 across several sectors. The COVID-19 pandemic has shed useful insights into the resilience of each nation, and the development of robotic solutions to isolate patients and practise social distancing has made good contribution. Clearly, the adoption of robotics across sectors like manufacturing and services can boost productivity whereas their integration in traditional sectors such as agriculture and food processing can improve food security.

An important barrier for developing countries is the need to import these systems at considerably higher costs, especially with the freight charges skyrocketing in the aftermath of the pandemic and maintaining high levels since. Coupled with this, a lack of local or regional suppliers of robot systems leads to a logical lack of local expertise in maintaining and troubleshooting problems with the robots, needing the intervention of professionals from abroad, which is a non-negligible cost element to factor in the equation. The highly multidisciplinary nature of robotic systems means that the capacity building of practising engineers or fresh graduates needs to be brought about through specialized vocational and university programmes.

As noted from the literature findings, the adoption of robotic systems requires a good understanding of the local economy, and the influence is heterogeneous across sectors and countries. This places uncertainties on the national policy-makers to develop strategies for I4.0 based on robotics. Still, clear trends have been reported in the literature for successful adoption and positive influence on working conditions and employment in certain sectors and levels of skill. Therefore, it is crucial that governments work in collaboration with universities and research institutes to develop a better understanding of the local sectors and their affinity and leverage points for robotic system integration. The adoption of a state-of-the-art automated system can be used as a powerful lever to encourage our youth to take to traditional sectors such as farming and agriculture, crucial for developing local capacity in essential areas. With the challenge currently being faced by Mauritius to retain its talented workforce with migration, investment in technological solutions such as AI-based robotic systems can lead to the desired effect of attracting our young generation to nurture a career locally and contribute to the economic growth of the country.

4.3 APPLICATION AREAS FOR ROBOTICS

This section describes the main sectors where robotics has been applied and where there is growing evidence reported in the literature for the integration of digital technologies. The goal for technological advancement in these application areas is to target enhanced performance of existing systems or to offer solutions to problems where robotic systems can offer a better outcome compared to current practice, or even for which no solution currently exists. As covered in the section earlier where the advantages of robotics were discussed, the unique ability of robots to work in challenging, if not inaccessible environments, as well as compatibility to work with a whole range of emerging technologies such as IoT, machine learning, cloud computing, and multi-modal sensors, makes them excellent candidates for working in an I4.0 framework. The strong body of knowledge documenting the design, testing, and improvement of various types of robots, able to operate in aerial, terrestrial, and marine environments, is yet another strong point supporting robotic technology. The advent of I4.0 has opened opportunities to further equip robots with added capabilities as presented below.

Javaid et al. [42] discussed the extensive application of robotics in various sectors of the world economy. The following main areas were reported, with the use of AI tools to improve performance:

- Manufacturing industry, by adopting smart manufacturing approaches with reconfigurable systems for a wide range of processes. This has been based on the technological advancement of Computerised Numerical Control (CNC) technologies and recently complemented with additive manufacturing methods. The adoption of AI-based robotic systems further supports key paradigms such as lean and just-in-time manufacturing ideologies that have been the backbone of a competitive and productive manufacturing sector.
- Agriculture, which can benefit immensely from mechanization in its basic form to remove the strenuous tasks from the farmers and further extended through an I4.0 paradigm to allow the farmers to focus more on the throughput, facilitated with the use of sensor networks to collect data on variables affecting yield, and using same to improve production. These knowledge bases on best practices for a given country across the various climate zones have been captured in mobile apps for use by farmers to understand the dynamics of crop yield over a year;
- Smart homes, forming part of the built environment, where we spend the majority of our time, and unfortunately, have been a major driver behind environmental degradation and lack of resilience to climate change due to

the disruption of natural ecosystems. As proposed by Gooroochurn [43], a passive design approach coupled with mechatronic systems can be considered to decouple the state of the interior conditions from the ambient external conditions by intelligently controlling the influence of the very factors impoverishing indoor conditions at the source. In its simplest form, shading devices (see Figure 4.2) can be modulated to control heat gains at source at the interface of the building envelope with the external environment, and more advanced systems can be considered to make use of natural resources available on site to regulate the indoor conditions such as the solar chimney prototype for fostering indoor air quality by air exchanges between the inside and outside (see Figure 4.3) proposed by Gooroochurn [44] for the tropical context of Mauritius. The solar chimney can be controlled by opening suitable louvres to shift between the heating and cooling modes by using a fuzzy logic system based on external and indoor conditions.

- Healthcare industry, which has been a rich area of research for robotics in conjunction with several imaging modalities such as MRI, CT, and X-ray to provide robotic-assisted devices for performing a wide range of surgical procedures, including in neurosurgery, orthopaedics, and spinal interventions. Robotics has also played a vital role in the rehabilitation of patients with impaired functions in the aftermath of traumatic situations. For example, a robot-assisted system has been applied for the rehabilitation of cerebral palsy patients to retrain their brains to develop prior motor skills by allowing the patient to implement the normal walking gait through robotic assistance. Furthermore, robotic systems have been used in cases of limb loss to provide back some level of limb function, and by interfacing with a range of sensors collecting biological signatures such as heart rate and muscle action, automated robotic prosthesis pave to way to returning as much as possible of the full functionality of the lost limb. Overall, robots have been at the forefront of medical treatment innovation for minimally invasive surgery, reduced recovery time, lower infection, and less morbidity.

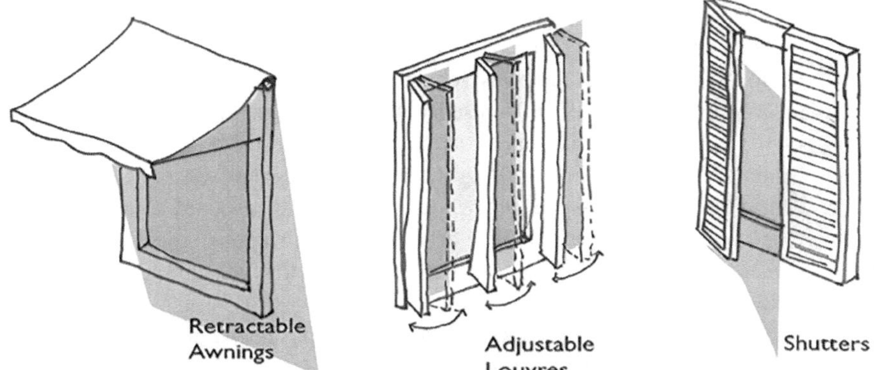

FIGURE 4.2 External shading devices for low elevation sun control using machine learning.

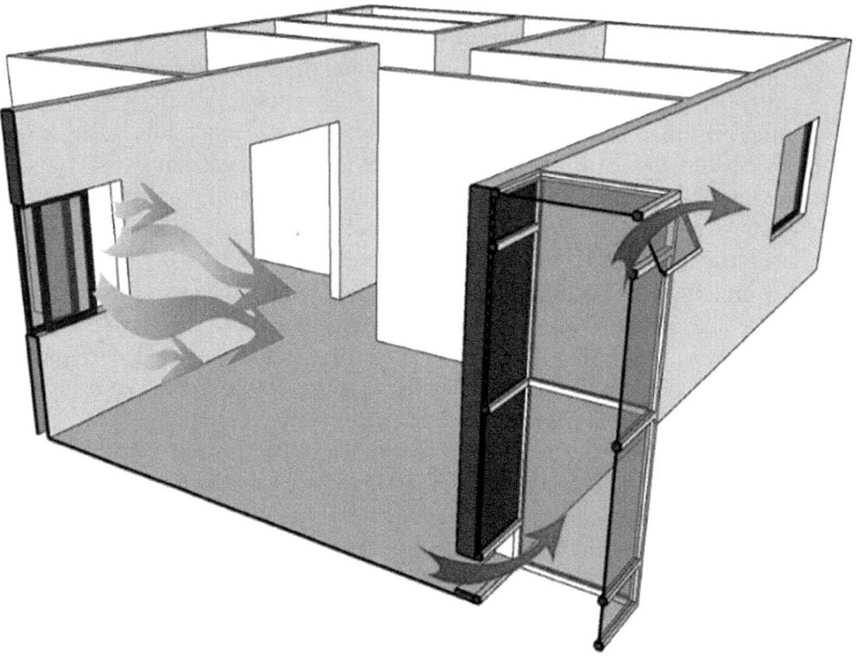

FIGURE 4.3 Aluminium frame and glass-based solar chimney prototype with various operating modes using machine learning.

- The automotive industry, which has been one of the seminal sectors investing in robotic technology to automate production lines, led the way to several innovations in safety standards for robot use and end-effector design. Now, this industry is aiming for autonomous vehicles, with significant progress made so far to arrive at working prototypes, albeit testing in controlled environments. This field has been enabled by AI algorithms in conjunction with 5G connectivity and cloud computing to ensure minimal latency and massive coverage of sensors, vital for continuous mapping of the immediate environment. The autonomous vehicle can be likened to a mobile robotic system, integrated with built-in sensors for controlling the actuation of several systems such as fuel injection and ignition and the Anti-Lock Braking System (ABS) anti-lock braking system, and external sensors to provide continuous feed on happenings so that the vehicle on-board computer can take proper, safe decisions. The crucial issue of responsibility in cases of malfunction or accidents needs to be resolved before we can see automated vehicles in our midst.
- The logistics and warehousing industry, which has been boosted heavily by e-commerce platforms such as Amazon and eBay, has invested heavily in robotic technology. Given the huge volume of retail transactions, the planning for logistics for orders and inventory control cannot be done without

robot process automation to ensure efficient delivery for customer satisfaction. Moreover, the need for rapid and correct retrieval and storage of items has been achieved by having exclusively fully automated robotic systems, which can operate in highly congested areas, and vertically expanded to maximize storage capacity.

4.4 EMERGING TRENDS AND RESEARCH DIRECTIONS IN ROBOTICS FOR I4.0

Based on the wide literature review presented in the earlier part of this chapter, setting the background for the ensuing discussion of the various facets of the use of robotic and automated systems in an I4.0 framework for monitoring, control, and decision-making ends, it is clear that developing countries have a lot to gain on the socio-economic and environmental stewardship fronts. The published literature showcased clear research directions, firstly in providing adaptive robot programming tools to align closer to the precepts of I4.0 as well as the development of more inclusive and safer environments and collaborative robots, where the skill set of both protagonists can be reaped to a maximum, as opposed to forcing them to work in siloed configurations. In so doing, it has become essential to develop frameworks to model and better understand this relationship between man and machines and how they can co-habit and co-create in the same environment by consolidating the strengths of each.

The need for accurate modelling was also reported in the use of robotics for minimally invasive surgery, where the non-linear behaviour of robots and the complex nature of the interaction between the robot's end-effector and the human tissue being operated on present highly difficult challenges. The operation of robotic systems in such complex and unstructured environments was also observed in manufacturing set-ups for HRC systems where AI techniques can be used to better understand the process, and for which reinforcement learning and vision cues are reported plausible research directions to develop solutions. Coming back to the medical field, the healthcare sector continues to attract research into the development of robotic systems for surgical and rehabilitation purposes, and research to advance and measure the efficacy of these robot-assisted procedures is anticipated as well as the use of biometric signals and operative environment data to improve the efficacy of medical interventions.

From the social viewpoint, ethical issues with the use of AI and robotics have been reported pertaining to job redundancy, data privacy and security, transparency, conflict in decision-making, acceptance, and trust among other aspects. Although there are several studies showing the acceptance of robotic systems for service delivery, there is a need for this type of research in different contexts for this robot–human interaction. The research directions reported to promote the vital aspect of ethics in I4.0 are cybersecurity, explainable AI, safety with cobot, fairness in decision-making and AI and robotics standards, and a responsibility framework. An area expected to further push the limits of soft robots is tactile sensing, with the goal of mimicking as much as possible the functioning of the human skin. The ability to integrate these sensors in the structure of soft robots and power them is further research directions in this area.

REFERENCES

1. Masood Fathi, Arash Sepehri, Morteza Ghobakhloo, Mohammad Iranmanesh, Ming-Lang Tsenge, Balancing assembly lines with industrial and collaborative robots: Current trends and future research directions, *Computers & Industrial Engineering*, Volume 193, 2024, 110254, ISSN 0360-8352, https://doi.org/10.1016/j.cie.2024.110254
2. Matteo Capponi, Riccardo Gervasi, Luca Mastrogiacomo, Fiorenzo Franceschini, Assembly complexity and physiological response in human-robot collaboration: Insights from a preliminary experimental analysis, *Robotics and Computer-Integrated Manufacturing*, Volume 89, 2024, 102789, ISSN 0736-5845, https://doi.org/10.1016/j.rcim.2024.102789
3. Pai Zheng, Chengxi Li, Junming Fan, Lihui Wang, A vision-language-guided and deep reinforcement learning-enabled approach for unstructured human-robot collaborative manufacturing task fulfilment, *CIRP Annals*, Volume 73, 2024, 341–344, ISSN 0007-8506, https://doi.org/10.1016/j.cirp.2024.04.003
4. Yuming Liu, Aidi Hizami Bin Alias, Nuzul Azam Haron, Nabilah Abu Bakar, Hao Wang, Exploring three pillars of construction robotics via dual-track quantitative analysis, *Automation in Construction*, Volume 162, 2024, 105391, ISSN 0926-5805, https://doi.org/10.1016/j.autcon.2024.105391
5. Qiang Guo, Zi Yang, Jinting Xu, Yan Jiang, Wenbo Wang, Zonglin Liu, Weisen Zhao, Yuwen Sun, Progress, challenges and trends on vision sensing technologies in automatic/intelligent robotic welding: State-of-the-art review, *Robotics and Computer-Integrated Manufacturing*, Volume 89, 2024, 102767, ISSN 0736-5845, https://doi.org/10.1016/j.rcim.2024.102767
6. Sufian Imam Wahidi, Selda Oterkus, Erkan Oterkus, Robotic welding techniques in marine structures and production processes: A systematic literature review, *Marine Structures*, Volume 95, 2024, 103608, ISSN 0951-8339, https://doi.org/10.1016/j.marstruc.2024.103608
7. Henry Stemmler, Automated deindustrialization: How global robotization affects emerging economies—Evidence from Brazil, *World Development*, Volume 171, 2023, 106349, ISSN 0305-750X, https://doi.org/10.1016/j.worlddev.2023.106349
8. Ronald Bachmann, Myrielle Gonschor, Piotr Lewandowski, Karol Madoń, The impact of robots on labour market transitions in Europe, *Structural Change and Economic Dynamics*, Volume 70, 2024, 422–441, ISSN 0954-349X, https://doi.org/10.1016/j.strueco.2024.05.005
9. Wenyuan Sun, Zhonghui Zhang, Yang Chen, Fushu Luan, Heterogeneous effects of robots on employment in agriculture, industry, and services sectors, *Technology in Society*, Volume 75, 2023, 102371, ISSN 0160-791X, https://doi.org/10.1016/j.techsoc.2023.102371
10. Ci-Jyun Liang, Thai-Hoa Le, Youngjib Ham, Bharadwaj R.K. Mantha, Marvin H. Cheng, Jacob J. Lin, Ethics of artificial intelligence and robotics in the architecture, engineering, and construction industry, *Automation in Construction*, Volume 162, 2024, 105369, ISSN 0926-5805, https://doi.org/10.1016/j.autcon.2024.105369
11. Ting Wang, Yi Zhang, Chun Liu, Robot adoption and employment adjustment: Firm-level evidence from China, *China Economic Review*, Volume 84, 2024, 102137, ISSN 1043-951X, https://doi.org/10.1016/j.chieco.2024.102137
12. Baogui Xin, Xiaopu Ye, Robotics applications, inclusive employment and income disparity, *Technology in Society*, Volume 78, 2024, 102621, ISSN 0160-791X, https://doi.org/10.1016/j.techsoc.2024.102621
13. Aminata Diouf, Bruno Belzile, Maarouf Saad, David St-Onge, Spherical rolling robots—Design, modeling, and control: A systematic literature review, *Robotics and Autonomous Systems*, Volume 175, 2024, 104657, ISSN 0921-8890, https://doi.org/10.1016/j.robot.2024.104657

14. Zhongbao Luo, Weiqi Cheng, Tianyu Zhao, Nan Xiang, Intelligent sensory systems toward soft robotics, *Applied Materials Today*, Volume 37, 2024, 102122, ISSN 2352-9407, https://doi.org/10.1016/j.apmt.2024.102122
15. Xidong Zhou, Hang Zhong, Hui Zhang, Wei He, Hean Hua, Yaonan Wang, Current status, challenges, and prospects for new types of aerial robots, *Engineering*, 2024, ISSN 2095-8099, https://doi.org/10.1016/j.eng.2024.05.008
16. Joseph C. Brinkman, Zachary K. Christopher, M. Lane Moore, Jordan R. Pollock, Jack M. Haglin, Joshua S. Bingham, Patient interest in robotic total joint arthroplasty is exponential: A 10-year Google trends analysis, *Arthroplasty Today*, Volume 15, 2022, 13–18, ISSN 2352-3441, https://doi.org/10.1016/j.artd.2022.02.015
17. Wei-Shang Li, Qi Yan, Wen-Ting Chen, Gao-Yu Li, Lin Cong, Global research trends in robotic applications in spinal medicine: A systematic bibliometric analysis, *World Neurosurgery*, Volume 155, 2021, e778-e785, ISSN 1878-8750, https://doi.org/10.1016/j.wneu.2021.08.139
18. Qingjie Zeng, Jin Wang, Global scientific production of robotic liver resection from 2003 to 2022: A bibliometric analysis, *Laparoscopic, Endoscopic and Robotic Surgery*, Volume 6, Issue 1, 2023, 16–23, ISSN 2468-9009, https://doi.org/10.1016/j.lers.2023.02.002
19. Kamil Erozkan, Emre Gorgun, Robotic colorectal surgery and future directions, *The American Journal of Surgery*, Volume 230, 2024, 91–98, ISSN 0002-9610, https://doi.org/10.1016/j.amjsurg.2023.10.046
20. Di Wu, Renchi Zhang, Ameya Pore, Diego Dall'Alba, Xuan Thao Ha, Zhen Li, Yao Zhang, Fernando Herrera, Mouloud Ourak, Wojtek Kowalczyk, Elena De Momi, Alícia Casals, Jenny Dankelman, Jens Kober, Arianna Menciassi, Paolo Fiorini, Emmanuel Vander Poorten, A review on machine learning in flexible surgical and interventional robots: Where we are and where we are going, *Biomedical Signal Processing and Control*, Volume 93, 2024, 106179, ISSN 1746-8094, https://doi.org/10.1016/j.bspc.2024.106179
21. Marcantonio V. Pinci, Norberto J. Torres-Lugo, David E. Deliz-Jimenez, Joseph Salem-Hernandez, Alexandra Claudio-Marcano, Norman Ramírez, Antonio Otero-López, Patient perception of robotic-assisted total joint arthroplasty in a hispanic population, *Arthroplasty Today*, Volume 25, 2024, 101286, ISSN 2352-3441, https://doi.org/10.1016/j.artd.2023.101286
22. Rebecca Stower, Arvid Kappas, Kristyn Sommer, When is it right for a robot to be wrong? Children trust a robot over a human in a selective trust task, *Computers in Human Behavior*, Volume 157, 2024, 108229, ISSN 0747-5632, https://doi.org/10.1016/j.chb.2024.108229
23. Xiaolong Wu, Shuhua Li, Yonglin Guo, Shujie Fang, Human or AI robot? Who is fairer on the service organizational frontline, *Journal of Business Research*, Volume 181, 2024, 114730, ISSN 0148-2963, https://doi.org/10.1016/j.jbusres.2024.114730
24. Laxmi Pandit Vishwakarma, Rajesh Kr Singh, Ruchi Mishra, Denizhan Demirkol, Tugrul Daim, The adoption of social robots in service operations: A comprehensive review, *Technology in Society*, Volume 76, 2024, 102441, ISSN 0160-791X, https://doi.org/10.1016/j.techsoc.2023.102441
25. Neethu Robinson, Ravikiran Mane, Tushar Chouhan, Cuntai Guan, Emerging trends in BCI-robotics for motor control and rehabilitation, *Current Opinion in Biomedical Engineering*, Volume 20, 2021, 100354, ISSN 2468-4511, https://doi.org/10.1016/j.cobme.2021.100354
26. Saad M. Sarhan, Mohammed Z. Al-Faiz, Ayad M. Takhakh, A review on EMG/EEG based control scheme of upper limb rehabilitation robots for stroke patients, *Heliyon*, Volume 9, Issue 8, 2023, e18308, ISSN 2405-8440, https://doi.org/10.1016/j.heliyon.2023.e18308

27. Xiaogang He, Ruifeng Teng, Dawei Feng, Jiahui Gai, Industrial robots and pollution: Evidence from Chinese enterprises, *Economic Analysis and Policy*, Volume 82, 2024, 629–650, ISSN 0313-5926, https://doi.org/10.1016/j.eap.2024.03.001
28. Simron J. Singh, Allison Elgie, Dominik Noll, Matthew J. Eckelman, The challenge of solid waste on Small Islands: Proposing a socio-metabolic research (SMR) framework, *Current Opinion in Environmental Sustainability*, Volume 62, 2023, 101274, ISSN 1877-3435, https://doi.org/10.1016/j.cosust.2023.101274
29. Shanuka Dodampegama, Lei Hou, Ehsan Asadi, Guomin Zhang, Sujeeva Setunge, Revolutionizing construction and demolition waste sorting: Insights from artificial intelligence and robotic applications, *Resources, Conservation and Recycling*, Volume 202, 2024, 107375, ISSN 0921-3449, https://doi.org/10.1016/j.resconrec.2023.107375
30. Haipeng Wang, Application of new features based on artificial intelligent robot technology in medium-scale urban design pedigree and intelligent management and control, *Intelligent Systems with Applications*, Volume 22, 2024, 200379, ISSN 2667-3053, https://doi.org/10.1016/j.iswa.2024.200379
31. Samuel Leder, Achim Menges, Architectural design in collective robotic construction, *Automation in Construction*, Volume 156, 2023, 105082, ISSN 0926-5805, https://doi.org/10.1016/j.autcon.2023.105082
32. Yongding Tian, Chao Chen, Kwesi Sagoe-Crentsil, Jian Zhang, Wenhui Duan, Intelligent robotic systems for structural health monitoring: Applications and future trends, *Automation in Construction*, Volume 139, 2022, 104273, ISSN 0926-5805, https://doi.org/10.1016/j.autcon.2022.104273
33. Jennifer A. Cardenas, Zahra Samadikhoshkho, Ateeq Ur Rehman, Alexander U. Valle-Pérez, Elena Herrera-Ponce de León, Charlotte A.E. Hauser, Eric M. Feron, Rafiq Ahmad, A systematic review of robotic efficacy in coral reef monitoring techniques, *Marine Pollution Bulletin*, Volume 202, 2024, 116273, ISSN 0025-326X, https://doi.org/10.1016/j.marpolbul.2024.116273
34. Hadi Jahanshahi, Zheng H. Zhu, Review of machine learning in robotic grasping control in space application, *Acta Astronautica*, Volume 220, 2024, 37–61, ISSN 0094-5765, https://doi.org/10.1016/j.actaastro.2024.04.012
35. Antar Mazumder, Md. Fahad Sahed, Zinat Tasneem, Prangon Das, Faisal R. Badal, Md. Firoj Ali, Md. Hafiz Ahamed, Sarafat Hussian Abhi, Subrata Kumar Sarker, Sajal Kumar Das, Md. Mehedi Hasan, Manirul Islam, Md. Robiul Islam, Towards next generation digital twin in robotics: Trends, scopes, challenges, and future, *Heliyon*, Volume 9, Issue 2, 2023, e13359, ISSN 2405-8440, https://doi.org/10.1016/j.heliyon.2023.e13359
36. Luige Vlădăreanu, Alexandru I. Gal, Octavian D. Melinte, Victor Vlădăreanu, Mihaiela Iliescu, Adrian Bruja, Yongfei Feng, Alexandra Ciocîrlan. Robot digital twin towards Industry 4.0, *IFAC-PapersOnLine*, Volume 53, Issue 2, 2020, 10867–10872, ISSN 2405-8963, https://doi.org/10.1016/j.ifacol.2020.12.2815
37. Moid Sandhu, David Silvera-Tawil, Paulo Borges, Qing Zhang, Brano Kusy, Internet of robotic things for independent living: Critical analysis and future directions, *Internet of Things*, Volume 25, 2024, 101120, ISSN 2542-6605, https://doi.org/10.1016/j.iot.2024.101120
38. Jorge Ribeiro, Rui Lima, Tiago Eckhardt, Sara Paiva. Robotic process automation and artificial intelligence in Industry 4.0 – A literature review, *Procedia Computer Science*, Volume 181, 2021, 51–58, ISSN 1877-0509, https://doi.org/10.1016/j.procs.2021.01.104
39. Ellen McArthur Foundation, https://www.ellenmacarthurfoundation.org/circular-economy-diagram The latest access date is: 18 November 2024.

40. Chandra Prakash Garg, Ömer F. Görçün, Pradip Kundu, Hande Küçükönder. An integrated fuzzy MCDM approach based on Bonferroni functions for selection and evaluation of industrial robots for the automobile manufacturing industry, *Expert Systems with Applications*, Volume 213, Part A, 2023, 118863, ISSN 0957-4174, https://doi.org/10.1016/j.eswa.2022.118863
41. Hamid Taheri, Nasser Mozayani, A study on quadruped mobile robots, *Mechanism and Machine Theory*, Volume 190, 2023, 105448, ISSN 0094-114X, https://doi.org/10.1016/j.mechmachtheory.2023.105448
42. Mohd Javaid, Abid Haleem, Ravi Pratap Singh, Rajiv Suman. Substantial capabilities of robotics in enhancing Industry 4.0 implementation, *Cognitive Robotics*, Volume 1, 2021, 58–75, ISSN 2667-2413, https://doi.org/10.1016/j.cogr.2021.06.001
43. Mahendra Gooroochurn, Adaptive passive measures for tropical climates – A case study for Mauritius. In: *Removing Barriers to Environmental Comfort in the Global South*. Laura Marín-Restrepo, Alexis Pérez-Fargallo, María Beatriz Piderit-Moreno, Maureen Paulina Wegertseder-Martínez (Eds.), 117–131. Springer Nature Switzerland AG, 2022, Published May 2023.
44. Mahendra Gooroochurn, A hybrid glass-based solar chimney to promote cross-ventilation and night flushing. *2022 International Conference on Electrical, Computer, Communications and Mechatronics Engineering (ICECCME)*, 16–18 November 2022, Maldives National University, IEEE Xplore, https://ieeexplore.ieee.org/document/9988027

5 A Paradigm Shift towards Sustainable Production with Additive Manufacturing in the Industry 4.0 Era

Raviduth Ramful

5.1 INTRODUCTION

In this day and age, advanced research, alternative methods and industrial improvements are required to tackle the radical societal and technical changes faced in the field of modern manufacturing. The manufacturing world is at its turning point in time, and new solutions are required to meet the ever-rising demand in terms of both the quality and quantity of new products. The challenges which are impacting global trade, namely climate change, rising freight costs and increased material costs, have a resounding effect on manufacturing industries. To curb this problem and keep pace with new product development and innovation, new alternative approaches are required to address the foregoing challenges. In addition, increased labor cost and tool cost in material removal processes are further driving the transition from traditional manufacturing processes toward automated and cost-effective processes.

With the new generation of consumers favoring customized products at affordable prices, the conventional method of product development and manufacturing has to be revisited and refined. The surge in products' customization to meet specific customer needs is forecasted to rise in the coming years [1]. Advanced manufacturing technology is regarded as a potential solution to address the increased competitiveness observed among companies involved in the production of customized consumer goods [1]. Moreover, the optimization of production processes and cost reductions on supply chain (SC) inventories is enhancing the value chain business model, thereby leading to added competitiveness and increased profit margins [2]. The improved and concise value chain models are localized, and their collaborative strategy further promotes sustainability [3].

One proven method to boost productivity and increase the competitiveness of modern enterprises is via high-end automation technologies. The fast-evolving technological advancements in modern manufacturing such as increased connectivity, advanced robot-assisted technology and automated monitoring capabilities have

DOI: 10.1201/9781003511298-5

revolutionized the conventional manufacturing world. Besides no economic benefits, this transformation has greatly enhanced production lines and optimized resource utilization while promoting sustainability. The attainment of sustainable targets via carbon offsetting is key for many industries due to increased rigorous policies imposed by governments worldwide. One technology which can make a significant impact from both the production and environmental perspective is additive manufacturing (AM) technology, also commonly known as 3D-printing [4,5].

5.2 ADDITIVE MANUFACTURING

AM is a computer-driven process which utilizes a three-dimensional system of coordinates to generate 3D models from Computer Aided Design (CAD) files in a layer-by-layer approach. The customization and design possibilities through this advanced technique of manufacturing are immense. The capabilities of present-day matured AM technologies to generate 3D models from a diverse range of materials such as polymeric, metallic and ceramic among others has extended its use in numerous applications ranging from aerospace to medical science [6]. In contrast to conventional means, AM also provides the capabilities to enhance the durability and life cycles of 3D printed components [2]. The AM process has the capability to make complex parts in batch quantities as well as a single part [7]. Among its numerous benefits, the ability for rapid prototyping through AM is giving an edge to designers.

The rapid translation from ideas generated on drawing boards to real-life tangible-object is transforming the design world. The key steps required to generate a product from a bare idea via the AM process are minimal. The main steps from sketching to 3D-printing process are outlined in Figure 5.1. From sketch to 3D-print, the process consists of CAD file generation, conversion to STL file and G-code generation. The CAD files are normally designed in commercially available CAD software such as AutoCAD, CATIA or Fusion 360. Prior to the 3D-printing stage, there is a requirement to render the CAD file into a format that can be interpreted by the 3D-printer. For that purpose, slicer software is generally used even though some latest CAD software offers an additional toolbar for rendering the CAD file for 3D-printing. Examples of slicer software include Cura, 3D Slicer and Simplify3D among others.

Most slicing software is compatible with specific file formats, namely 3MF, STL, STEP and OBJ. In the slicer software, the compatible CAD files are transformed into a G-code format by considering the user specification and 3D-printer settings. The user specifications in terms of quality of print, infill percentage and amount of raw material consumed can be directly defined while the 3D-printer settings are bounded by the hardware capabilities. Some of the hardware capabilities include the print

FIGURE 5.1 Main steps in the AM pre-processing stage from sketch to 3D-printing process.

volume, which is predominantly affected by the size of the 3D-printer, the extruder nozzle specification and the build-plate surface.

5.2.1 Types of AM Processes

In recent decades, the popularization of AM technologies has led to the breakthrough development and innovative creation of distinctive variants of AM equipment applicable to a wide variety of engineering applications. Some of these specific technologies, namely vat photo polymerization, binder jetting process, material jetting, powder bed fusion, material extrusion, sheet lamination and directed energy deposition, among others, are further enabling AM processes to penetrate the global manufacturing scene, which is largely dominated by conventional manufacturing operations [8]. The main types of materials used in these AM processes are generally in the form of liquid, filament/paste, powder and solid sheet [9]. Some of the commonly used AM technologies are displayed in Figure 5.2. The processes which are illustrated on the right-hand side of Figure 5.2c, namely powder bed fusion and direct energy deposition, are normally applicable to metal additive manufacturing (MAM).

5.2.2 The Benefits and Needs of AM in the 21st Century

From an economic viewpoint, the advent of the emerging field of AM can be advantageous in numerous ways in contrast to the conventional way of operation. The economic advantages provided by AM are diverse ranging from reduced material waste and energy consumption to shortened time-to-market and just-in-time production [1]. In contrast to assembly lines which mostly favor the mass production of identical products, one of the standing out features of AM is their capacity to produce customized products. The AM process is causing a paradigm shift in conventional production practice by offering new manufacturing possibilities. These manufacturing possibilities provide mutual benefits to both the producers and consumers alike. Examples of some of the benefits provided by AM are reduced lead time, faster introduction of new designs to the market and meeting customers' demands more efficiently [10].

AM is classified as a Near Net Shape (NNS) manufacturing process. In NNS manufacturing, the aim of the initial fabrication process of a component is to achieve the nearest form of the finished product in terms of both shape and size.

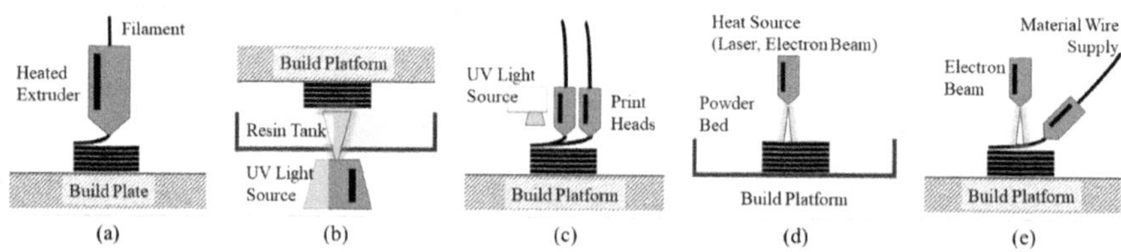

FIGURE 5.2 Commonly used AM technologies, namely: (a) fused deposition modelling (FDM), (b) vat photo polymerization, (c) material jetting, (d) powder bed fusion and (e) directed energy deposition (DED).

NNS manufacturing is highly regarded in the industry given its noteworthy benefits and is generally associated with lean manufacturing methods. With NNS, the cost of production is greatly reduced as the production time of downstream processes, namely finishing and additional machining, is minimized. In addition, the reduction in the utilization of resources in terms of raw materials and energy requirements further improves the cost of production. Moreover, being an NNS-type manufacturing method, AM can be considered for the manufacturing of high-value materials such as high-grade titanium as one of its core objectives via lean manufacturing is to reduce waste [11].

5.2.3 The Adoption of AM Technologies and Their Current Applications

The reasons for adopting AM technologies in industries in the 21st century are diverse. The responses obtained from a survey study conducted with adopters of AM technologies around the world showed that the driving reasons were mainly associated with the ease of new product development, innovation, rapid prototyping, supporting complex design creations, customization time and cost saving among other factors [12]. Moreover, the advent and industry-wide implementation of advanced technologies in Industry 4.0, such as artificial intelligence (AI), cloud computing and big data analytics among others, are prompting the consideration for compatible and sophisticated manufacturing technologies along with specialized workforce [4]. AM technology was considered in industries in the early 1990s, and specific job creation in these industries comprised process engineers, application engineers, rapid prototyping engineers, AM technicians and R&D staff among others [2]. Some of the ground-breaking applications of 3D-printing technology have been achieved in emerging fields of engineering, such as aerospace engineering, automotive engineering, manufacturing and processing and healthcare [2].

5.2.3.1 Aerospace Industry

Another field whereby high-end components are manufactured by 3D-printing technology is the aerospace industry. In recent years, full-scale rocket enclosures have been manufactured using AM technology [5]. The implementation of AM technologies in the aerospace industry can be impactful and sustainable given the sector's requirement for high needs and low production scale [2,13]. Further optimization in the design and manufacturing of high-performing and material-efficient parts can be by and large advantageous to the aerospace industry [10,14].

In the aerospace industry, the cost of manufacturing critical parts such as landing gear, engine mounts and other aircraft structures can be overly high due to the high buy-to-fly ratio. The high buy-to-fly ratios are normally encountered when a large billet of material is required due to unique part geometry and dimension which need to be machined in one piece. This process generally leads to high material wastage and high tool cost and can be time consuming in contrast to NNS manufacturing. The implementation of AM can address some of these issues [15,16].

The future of AM in the aeronautical sector is promising as it can specifically address issues in the field of maintenance [6]. The establishment of future regulations and certification of AM metal parts is forecasted to solve logistical problems faced in the maintenance of aircraft. For instance, extensive logistic items in terms of spare parts could be replaced by AM machines and powder magazines [6]. Moreover, the in-situ production of parts is also postulated to shorten the SC provided that aircraft manufacturers have the willingness to decentralize their production and disposition to share the digital models of parts [2,6].

5.2.3.2 Automotive Industry

In the automotive industry, the goal is to optimize vehicle production by considering lightweight structures while striving harder to produce greener and more energy-efficient vehicles to limit their carbon footprint. AM processes provide alternative solutions to address the aforementioned challenges in the automotive industry [10]. AM has the capability to reproduce high-quality and complex parts for the automotive sector [2]. These parts, which can be further optimized through CAD software by considering a generative design toolbox, are structurally viable and lightweight, thereby minimizing the resources and cost required for manufacturing.

5.2.3.3 Architecture

Moreover, in architecture, AM is being used by architects to translate building designs and project ideas into tangible models for ease of understanding and perception prior to the realization of full-fledged architecture [17]. The conventional way of building architectural concepts, which entails attention to details and accurate representations, can be a tedious and time-consuming process. AM is transforming the traditional way of creating architectural models by bypassing the need for tool and mold making, thereby drastically reducing manufacturing time [10,17,18].

5.2.3.4 Medical Field

In the medical field, for instance, 3D-printed components are being used in numerous areas ranging from orthopedics to dentistry. One of the notable benefits of implementing AM technology in medical applications is the possibility and ease of producing customized parts [10]. In addition, customized medical devices, such as orthotic and prosthetic devices in the orthopedic field, are required to cater to specific individual needs. AM technology is also reshaping the field of dentistry, which is a branch of medicine whereby customized products are required to suit the needs of individual patients. Examples of customized products in dentistry, which can be manufactured by more efficient means via AM, are elaborate dental crowns, bridges and orthodontic braces [19,20].

5.2.4 Tendency in AM Technology and Industry Changes

5.2.4.1 Research & Development in AM Technology

With the surge in interest from manufacturers and consumers alike, the development of AM technology in the past decades has seen a drastic evolution. Innovation is one of the key factors behind the remarkable ascension observed in the AM field. In recent years, developers have been focusing on key attributes which are rendering AM devices more usable and efficient [5]. Some of the high-end attributes of AM, namely automation, ease of use and reliability are fundamental to enable the 3D-printing technology to attain the high standards sustained by conventional methods of manufacturing in the modern world. Future innovation in 3D-printing technology is focused on meeting the global demand of large-scale production and reduced lead time by rendering the process more efficient and affordable [5].

5.2.4.2 Production Time with Current AM Technology

The production time in the field of AM remains a popular subject of discussion despite AM technology being much more efficient and speedier in its operation in comparison to standard manufacturing methods. For instance, a typical 3D-print can still take a lengthy time in terms of hours or even days to be fully manufactured. This further complicates the capacity of such 3D-printers for large-scale commercialization. Innovation in the present day and age is enabling faster print time in modern 3D-printers. Some of the explored solutions are the inclusion of multiple print heads and the ability to print more than one material type simultaneously [5]. Further research is being conducted to explore the possibility of using larger nozzles for faster polymer deposition and high-speed motors for faster movement of the printer heads [5].

5.2.4.3 Autonomous Capabilities of AM Technology

The full autonomous capacity of modern 3D-printers remains one of the most engrossing factors in luring future investors and consumers. Fully autonomous 3D-printers have enormous potential to substitute labor-intensive manufacturing operations while providing the prospects of delivering products to a standard with high consistency [5]. Automation in 3D-printing technology is forecasted to drive businesses and productivity at both ends of the spectrum. Many of the current 3D-printers, mainly designed for the mass market, still require considerable human supervision to ensure accurate and seamless printing operations.

The practical and economical aspects of fully autonomous 3D-printers will be more appealing to the masses including to Small Medium Enterprises. For instance, smart features such as auto-leveling or intelligent fault diagnosis, which are fitted in modern 3D-printers specifically designed for the mass market, are further propelling the interest and growth in the AM sector [5]. The aim of implementing AM in conventional industrial operations is to ensure accurate production with high-quality standards by shifting toward a fully autonomous production system and by minimizing human intervention.

5.2.5 Limitations to be Addressed to Scale up the AM Industry

In industrial or high-end engineering applications, where metallic components are predominantly used, there is a demand for innovative MAM processes to address some of the challenges faced in the current manufacturing processes. Among the numerous factors affecting the manufacturing world such as scarcity of labor, overhead cost and increasing raw material cost, efficiency is of the key essence. Fully automated processes such as the MAM process or even hybrid manufacturing are potential solutions which could be considered to address some of the current challenges. In addition, the adoption of such an approach will not only boost productivity but also minimize waste.

Moreover, being a relatively new technology, the lack of standards in AM is restricting its use for parts production in numerous manufacturing sectors. Advanced engineering sectors such as the automotive and aerospace industries rely on production parts with high precision and accuracy. The shortfall of relevant and established standards to support AM-produced parts affects the production process and material selection to ensure high consistency and quality [21]. A summary of the main challenges observed in the implementation of AM technologies is summarized in Table 5.1.

TABLE 5.1
Summary of the Main Challenges Observed in the Implementation of AM Technologies

Barrier Type	Challenges of AM	References
AM process	Production time	[12]
	Throughput – number of units which can be processed in a given amount of time	[12]
	Output quality: accuracy, surface finish	[12]
	Requirement for post-processing machining	[12]
	In-process monitoring and inspection	[12]
Equipment/ machinery	Low-volume – limitation for mass production on a large scale	[12]
	Restricted to perform production processes for large-sized objects	[12]
	Cost of production requires high capital investment in the inception phase	[12]
	Affordability	[12]
Material	Validation of the mechanical/thermal performance of 3D-printed AM parts	[9]
	Produced parts' strength	[9]
	Recycling of AM material and part	[22]
Design	Size restrictions	[12]
	Lack of professional and integrated software	[12]
Policy	Lack of standards pertaining to equipment, materials and processes	[21]
	Protecting intellectual property (IP)	[21]
Awareness	Perception of being suitable for rapid prototyping rather than product manufacture	[12]
	Resistance to shifting to high-tech	[12]

5.2.6 AM – An Indispensable Tool in Industry 4.0

The rivalry among manufacturing enterprises is ever so high on the global scale to meet the pressing demand of customers while striving hard to maintain their competitiveness. Reforms and research and development are some of the key aspects required to maintain competitiveness which can partially be achieved in the Industry 4.0 era by considering the breakthrough technologies. The stimulation provided by Industry 4.0 technologies can lead to the creation of numerous opportunities to facilitate the development of new products and services in enterprises while streamlining the means to address customer's demands [10]. AM is one such breakthrough technology and is regarded as the driving force and an important tool in the fourth industrial revolution [10]. When incorporated with other technological innovations provided by Industry 4.0, namely Internet of Things (IoT), big data and AI, AM can promote remote, on-site and flexible manufacturing operations [10,23,24].

5.3 THE REALM OF INDUSTRY 4.0

Industry 4.0 is considered a new paradigm to enhance the process performance in modern manufacturing enterprises through large-scale digitization and vertical, horizontal and end-to-end integration concepts [25]. The manufacturing sector has gone through a series of drastic changes and evolution in the course of the past few centuries with industrial developments ranging from Industry 1.0 in the 18th century to Industry 4.0 in the present day and age. From the first assembly line introduced by Henry Ford for the mass production of automobiles till now, the surge in technological innovations ranging from electric-powered assembly lines to advanced automation and robotics has transformed the modern manufacturing world. The modern manufacturing processes can only get better when coupled with Industry 4.0 technology.

The transition from the third industrial revolution to Industry 4.0 has opened a new era of manufacturing with the introduction of the concept of mass customization. Mass customization of products which was unthinkable in the recent past and considered as being unsustainable was only made possible with the advent of the cloud manufacturing model and smart manufacturing. The cloud manufacturing model involves the cloud-based computing system which is paired with automated technologies and AI to streamline and improve the efficiency of existing manufacturing operations [4]. On the contrary, smart manufacturing systems, which capitalizes on state-of-the-art technologies such as edge computing, blockchain-IoT and AI, have yielded successful results in Industry 4.0 [4]. The use of these tools in smart manufacturing is to improve operational efficiency and avoid downtime [4].

5.3.1 The Advantages of Adopting Industry 4.0 in Modern Manufacturing Operations

The advantages of adopting Industry 4.0 in modern manufacturing operations are extensive as it incorporates the strength of numerous technologies. The benefits provided by the incorporation of such stand-alone or combined technologies in a manufacturing environment are sizeable as they can lead to the reduction of processes, improve the work environment, reduce processing time and facilitate decision making among other factors [26–28]. Moreover, besides directly impacting the work environment and manufacturing operations, the technologies in Industry 4.0 can markedly improve customer satisfaction. Some of the noteworthy features of this development which have transformed mass consumerism in modern times are the capacity for customization on a large-scale, reduced lead times, virtual customer assistance and overall increase in productivity [26,29,30].

5.3.1.1 The Benefits of Virtual Manufacturing in Industry 4.0

The concept of virtual manufacturing is further improving manufacturing systems in the Industry 4.0 era. It is a key component which enables the creation of highly integrated and efficient manufacturing systems by capitalizing on advanced technologies such as AI, IoT and big data [31,32]. The concept of virtual manufacturing offers a wide range of advantages to manufacturing companies. Some of the notable ones are reduced cost through simulation and optimization of production processes, improved efficiency, faster time to market, added flexibility and improved product quality [31,32].

5.3.1.2 The Influence of Industry 4.0 Technologies on Occupational Health and Safety

Past studies have shown that the implementation of Industry 4.0 and its associated technologies have a positive impact on the occupational health and safety risks observed in manufacturing industries. Some of these related technologies, namely AM, AI, machine vision, IoT and robot technology were found to lead to a reduction in physical, ergonomic and mechanical risks among others in the work environment. Some of these technologies provide a framework for real-time monitoring and risk assessment of workers in various working environments [26]. In addition, hazards at the workplace and operations with high physical risk can be continuously and effectively detected with advanced wireless communications and information technologies [26]. Other types of risks such as psychosocial risks which relate to the interrelation of social factors and individual thought and behavior are also being addressed by AI, big data and cybersecurity among others [26].

5.3.2 The Integration of Industry 4.0 Technologies with AM

The Industry 4.0 era has seen the breakthrough and development of a wide range of automated technology for industries through the combination of AI, IoT and advanced robotics in processes such as robot-assisted production lines and automated defect detection among others. A schematic outline of the main technological breakthroughs in the Industry 4.0 era is illustrated in Figure 5.3. These technological advancements are harmonized to work in unison with the ultimate goal of providing enhanced solutions to conventional problems. This approach has led to a restructuring of the entire operation of our society and that of the manufacturing world [7,10].

An increase in the demand for customized products has also been observed in the Industry 4.0 era. With the integration of AM, the demand for mass customization can

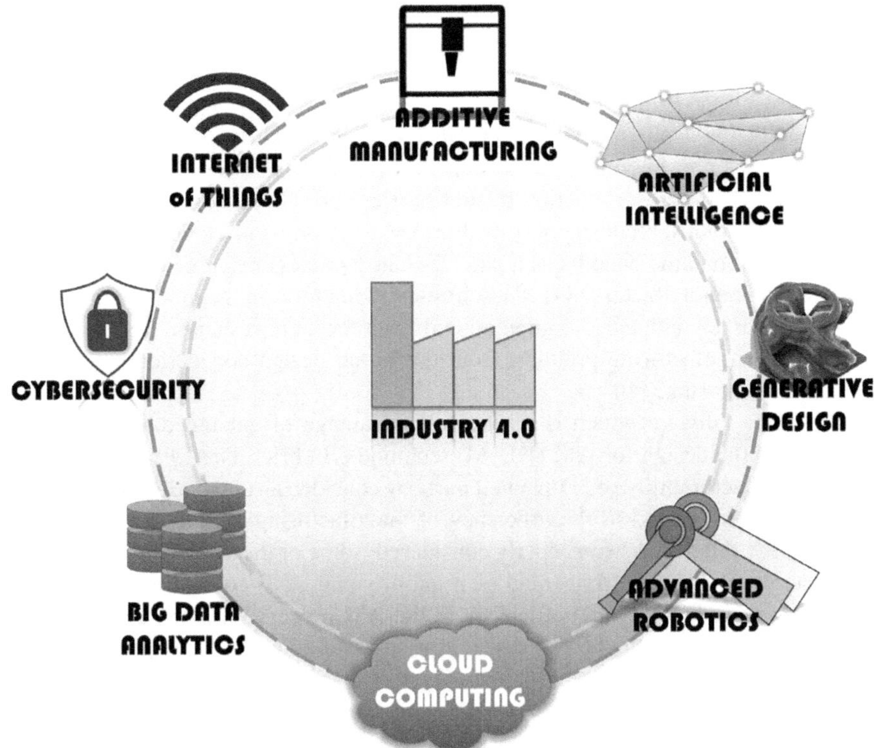

FIGURE 5.3 The main technological breakthroughs in the Industry 4.0 era.

be easily addressed as AM can expedite the development of products through quick prototyping and produce sophisticated products with advanced attributes in a more efficient manner [33]. Changes in product design or product mix normally require modification and adaptation of the production set-up. AM enables the seamless incorporation of changes in product design during manufacturing without requiring major modifications of the production set-up [7]. To assist AM processes, advanced robotics are used for numerous purposes ranging from inspection in AM's quality assurance to flexibility and efficiency in part and material handling [7].

5.3.2.1 IoT and AM

Industry 4.0 is the convergence of numerous technological innovations such as big data, cloud computing, advanced robotics, AI and AM. The IoT is one of the prominent technological innovations in Industry 4.0. The aim of integrating IoT technology in modern manufacturing operations involving AM is to create smart factories. This integration can lead to a notable increase in the efficiency and productivity of the operations [7]. IoT technology comprises sensors and sophisticated devices which are used to gather data and optimize processes while enabling the detection and offsetting of potential problems. With Industry 4.0, the conventional IoT systems are transformed into the Industrial IoT or IIoT, which consist of an ecosystem of devices, sensors and a network of equipment that work in unison to gather, monitor and process data from industrial operations [7].

5.3.2.2 Cloud Computing and AM

Additionally, the deployment of the aforelisted tools in Industry 4.0, conjointly with novel production technologies such as the AM process, has created opportunities to overcome design- and production-related barriers [4]. Cloud computing is another indispensable tool in Industry 4.0. In the AM environment, cloud computing can provide key economic benefits as it can efficiently process resource allocation without affecting quality factors [34]. These fast-evolving tools in the modern era of digitalization, namely enhanced computational resources, are reshaping the traditional method of manufacturing products from the initial design conception stage to the final fabrication stage [4].

One such software which is taking full advantage of enhanced computational resources is the design for AM (DfAM) technique. In brief, DfAM mainly focuses on the manufacturability of 3D-printed parts by considering and optimizing key process parameters related to the efficiency of manufacturing of fully functional and assembled AM parts. When properly considered, some of the key optimized process parameters, namely cost, time, quality, reliability and CAD constraints, can significantly boost productivity and efficiency in the AM process. The implementation of DfAM techniques and its best practices can however be challenging given its complexity as well as the fast-evolving environment of additive technologies and engineering design software [4].

5.3.2.3 Big Data and AM

With the advent of IIoT, data at every stage of manufacturing from design to production is collected, stored and processed. This data collection and analysis enables the identification of inefficient processes as well as discrepancies in the various stages of manufacturing. In AM, the data which can be generated from its pre-processing to post-processing stages can be diverse. To improve the performance, remedial measures based on the system response are swiftly applied to address those deficiencies [35].

In the pre-processing stage, data are mainly obtained from the generated CAD model, while in the post-processing stage, data are collected to assess the final print quality. Data are also generated in the processing stage and via instantaneous measurement of the part being manufactured as part of the in-situ monitoring, quality assurance and detection of irregularities. Irregularities such as surface defects in deposited layers, over-extrusion or under-extrusion can be obtained by using a high-resolution camera [7] and via advanced optical technology such as the optical tomography technique [36]. Optical tomography is a form of computed tomography technique which uses a high-resolution sensor operating in the visible light spectrum to identify porosity and defects in additively manufactured parts in a non-destructive approach.

5.3.2.4 The Advent of Digital Twins and Their Significance in AM

The implementation of digital twin technologies will enable to predict and correct errors prior to manufacturing the part. Digital twin technologies comprise the comprehensive physical and functional description of a component or product which incorporates all the information useful throughout its life cycle. This useful information will be used throughout the product life cycle as part of monitoring, prediction and optimization processes during manufacturing [37]. Other technologies such as the IoT are used to establish a connection between the real model and the replicated version to form a digital thread.

Given the numerous process variables affecting the quality of manufactured parts via AM, such as defective parts, dimensional and geometrical inaccuracy and surface roughness, the digital twin technology can be considered and applied from the design stage all the way up to the SC. This will also avoid the tedious trial and error methods to find optimum parameters for the AM process which can be time consuming and costly. In addition, with the digital twin technology, any issues and challenges encountered in the production process or in the design can be duly predicted in advance and subsequently solved, thereby limiting major disruptions or defects, respectively [38].

5.3.2.5 AI in AM

The implementation and consideration of AI Technology as part of Industry 4.0 has led to the breakthrough development and transformation of traditional manufacturing processes into smart manufacturing operations [7]. In AM in particular, AI can be used at various stages of manufacturing namely in the pre-processing and processing stages [39]. In the pre-processing stage of AM, AI can be implemented to determine the optimized design features from iterated designs via AI-based computer algorithms. One of the ground-breaking software add-ons which is pushing the limit of the AM industry is the AI-powered tool of generative design. Generative design is an iterative design method which is powered by advanced AI-based computer algorithms to generate outputs that satisfy a set of constraints iteratively. Moreover, AI techniques are employed to optimize the slicing process of CAD models in the pre-processing stage.

5.3.3 How Can Industry 4.0 Improve the AM Processes?

5.3.3.1 Defect Detection and Fault Diagnosis

Advanced sensor technology and live data sharing are enabling the in-situ monitoring of AM processes, thereby allowing the detection of AM defects [8]. From past literature, computer vision and AI have both been applied to autonomously detect and analyze defects in the AM process [40,41]. Defects in the AM process normally arise during printing and in the post-processing stage. Some of the typical defects observed in AM including their descriptions are summarized in Table 5.2.

Another means through which Industry 4.0 can improve the AM process is via advanced fault diagnosis of hardware systems. Unsupervised faults in the hardware operation can have a knock-on effect on the performance and productivity of 3D-printers. To reduce the lead time and improve the efficiency of the AM process, preventive maintenance, advance detection and replacement of faulty components are recommended. Typical hardware which are vulnerable to failure, improper functioning and prone to external disturbances are the AM actuators and filament extruders [8].

TABLE 5.2
Typical Defects Observed in Two Main Types of AM Processes, Namely in Fused Deposition Modelling (FDM) and in Metal Additive Manufacturing (MAM)

No.	Type of Defect	Description	References
		Fused deposition modelling (FDM)	
1.	Over- or under-fill	Excess or lack of local volumetric material	[42]
2.	Delamination	Weak bonding between deposited layers	[42]
3.	Cracking	Resulting from thermal stress	[42]
4.	Distortion	Residual stresses trapped in the printed parts due to non-uniform temperature gradients	[42]
5.	Over- or under-melting	Excessive or insufficient temperature	[42]
6.	Porosity	Due to insufficient fusion, inconsistency in input material	[42]
		Metal additive manufacturing (MAM)	
7.	Residual stress	Due to localized heating and cooling	[43]
8.	Lack of fusion	Insufficient energy density during melting	[43]
9.	Surface roughness	Depends on the quality of the material feedstock	[43]
10.	Spherical micropores	Arising due to trapped gas in the raw metal powder particles	[43]

5.3.3.2 Open-Loop and Closed-Loop Control Systems

The open-loop control system is mainly considered in the pre-processing stage of the AM process to generate the print path and support for the design structure. The open-loop control system also utilizes the machine learning (ML) technique in the pre-processing stage of the AM process to predict and optimize the printing process [8]. The implementation of a closed-loop control system in the AM process enables advanced monitoring and optimization techniques. For instance, components of Industry 4.0 and AM enable in-situ monitoring in a closed-loop system during manufacturing, hence enabling the production of high-quality products. Traditional closed-loop control system was based on PID (Proportional – Integral – Derivative) controllers and fuzzy inference systems [8]. Modern closed-loop control system operates on ML-based monitoring systems. Some of the commonly used ML algorithms to perform control tasks include the support vector machine, neural network, linear regression, decision tree and random forest [8].

5.4 ROLE OF AM IN THE INTEGRATED MANUFACTURING PROCESS CHAIN

The advent of AM technology has provided numerous benefits to the manufacturing industries in the Industry 4.0 Era. The decentralization of SCs and their shortening is one example which has led to the notable transformation of the modern manufacturing world. Moreover, coupled with Industry 4.0, AM is fostering the digitization of SCs by shifting to the digital inventory system. Besides AM technologies, AI and ML, big data, cloud computing, IoT and augmented reality are among the noteworthy technologies which have additionally boosted the rapid growth and implementation of Industry 4.0.

AM, which is a relatively new field of advanced manufacturing, is still undergoing considerable research and development. In recent years, the emergence of advanced AM technologies has displayed the full potential and capacity to be further scaled up and developed to compete with conventional manufacturing processes in the near future [44]. In contrast to conventional subtractive manufacturing or metal-casting or forming processes, the AM process provides the added benefits of lean production, reduced tool cost, reduced labor cost and rapid product development. The relevance and significance of AM among Industry 4.0 technologies rest upon the paradigm shift that is observed in the current manufacturing world whereby the increasing demand for mass-customized goods is forecasted to outclass the conventionally mass-produced goods. Further research is ongoing to scale up the throughput in AM to match volume production where the cycle time of manufactured products can be in seconds [7].

To further promote their integration in the modern manufacturing process chain, new policies and standards pertaining to AM equipment, materials and processes need to be developed. AM process has been classified under the same category for nonconventional manufacturing operations such as casting, die casting, drilling and cutting in accordance with the DIN EN ISO 286-1, which defines the standard for international tolerance classes [7,45]. New standards are being developed by the American Society for Testing Materials for AM technology to promote knowledge of the industry, stimulate research and boost their implementation in further applications [21]. The developed standards are meant to cover aspects of measuring the performance of various production processes in AM, assessing the quality of manufactured products and defining the calibration procedures for AM machines [21].

To improve the accuracy of manufactured parts via the MAM process, other approaches to optimize the printing parameters are being considered, such as uniform powder spreading, enhanced optical system, improved focal positioning control of laser spot size, reducing layer thickness and reducing particle size among other factors [7]. Moreover, AI-based ML techniques, which can be employed in predicting and optimizing the process parameters in MAM, have been proposed to further improve the accuracy of manufactured parts [7,46]. In another study, a smart and holistic production system for MAM was implemented by considering key elements of Industry 4.0 to enable complete monitoring of the production line while collecting manufacturing data to record responses and control the performance through continuous process optimization. The integrated manufacturing process chain was thought to enhance production flow from the raw material to the final product stage through the smart manufacturing and digitalization processes [47] (Table 5.3).

TABLE 5.3
The Notable Benefits of AM over Traditional Manufacturing and Their Corresponding Effect on Supply Chain

No.	Benefits of AM over Traditional Manufacturing	Effect on Conventional Manufacturing and on Supply Chain	References
	To achieve efficient manufacturing		
1.	Reduced manufacturing waste and promote recycling.	Promoting sustainable manufacturing	[22,48]
2.	Ability to manufacture/replicate complex geometries/intricate parts	Efficient and versatile manufacturing process	[9,12,48]
3.	Ability to create functional parts in one piece	More efficient manufacturing – enabling to promptly address customer's demand	[48]
4.	Decentralized manufacturing	Reduced lot size	[12]
5.	Ease of operation	Automated process	[7]
6.	Direct production from CAD models	More efficient manufacturing – enabling to promptly address customer's demand	[9]
7.	Near net shape (NNS) manufacturing process	Reduced cost of production Fewer downstream processes Efficient use of resources – energy/ material	[48]
	To promote cost-effective production		
8.	Small volume manufacturing – small batches of customized products	Cost-efficient	[7]
9.	Faster batch customization	Enable mass customization at a low cost	[12]
10.	No requirement for special tooling	Reduced tooling cost requirement	[12]
11.	Material savings	Efficient resource utilization	[7,12,22]
	To meet customer demands		
12.	One-site manufacturing of replacement part	Minimize storage and transportation cost Shorten supply chain Reduced repair/labor costs	[7,48]
13.	Promotes modification and customization of components through sharing of digital files	More efficient manufacturing – enabling to promptly address customer's demand	[9,48]
14.	Mass customization	Reduced manufacturing lead time	[12]
15.	Rapid prototyping	More affordable prototypes and reduced development costs. Reduced time to market Increased competitiveness	[12]
16.	Fosters make-to-order concept/ on-demand manufacturing	Reduces the need for high inventory	[48]

5.4.1 Advantages and Challenges of the Integrated Manufacturing Process Chain Involving AM Technologies

The implementation of the integrated manufacturing process chain has both benefits and disadvantages at the same time. In terms of notable benefits, the inclusion of integrated manufacturing technology such as the MAM provides the added flexibility to meet the request for innovative and customized production [2]. This added flexibility will enable the integrated manufacturing process chain to achieve the performance of a flexible manufacturing system by smoothly adapting to changes in the product being made and its quantity. Moreover, an integrated manufacturing process chain has been reported to provide full traceability of products and tools used for manufacturing, hence allowing better inventory control and management [47]. The high level of automation in an IoT environment adds to the robustness and reliability of the manufacturing process through high-end connectivity and data transfer, thereby enabling the monitoring of the performance metrics of various manufacturing systems [47].

5.4.1.1 Re-skilling Program for Workforce

The successful implementation of Industry 4.0 and its related technologies requires significant training and the formation of the company's workforce [47]. This crucial step will smoothen the company's transformation, thereby enabling the workforce to familiarize and adapt to the tools that come with digitalization. The new skill sets, change in working habits and mindset acquired through training by the workforce will be beneficial to the company, rendering it more efficient and productive. Such re-skilling programs will also provide operators with an increasing awareness about the health and safety prevailing in an automated environment, which can be hazardous as it involves a multitude of robotic arms, machines and automated guided vehicles [47]. Moreover, with the advent of digitalization and transformation of the conventional manufacturing environment, the role of the operator is anticipated to change and shift from the direct operation of the machine to remotely controlling and monitoring the full chain. Highly skilled job positions for control engineers and software programmers to maintain, service and set up advanced robotics in automated plants will also be on the rise with the emergence of this new manufacturing age [47].

The full implementation of an integrated manufacturing process chain by the latest technological advancements such as cloud computing and big data management has its drawbacks. The inability to handle the massive amount of data coming from a fully monitored process chain and the inefficient means to interpret data can be meaningless to the manufacturing concern. Moreover, the leakage of intellectual property and confidential and sensitive company information is an ever-growing concern in this era of high connectivity due to rising cybersecurity threats. Another downside of the integrated manufacturing process chain involving AM technologies is the requirement for high capital investments in the initial phase [44].

5.4.2 Furthering the Progression toward Industry 5.0

Some of the challenges which have been encountered during the implementation of Industry 4.0, namely technical integration, lack of skilled workforce and SC issues among others, can be addressed with the emergence of Industry 5.0. The emergence of Industry 5.0 is assumed to create a more resilient and sustainable work environment between robot-assisted processes and human workforce. A few of the breakthrough technologies which will expedite these tasks are hyper customization techniques, collaborative robots and predictive maintenance among others [49].

The growing trend toward more tailor-made products in Industry 4.0 has given rise to a new era of customization known as hyper customization. This means of customization makes use of real-time data to manufacture specific customers' products by considering AI and big data. One of the key requirements to achieve hyper customization is the integration of agile manufacturing [50]. Agile manufacturing, which is closely related to the concept of lean manufacturing, enables a manufacturing organization to have the right processes, tools and training to swiftly respond to customers' needs and market changes without compromising on costs and quality. Flexible systems of manufacturing and modularization in the production line can also help to achieve hyper customization in the fifth industrial revolution [49].

The implementation of collaborative robots is also believed to maximize productivity while ensuring a safe and favorable work environment for humans. These robots are implemented with the aim to cater to repetitive and dangerous tasks, thereby helping manufacturers to increase their production while alleviating the monotonous and risky tasks from human operators, respectively. Unlike conventional robots, the implementation of collaborative robots for automating the parts of a production line requires minimal changes [49,51]. Another technological advancement that is believed to lead to further improvement in modern manufacturing is predictive maintenance. The prediction of assets' life in manufacturing enterprises by collecting data from various sensors can contribute to the reduction of maintenance costs while improving productivity and quality of manufactured products [49].

5.5 ATTAINING SUSTAINABILITY WITH AM IN INDUSTRY 4.0

5.5.1 Adopting Sustainable Production for Economic Reasons

There is a worldwide interest in sustainable production which establishes a direct relationship between product and process design by addressing production issues. The enticing option of sustainable production enables enterprises to enhance the company's image while improving customer loyalty [48]. With the concept of sustainability gaining momentum among manufacturing firms, there is a need to further strengthen their eco-footprint by considering alternative approaches such as the AM process [48]. Attaining sustainable production with AM can be highly economical. Besides being a green production method, AM promotes lean production by reducing waste generation and fostering the recycling of materials. The factors enumerated and discussed in this section cover the key sustainable aspects of AM which in turn has direct implications on its economic viability. The attainment of sustainability through AM can be achieved in numerous ways, namely by enhancing the value chain model, by considering efficient resource utilization techniques and by adopting measures to preserve the environment among other factors [48].

5.5.2 Enhancing the Value Chain

Sustainability in AM can be achieved by improving resource efficiency, by extending product life and by reconfiguring value chains [13]. The reconfiguration of value chains by implementing the make-to-order concept is also enhancing the efficiency of product manufacturing via AM. With the make-to-order concept, goods are produced to exactly match the demand, thereby minimizing the inventory, producing less waste and resulting in fewer unsold finished goods [13]. AM, which fosters the concept of on-demand manufacturing and direct delivery to customers, eliminates high inventory levels by reducing the need for large amounts of raw materials within the SC and transportation. AM also promotes the fabrication of lightweight, optimized products and readily assembled objects, thereby shortening the SC and reducing operating costs [48].

5.5.3 Efficient Resource Utilization

Moreover, given its automated nature, AM is classified as a less resource-intensive process or green production method. AM provides the advantage of having reduced resource consumption in terms of raw materials during manufacturing [52,53]. In addition, the high efficiency of this specific manufacturing method has also resulted in reduced resource utilization in terms of energy, human workforce and factory space among other factors [12]. Additionally, with its autonomous and distant manipulation capabilities, AM alleviates the risk posed to the human workforce by harsh and hazardous working environments [12].

5.5.3.1 Reduced Energy Consumption

The transition toward decarbonization is prompting many manufacturing enterprises to review their energy consumption policies [48]. The adoption of AM in contrast to conventional machining in part manufacturing results in a more efficient process with reduced energy consumption and hence lower carbon footprint and less embodied energy [21]. AM processes normally require less energy to produce parts in contrast to other manufacturing processes as it is a highly efficient process. Design and optimization of complex parts with the layer-by-layer method of fabrication in AM was found to significantly reduce energy consumption [12]. On the contrary, further research is still required in MAM processes to optimize their energy consumption, which involves the utilization of high-intensity energy beams from laser sources [54]. Besides strengthening the economic vitality and improving the competitiveness of manufacturing companies, the adoption of alternative and energy-efficient measures in AM also helps to protect the environment [21].

5.5.3.2 Lower Tooling Cost

The enhancement in the efficiency of AM processes, which lessens the need for specialized tools and tool replacement, improves the profitability margin of companies. Moreover, besides reduced tool usage, AM processes generally operate in stand-alone mode with a minimum requirement for additional support equipment, fixtures and assembly parts [48]. The tooling cost requirement in the AM process is also insignificant as the method of manufacturing does not involve any material removal processes. The unrequired tools and molds in AM, which is normally considered tool-less manufacturing, enable on-demand production systems while leading to a reduction in inventory costs [12].

5.5.4 Preserving the Environment

There is an ever-increasing awareness about the use of alternative and greener technologies to minimize the impact of industrialization and exploitation of resources on the environment and its biodiversity. To achieve compliance with the sustainable development goals set forth by the United Nations, various stakeholders involving the government, industries and scientists and considering alternative and greener approaches to protect the environment. The concept of sustainability, which is established on the ideology of minimizing the impact of human development on its environment and on its natural resources, can be well attained with AM in Industry 4.0 [7].

5.5.4.1 Promoting Circular Economy

The green approach of considering AM technologies aligns with the concept of the circular economy. The circular economy model is founded on the basis of resource production and consumption. It principally considers the factors affecting the life cycle of developed products such as reuse, repair and recycling among other factors [22]. The goal of the circular economy, which is to obtain the maximum out of the developed product from inception to the disposal stage, is boosted by the implementation of AM technologies [22]. For instance, the transition of modern manufacturing toward green materials and processes in the future will instill the reuse, recyclability and circularity of materials in AM [22]. Moreover, the integration of AM technologies in a modern production environment was reported to achieve greater production efficiency in terms of manufacturing costs within the framework of the circular economy [22,55].

5.5.4.2 Eco-Friendly and Durable Products

To address the environmental issues associated with modern manufacturing and to further promote the concept of circular economy, the creation of eco-friendly and durable products is being envisaged via the AM technique. New research studies are focusing on the development of greener input materials for AM and means to enhance their processing stage are also being looked at. The adoption of greener AM processes is also promoting the generation of eco-friendly products. One notable example is in the fused deposition modelling (FDM) process which utilizes polylactic acid (PLA) filament as input material. PLA is a biodegradable, sustainable and non-toxic material which is suitable for large-scale production of parts in numerous applications given its low melting point [48]. Moreover, the optimization of assembly design via AM enables the development of products with greater operational efficiency, functionality and durability [13].

5.5.4.3 Waste Reduction

Sustainability can also be achieved by reducing waste during production. AM is an NNS-type manufacturing process, and the material waste generated by this process is substantially lower than conventional manufacturing processes. This approach of reducing waste during manufacturing further promotes the concept of lean production. The risk of overproduction possibilities is also addressed in AM, as it enables the quick production of personalized products by considering improved printing parameters such as printing speed and optimized infill percentage [48].

5.5.5 Means to Further Promote Sustainability by Considering AM Technologies in Industry 4.0

Sustainability can also be well achieved via the emergence of smart factories whereby the incorporation of various Industry 4.0 technologies conjointly with AM can significantly boost productivity. The factory based on AM systems coupled with Industry 4.0 is foreseen to be fully autonomous from start to end of operation in the future. The integration of AM with Industry 4.0 can be more rewarding and favorable toward the sustainability goals as several key aspects of the manufacturing operation can be further optimized [13]. For instance, the modularization in the production line of smart factories, and their quick response to customer needs and product changes, can beneficially enhance the sustainable traits and profitability of companies [7].

Besides the generation of less wasteful material through efficient resource consumption and monitoring, the integration of AM with Industry 4.0 was found to lead to shorter SC, efficient movement of materials and reduced inventory among other factors [13]. Moreover, there are key areas in the manufacturing operations involving AM technology under the Industry 4.0 era which still need to be further addressed in order to meet the sustainable development goals. For instance, besides the optimization of energy-intensive processes such as in MAM processes, advanced research is focused toward the inclusion of greener raw materials, the re-utilization of input materials and addressing the need for further post-machining and finishing operations in AM parts [13].

Moreover, findings from a recent survey conducted with AM adopters around the world revealed that companies tend to mostly favor the economic benefits at the expense of social and environmental sustainability benefits provided by the implementation of AM technologies [12]. The developers of AM technologies will eventually need to emphasize further the beneficial and sustainability attributes of green production that can be achieved with the AM process. This could further instigate more companies to shift toward this advanced manufacturing method by considering their sustainable features as a determinant factor.

5.6 CONCLUSION

Recent advancement and breakthrough in AM technologies have led to its implementation acceptance in a wide range of engineering applications including automated production lines. Coupled with the tools of Industry 4.0, namely AI, machine vision, IoT, big data and cybersecurity among others, the full potential of AM is being exploited and utilized to transform the modern manufacturing world. The exigency to meet the demand for increasingly customized products from mass consumers while remaining competitive and profitable is a major challenge for many enterprises. AM addresses these product requirements with relative ease and in contrast to conventional machining, it can provide seamless and automated solutions to complex designs and enables mass customization. Moreover, the advent of digitization through Industry 4.0 and the implementation of the integrated manufacturing process chain through AM have resulted in the decentralization of production lines and have led to a boost in the productivity of enterprises. For instance, the incorporation of such technologies in manufacturing industries has shortened the SCs improved the movement of materials and reduced inventory among other factors. Besides their numerous economic benefits and competitive edge, the proposed integration of AM into modern manufacturing processes is enabling the attainment of sustainable manufacturing. In addition to having a more efficient utilization of raw materials, the energy requirement by AM processes to produce parts is relatively lower in contrast to conventional machining. Being an NNS-type of manufacturing, AM has at its core a focus on reducing waste, thereby further promoting the concept of lean production. Other factors which have notably contributed to greater sustainability are reduced tool costs, reduced labor cost and rapid product development.

REFERENCES

1. Y. Huang, M. C. Leu, J. Mazumder, and A. Donmez, "Additive manufacturing: Current state, future potential, gaps and needs, and recommendations," *The Journal of Manufacturing Science and Engineering*, vol. 137, no. 1, p. 014001, 2015, doi: 10.1115/1.4028725
2. B. Elhazmiri, N. Naveed, M. N. Anwar, and M. I. U. Haq, "The role of additive manufacturing in Industry 4.0: An exploration of different business models," *Sustainable Operations and Computers*, vol. 3, pp. 317–329, 2022, doi: 10.1016/j.susoc.2022.07.001
3. M. Gebler, A. J. M. Schoot Uiterkamp, and C. Visser, "A global sustainability perspective on 3D printing technologies," *Energy Policy*, vol. 74, pp. 158–167, Nov. 2014, doi: 10.1016/j.enpol.2014.08.033
4. R. Ashima, A. Haleem, S. Bahl, M. Javaid, S. Kumar Mahla, and S. Singh, "Automation and manufacturing of smart materials in additive manufacturing technologies using Internet of Things towards the adoption of Industry 4.0," *Materials Today: Proceedings*, vol. 45, pp. 5081–5088, 2021, doi: 10.1016/j.matpr.2021.01.583
5. M. Attaran, "The rise of 3-D printing: The advantages of additive manufacturing over traditional manufacturing," *Business Horizons*, vol. 60, no. 5, pp. 677–688, Sep. 2017, doi: 10.1016/j.bushor.2017.05.011

6. A. Ceruti, P. Marzocca, A. Liverani, and C. Bil, "Maintenance in aeronautics in an Industry 4.0 context: The role of augmented reality and additive manufacturing," *Journal of Computational Design and Engineering*, vol. 6, no. 4, pp. 516–526, Oct. 2019, doi: 10.1016/j.jcde.2019.02.001
7. N. Balashanmugam, "Perspectives on additive manufacturing in Industry 4.0," In: M. Manjaiah, K. Raghavendra, N. Balashanmugam, and J. Paulo Davim (eds.), *Additive Manufacturing*, Elsevier, 2021, pp. 127–150. doi: 10.1016/B978-0-12-822056-6.00001-1.
8. T. S. Tamir et al., "3D printing in materials manufacturing industry: A realm of Industry 4.0," *Heliyon*, vol. 9, no. 9, p. e19689, Sep. 2023, doi: 10.1016/j.heliyon.2023.e19689
9. N. Guo and M. C. Leu, "Additive manufacturing: Technology, applications and research needs," *Frontiers of Mechanical Engineering*, vol. 8, no. 3, pp. 215–243, Sep. 2013, doi: 10.1007/s11465-013-0248-8
10. R. M. Mahamood, T. C. Jen, S. A. Akinlabi, S. Hassan, K. O. Abdulrahman, and E. T. Akinlabi, "Role of additive manufacturing in the era of Industry 4.0," In: M. Manjaiah, K. Raghavendra, N. Balashanmugam, and J. Paulo Davim (eds.), *Additive Manufacturing*, Elsevier, 2021, pp. 107–126. doi: 10.1016/B978-0-12-822056-6.00003-5
11. M. S. Safavi, A. Bordbar-Khiabani, J. Khalil-Allafi, M. Mozafari, and L. Visai, "Additive manufacturing: An opportunity for the fabrication of near-net-shape NiTi implants," *Journal of Manufacturing and Materials Processing*, vol. 6, no. 3, p. 65, Jun. 2022, doi: 10.3390/jmmp6030065
12. M. K. Niaki, S. A. Torabi, and F. Nonino, "Why manufacturers adopt additive manufacturing technologies: The role of sustainability," *Journal of Cleaner Production*, vol. 222, pp. 381–392, Jun. 2019, doi: 10.1016/j.jclepro.2019.03.019
13. S. Ford and M. Despeisse, "Additive manufacturing and sustainability: An exploratory study of the advantages and challenges," *Journal of Cleaner Production*, vol. 137, pp. 1573–1587, Nov. 2016, doi: 10.1016/j.jclepro.2016.04.150
14. P. Liu, S. H. Huang, A. Mokasdar, H. Zhou, and L. Hou, "The impact of additive manufacturing in the aircraft spare parts supply chain: Supply chain operation reference (scor) model based analysis," *Production Planning & Control*, vol. 25, no. 13–14, pp. 1169–1181, Oct. 2014, doi: 10.1080/09537287.2013.808835
15. S. M. Wagner and R. O. Walton, "Additive manufacturing's impact and future in the aviation industry," *Production Planning & Control*, vol. 27, no. 13, pp. 1124–1130, Oct. 2016, doi: 10.1080/09537287.2016.1199824
16. R. Huang et al., "Energy and emissions saving potential of additive manufacturing: The case of lightweight aircraft components," *Journal of Cleaner Production*, vol. 135, pp. 1559–1570, Nov. 2016, doi: 10.1016/j.jclepro.2015.04.109
17. H. Turunen, "Additive manufacturing and value creation - In architectural design, design process and end-products," In: A. Herneoja, Ö. Toni, and M. Piia (eds.), *Complexity & Simplicity - Proceedings of the 34th eCAADe Conference - Volume 1, University of Oulu, Oulu, Finland, 22–26 August 2016*, pp. 103–111. doi: 10.52842/conf.ecaade.2016.1.103
18. A. Pajonk, A. Prieto, U. Blum, and U. Knaack, "Multi-material additive manufacturing in architecture and construction: A review," *Journal of Building Engineering*, vol. 45, p. 103603, Jan. 2022, doi: 10.1016/j.jobe.2021.103603.
19. A. Haleem and M. Javaid, "Additive manufacturing applications in Industry 4.0: A review," *Journal of Industrial Integration and Management*, vol. 04, no. 04, p. 1930001, Dec. 2019, doi: 10.1142/S2424862219300011
20. M. Javaid and A. Haleem, "Current status and applications of additive manufacturing in dentistry: A literature-based review," *Journal of Oral Biology and Craniofacial Research*, vol. 9, no. 3, pp. 179–185, Jul. 2019, doi: 10.1016/j.jobcr.2019.04.004

21. M. Mani, K. W. Lyons, and S. K. Gupta, "Sustainability characterization for additive manufacturing," *Journal of Research of the National Institute of Standards and Technology*, vol. 119, p. 419, Oct. 2014, doi: 10.6028/jres.119.016
22. H. A. Colorado, E. I. G. Velásquez, and S. N. Monteiro, "Sustainability of additive manufacturing: The circular economy of materials and environmental perspectives," *Journal of Materials Research and Technology*, vol. 9, no. 4, pp. 8221–8234, Jul. 2020, doi: 10.1016/j.jmrt.2020.04.062
23. J. Moavenzadeh, "*The 4th Industrial Revolution: Reshaping the Future of Production.*" Accessed: May 27, 2024. [Online]. Available: https://na.eventscloud.com/file_uploads/fe238270f05e2dbf187e2a60cbcdd68e_2_Keynote_John_Moavenzadeh_World_Economic_Forum.pdf
24. T. Pereira, J. V Kennedy, and J. Potgieter, "A comparison of traditional manufacturing vs additive manufacturing, the best method for the job," *Procedia Manufacturing*, vol. 30, pp. 11–18, 2019, doi: 10.1016/j.promfg.2019.02.003
25. M. Ghobakhloo, "The future of manufacturing industry: A strategic roadmap toward Industry 4.0," *Journal of Manufacturing Technology Management*, vol. 29, no. 6, pp. 910–936, Jul. 2018, doi: 10.1108/JMTM-02-2018-0057
26. G. Arana-Landín, I. Laskurain-Iturbe, M. Iturrate, and B. Landeta-Manzano, "Assessing the influence of Industry 4.0 technologies on occupational health and safety," *Heliyon*, vol. 9, no. 3, p. e13720, Mar. 2023, doi: 10.1016/j.heliyon.2023.e13720
27. H. M. Elhusseiny and J. Crispim, "SMEs, barriers and opportunities on adopting Industry 4.0: A review," *Procedia Computer Science*, vol. 196, pp. 864–871, 2022, doi: 10.1016/j.procs.2021.12.086
28. L. S. Dalenogare, G. B. Benitez, N. F. Ayala, and A. G. Frank, "The expected contribution of Industry 4.0 technologies for industrial performance," *International Journal of Production Economics*, vol. 204, pp. 383–394, Oct. 2018, doi: 10.1016/j.ijpe.2018.08.019
29. D. Preuveneers and E. Ilie-Zudor, "The intelligent industry of the future: A survey on emerging trends, research challenges and opportunities in Industry 4.0," *Journal of Ambient Intelligence and Smart Environments*, vol. 9, no. 3, pp. 287–298, Apr. 2017, doi: 10.3233/AIS-170432
30. I. Gibson, D. Rosen, and B. Stucker, *Additive Manufacturing Technologies*. New York: Springer, 2015. doi: 10.1007/978-1-4939-2113-3
31. J. J. Roldán, E. Crespo, A. Martín-Barrio, E. Peña-Tapia, and A. Barrientos, "A training system for Industry 4.0 operators in complex assemblies based on virtual reality and process mining," *Robotics and Computer-Integrated Manufacturing*, vol. 59, pp. 305–316, Oct. 2019, doi: 10.1016/j.rcim.2019.05.004
32. M. Soori, B. Arezoo, and R. Dastres, "Virtual manufacturing in Industry 4.0: A review," *Data Science and Management*, vol. 7, no. 1, pp. 47–63, Mar. 2024, doi: 10.1016/j.dsm.2023.10.006
33. U. M. Dilberoglu, B. Gharehpapagh, U. Yaman, and M. Dolen, "The role of additive manufacturing in the era of Industry 4.0," *Procedia Manufacturing*, vol. 11, pp. 545–554, 2017, doi: 10.1016/j.promfg.2017.07.148
34. G. Senthilkumar, K. Tamilarasi, N. Velmurugan, and J. K. Periasamy, "Resource allocation in cloud computing," *Journal of Advances in Information Technology*, vol. 14, no. 5, pp. 1063–1072, 2023, doi: 10.12720/jait.14.5.1063-1072
35. M. Adnan, Y. Lu, A. Jones, and F. T. Cheng, "Application of the fog computing paradigm to additive manufacturing process monitoring and control," *SSRN Electronic Journal*, 2021, doi: 10.2139/ssrn.3785854
36. M. Dharnidharka, U. Chadha, L. M. Dasari, A. Paliwal, Y. Surya, and S. K. Selvaraj, "Optical tomography in additive manufacturing: A review, processes, open problems, and new opportunities," *The European Physical Journal Plus*, vol. 136, no. 11, p. 1133, Nov. 2021, doi: 10.1140/epjp/s13360-021-02108-1

37. B. Schleich, N. Anwer, L. Mathieu, and S. Wartzack, "Shaping the digital twin for design and production engineering," *CIRP Annals*, vol. 66, no. 1, pp. 141–144, 2017, doi: 10.1016/j.cirp.2017.04.040
38. G. L. Knapp et al., "Building blocks for a digital twin of additive manufacturing," *Acta Materialia*, vol. 135, pp. 390–399, Aug. 2017, doi: 10.1016/j.actamat.2017.06.039
39. J. Yang, Y. Chen, W. Huang, and Y. Li, "Survey on artificial intelligence for additive manufacturing," In *2017 23rd International Conference on Automation and Computing (ICAC)*, Huddersfield, United Kingdom, IEEE, Sep. 2017, pp. 1–6. doi: 10.23919/IConAC.2017.8082053
40. M. F. Khan et al., "Real-time defect detection in 3D printing using machine learning," *Materials Today: Proceedings*, vol. 42, pp. 521–528, 2021, doi: 10.1016/j.matpr.2020.10.482
41. K. Paraskevoudis, P. Karayannis, and E. P. Koumoulos, "Real-time 3D printing remote defect detection (stringing) with computer vision and artificial intelligence," *Processes*, vol. 8, no. 11, p. 1464, Nov. 2020, doi: 10.3390/pr8111464
42. M.-A. de Pastre, Y. Quinsat, and C. Lartigue, "Effects of additive manufacturing processes on part defects and properties: A classification review," *International Journal on Interactive Design and Manufacturing (IJIDeM)*, vol. 16, no. 4, pp. 1471–1496, Dec. 2022, doi: 10.1007/s12008-022-00839-8
43. C. Bellini et al., "Effects of recycling on defects and microstructure in Ti-6Al-4V powder particles and samples fabricated by electron beam melting process," *Procedia Structural Integrity*, vol. 47, pp. 359–369, 2023, doi: 10.1016/j.prostr.2023.07.088
44. S. Dias, P. Espadinha-Cruz, and F. Matos, "A Porter's Five Forces model proposal for additive manufacturing technology: A case study in Portuguese industry," *Procedia Computer Science*, vol. 217, pp. 165–176, 2023, doi: 10.1016/j.procs.2022.12.212
45. T. Lieneke et al., "Systematical determination of tolerances for additive manufacturing by measuring linear dimensions." In *26th International Solid Freeform Fabrication Symposium - An Additive Manufacturing Conference*, Austin, Texas, USA.
46. G. Masinelli, S. A. Shevchik, V. Pandiyan, T. Quang-Le, and K. Wasmer, "Artificial intelligence for monitoring and control of metal additive manufacturing," In: P. M. Pandey, N. K. Singh, and Y. Singh (eds.), *Industrializing Additive Manufacturing*, Cham: Springer International Publishing, 2021, pp. 205–220. doi: 10.1007/978-3-030-54334-1_15
47. M. Moshiri, A. Charles, A. Elkaseer, S. Scholz, S. Mohanty, and G. Tosello, "An Industry 4.0 framework for tooling production using metal additive manufacturing-based first-time-right smart manufacturing system," *Procedia CIRP*, vol. 93, pp. 32–37, 2020, doi: 10.1016/j.procir.2020.04.151
48. M. Javaid, A. Haleem, R. P. Singh, R. Suman, and S. Rab, "Role of additive manufacturing applications towards environmental sustainability," *Advanced Industrial and Engineering Polymer Research*, vol. 4, no. 4, pp. 312–322, Oct. 2021, doi: 10.1016/j.aiepr.2021.07.005
49. M. Khan, A. Haleem, and M. Javaid, "Changes and improvements in Industry 5.0: A strategic approach to overcome the challenges of Industry 4.0," *Green Technologies and Sustainability*, vol. 1, no. 2, p. 100020, May 2023, doi: 10.1016/j.grets.2023.100020
50. P. K. R. Maddikunta et al., "Industry 5.0: A survey on enabling technologies and potential applications," *Journal of Industrial Information Integration*, vol. 26, p. 100257, Mar. 2022, doi: 10.1016/j.jii.2021.100257
51. M. Doyle-Kent, "*Collaborative Robotics in Industry 5.0*."
52. M. Mehrpouya, A. Dehghanghadikolaei, B. Fotovvati, A. Vosooghnia, S. S. Emamian, and A. Gisario, "The potential of additive manufacturing in the smart factory Industrial 4.0: A review," *Applied Sciences*, vol. 9, no. 18, p. 3865, Sep. 2019, doi: 10.3390/app9183865

53. T. Peng, K. Kellens, R. Tang, C. Chen, and G. Chen, "Sustainability of additive manufacturing: An overview on its energy demand and environmental impact," *Additive Manufacturing*, vol. 21, pp. 694–704, May 2018, doi: 10.1016/j.addma.2018.04.022
54. M. Zavala-Arredondo et al., "Laser diode area melting for high speed additive manufacturing of metallic components," *Materials & Design*, vol. 117, pp. 305–315, Mar. 2017, doi: 10.1016/j.matdes.2016.12.095
55. C. M. Angioletti, F. Sisca, R. Luglietti, M. Taisch, and R. Rocca, "Additive manufacturing as an opportunity for supporting sustainability through implementation of circular economies. In *Proceedings of the Summer School Francesco Turco*, pp. 25–25. AIDI-Italian Association of Industrial Operations Professors, 2016".

6 Predictive Maintenance for Industry 4.0

Satyadev Rosunee and Roshan Unmar

6.1 INTRODUCTION

The systematic maintenance of machinery in a production plant ensures that machines operate efficiently, produce quality goods at competitive prices and give good return on investment (ROI) [1–3]. Maintenance entails significant costs. On the contrary, machine downtime due to lack of maintenance disrupts production flow, reduces productivity and obviously costs a lot more money.

The severity of financial losses owing to downtime is more consequent for some industries than others. Aviation, automotive, oil and gas and pharma industries are those that are most highly impacted. Some studies estimate that unplanned downtimes cost global manufacturing almost $1.5 trillion a year. Machine failures may also impact product quality and generate unnecessary waste [4,5]. Stricter sustainability standards (Higg Index Tools) and governments' push toward sustainable development goals are gradually "greening" the manufacturing sector. Therefore, the overarching objective of plant maintenance is to minimize breakdowns and stoppages and boost quality and productivity. Production should go on without interruption at the lowest possible cost.

Over the years, the pace of digitalization in manufacturing has been increasing. Companies are recognizing that machine learning (ML) and artificial intelligence (AI) can foster competitiveness and better ROI. Detecting machine failures before major manufacturing lines are affected is crucial in streamlining operations. However, the comprehensive integration of ML and AI tools is a major challenge as the aging workforce is finding it difficult to adapt to change. It requires an active and holistic involvement from every facet of the enterprise, and that includes empowerment of workers and top management. About 60% of the plant managers interviewed in this study said that AI would be a "game changer". However, they emphasized the need to rethink data security and privacy policies, ethical considerations, regulatory compliance, talent acquisition and proper management of user access to AI while transitioning to smart manufacturing.

6.2 INDUSTRY 4.0 AND ENABLING TECHNOLOGIES

Almada-Lobo [6] has elaborated on the Industry 4.0 revolution and the future of manufacturing systems. The myriad of technologies and smart systems underpinning the future progress of Industry 4.0 is illustrated in Figure 6.1. In this context, the integration of predictive maintenance (PdM) is a tangible enabler. It empowers

DOI: 10.1201/9781003511298-6

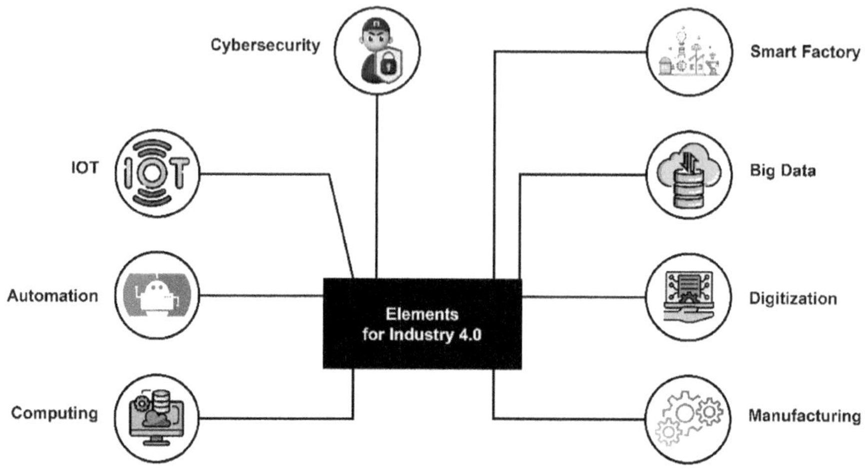

FIGURE 6.1 Industry 4.0 technologies in the context of smart manufacturing.

sustainable manufacturing through the digitalization of machine maintenance. Data capture extracted from a diverse range of manufacturing processes has increased exponentially due to the ubiquity of sensors and Internet of Things (IoT) technologies.

Even though PdM faces organizational, financial or even data sourcing and machine repair challenges, it remains a strong point for the companies that use it. Indeed, it allows for minimizing machine downtime and associated costs, maximizing the lifespan of the machine and maintaining the quality of output and speed of production.

6.3 MAINTENANCE MANAGEMENT METHODS

Industrial plants typically employ two types of maintenance management: reactive or preventive maintenance (PV). Maintenance can be classified into four categories [7,8]:

I. Reactive or run-to-failure maintenance
II. Scheduled maintenance
III. Preventive maintenance
IV. Predictive maintenance

6.3.1 Reactive Maintenance

In reactive maintenance, the equipment is repaired only when it is out of order. This method of maintenance works only when dealing with inexpensive and non-critical assets/pieces of equipment. A plant using run-to-failure management does not spend any money on maintenance until the machine or system fails to operate. Reactive maintenance generally leads to poor maintenance, excessive delays in production, more spoiled material and lower profits. This type of maintenance is justified for small enterprises where there is weak financial justification for scheduled maintenance. One major drawback is the erratic maintenance record of machines.

6.3.2 Scheduled Maintenance

Logically, in a fast-paced manufacturing set-up, reactive maintenance would be unsound for critical pieces of equipment like compressors, forklifts, boilers and pressure vessels. Scheduled maintenance requires early inspection of equipment and taking necessary action(s) to avoid breakdown. As the saying goes: "Maintenance that is scheduled is maintenance that gets done".

6.3.3 Preventive Maintenance

The concept of PM dates back to the 1950s when Japanese engineers came up with a set of procedures to ensure the reliability of machines and associated systems. This foundation eventually evolved into the total productive maintenance (TPM) system implemented by Toyota Motors in the early 1970s. The TPM philosophy of continuous improvement is all-inclusive and engages operators, supervisors, engineers and managers in a common goal of minimizing downtime and maximizing productivity. PM management programs are proactive and based on the number of hours the machine has been running. The typical productive life of a machine is illustrated in Figure 6.2.

During installation and commissioning, a new machine has a high probability of failure – first few weeks of operation. After this initial period, the probability of failure is relatively low for an extended period. After this normal machine life period, the probability of failure increases sharply with elapsed time.

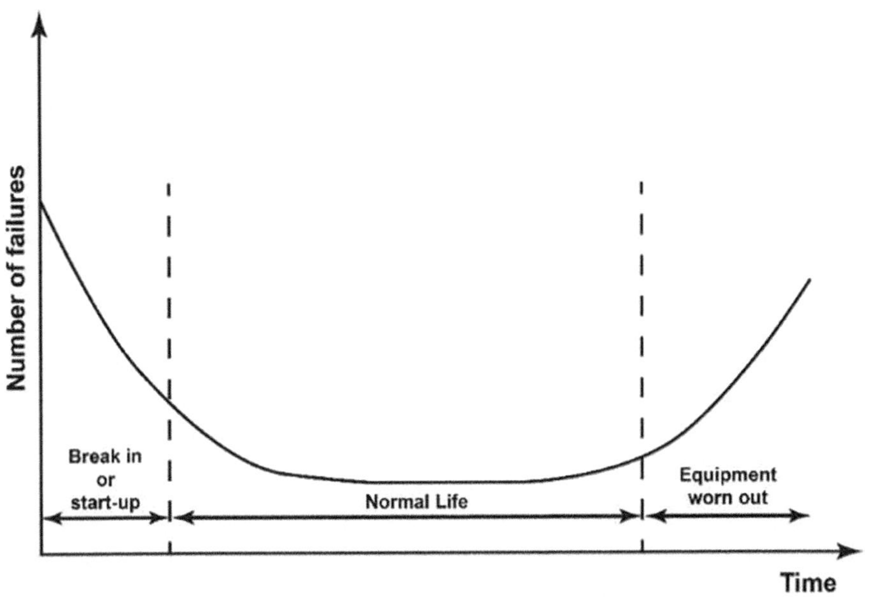

FIGURE 6.2 Time-dependent profile of preventive maintenance [8].

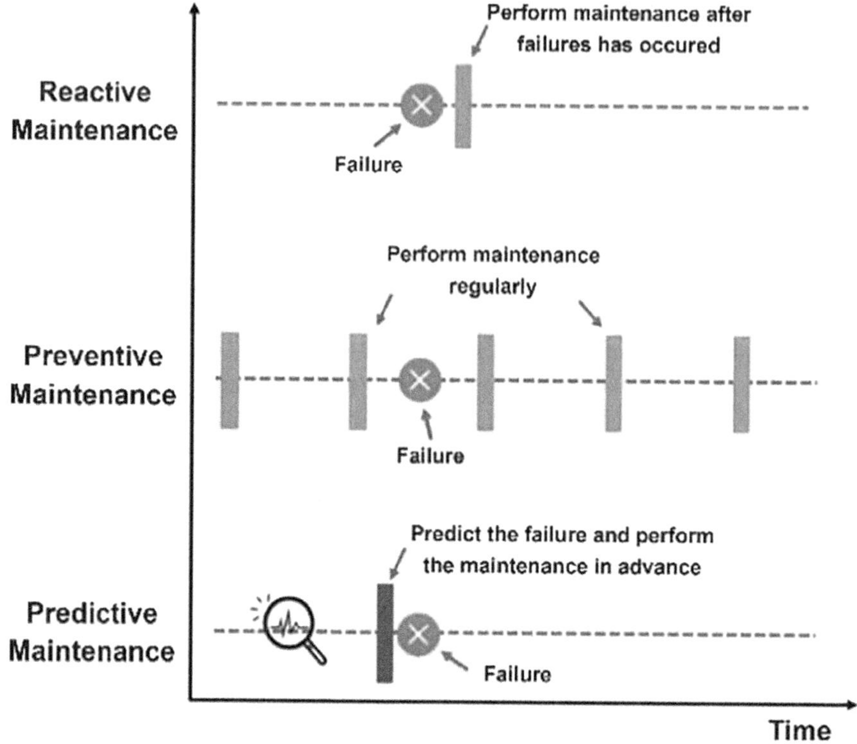

FIGURE 6.3 Summary of maintenance management methods and intervention points.

Figure 6.3 illustrates succinctly the different maintenance management methods and intervention points [9]. PM generally targets weak spots like gear trains, bearings, miscellaneous parts under excessive vibration and pressure vessels, which need regular inspection, thereby reducing the danger of breakdown. PM involves periodic inspection of the components mentioned above. Any fault detected is corrected while it is still in the minor stage. This ensures continuous production and higher plant efficiency at reasonable costs. Over 90% of managers interviewed mentioned that PM is an established feature in their plant.

6.3.3.1 PV and Quality Cost

The widely used prevention-appraisal-failure cost of quality model considers the trade-off between the cost of good quality (COGQ) and the cost of poor quality (COPQ) to determine the optimal quality cost. The COGQ includes prevention and appraisal costs, whereas COPQ involves both internal and external failure costs. Accordingly, PM and PdM add to the prevention component of COGQ.

As depicted in Figure 6.4, investment in PV helps reduce the total cost of quality as long as the optimal level is not exceeded. It is therefore necessary for enterprises to wisely work out maintenance plans based on a proper evaluation of the different quality cost factors.

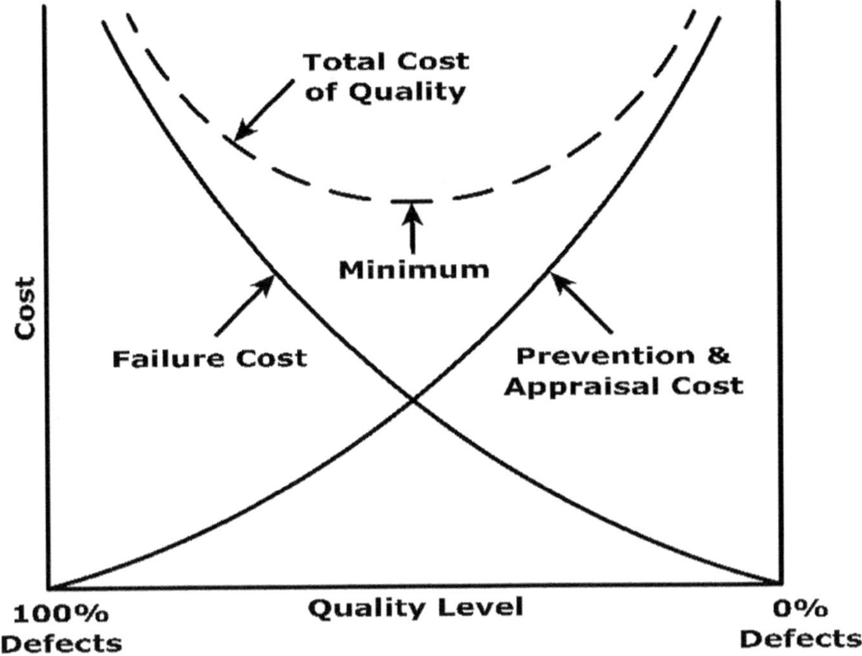

FIGURE 6.4 Preventive maintenance and the total cost of quality.

6.4 PREDICTIVE MAINTENANCE

The PdM of plant or equipment is data-driven. The availability of accurate and unbiased data is crucial when implementing PdM. The advent of powerful AI chips at increasingly affordable prices is facilitating the adoption of ML and AI tools and by extension enabling PdM, even in medium-tech manufacturing industries. PdM, similar to PV, is a proactive strategy that aims to predict when a machine might break down [8,9]. But, unlike PV, PdM tries to predict machine failure by integrating IoT devices and sensors [10]. PdM involves condition-based maintenance in which machines are monitored with sensor devices that supply data to an interface that diagnoses the asset's condition. ML algorithms use patterns in the data to predict when the asset will require maintenance to prevent equipment failure. It should be emphasized that a solid understanding of component failure greatly facilitates PdM.

The first step in implementing PdM is to establish reasonable machine condition baselines; i.e., the machine's condition baselines need to be set and data should be collected before installing sensors. This machine condition data is then used as "control" for diagnosis purposes. In PdM programs multiple IoT-enabled sensors and smart devices collect machine data and feed it to ML algorithms embedded in an appropriate system, namely a computerized maintenance management system (CMMS).

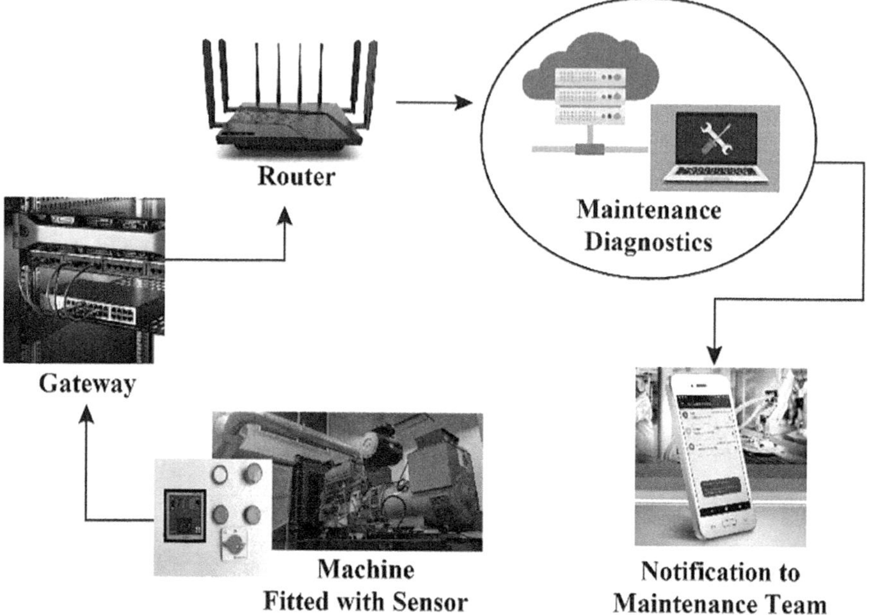

FIGURE 6.5 The components of an integrated predictive maintenance system.

When a piece of equipment is performing outside normal parameters, the CMMS will diagnose the issue and trigger a maintenance work order that is assigned to technicians so that they can perform any required repairs to address the anomaly (Figure 6.5). The information captured by PdM sensors can vary depending on the type of equipment and parts being monitored. The CMMS also reduces the need for close human supervision.

Private communication with maintenance managers in the local industry has highlighted the use of the following affordable tools, outlined in Table 6.1, for "predictive" maintenance purposes.

In fact, over 90% of manufacturers who deployed a PdM program saw a reduction in repair time as well as a 9% increase in equipment uptime and a 20% extension in the life cycle of aging assets.

PdM provides the data required to ensure the maximum interval between repairs and minimize the number and cost of unscheduled downtimes (Figure 6.6) [11].

The widely reported advantages of PdM are:

I. Fewer machine breakdowns, unplanned downtime and unexpected failures of other critical assets
II. Increased life expectancy of machines and equipment by about 20%
III. Reduced inventory of spare parts and capital tied to this item
IV. Improved safety and health throughout the workplace for technicians and operators

TABLE 6.1
Tools Used and Maintenance Diagnosis

Tool/Sensor	Maintenance Diagnosis
Digital camera	Digital camera is used to identify a defect or anomaly in a product or process in real time.
Infrared camera	Infrared camera is used to detect high temperatures owing to worn-out components in motors, stop mechanisms, etc.
Acoustic analysis	Acoustic analysis helps to detect liquid and air leaks and unusual sounds.
Vibration analysis	Sensors are used to determine an increase or decrease in the vibration of rotating components in pumps and compressors or machine parts that have a gear assembly.
Oil analysis	Proper lubrication extends a machine's useful life. Oil analysis can determine if the lubricant has been compromised by foreign particles and contaminants.

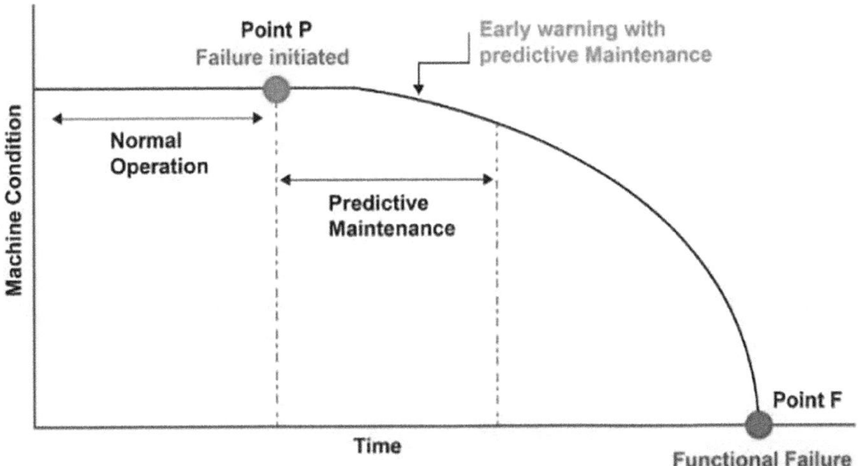

FIGURE 6.6 Machine condition and the optimum time for predictive maintenance.

6.5 THE PUSH FOR PdM

Industries such as aviation, railways, automotive, oil and gas, power generation and pharmaceuticals are generally at the forefront of innovation in key areas such as reliability, safety and dependability. Any system failure can have serious consequences on service levels, product quality and lead times. These industries have been integrating ML and AI in their day-to-day operations to predict possible failures. Notwithstanding ML and AI, it must be emphasized that deep human expertise is still required to understand machine failures as sometimes a single or simple observation can be used to determine machine conditions. As data storage or computing capacity becomes more affordable, lower-tech industries are rapidly catching up with the benefits of introducing ML and AI in their operations. In such industries, it is reported that it takes 3–6 months to build an ML model for a given machine.

In Mauritius, the immediate post-COVID years have been extremely difficult for the manufacturing sector, with steep rises in costs of raw materials, sea freight and other inputs. PdM is nascent in the Mauritian industry. The focus is on affordable, easily calibrated and reliable IoT devices and smart sensors that can be installed on production machinery for specific tasks. In addition, as the workforce ages, continuous improvement of production methods, overall plant maintenance as well as hiring and training of fresh recruits are becoming increasingly challenging. The re-skilling of staff is high on the agenda wherein the basics of ML and AI and the key importance of data are being introduced. Operators and supporting personnel are acutely aware that without correct data an ML model cannot be trained to identify the patterns in data that underpin potential triggers of PdM. ML is gradually assuming a transformative role in manufacturing and its influence is bound to spread.

6.6 CASE STUDY: PdM OF CIRCULAR KNITTING MACHINES

The process of knitting in textile manufacturing is highly versatile and can produce a wide range of fabrics such as jersey, purl, rib and pique. Circular knitting machines (CKMs) are ubiquitous in textile plants. High production speeds and fault-free knitted fabrics ensure productivity, quality, profitability and competitiveness.

The CKM is typically fitted with a variety of sensors, namely speed sensors and machine stop detection sensors to acquire data (Figure 6.7). The machine stop detection sensors are designed to detect different types of anomalies. The data captured is then used to train the ML model for PdM (Figure 6.8) [12]. The system provides real-time monitoring of the CKM, allowing timely detection of potential faults and

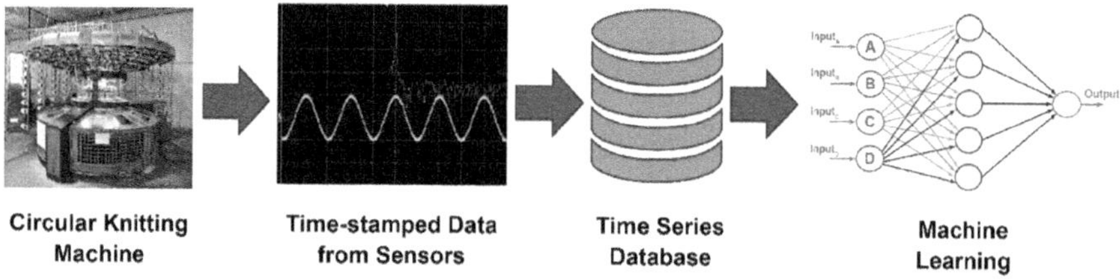

FIGURE 6.7 Data acquisition and training machine learning model.

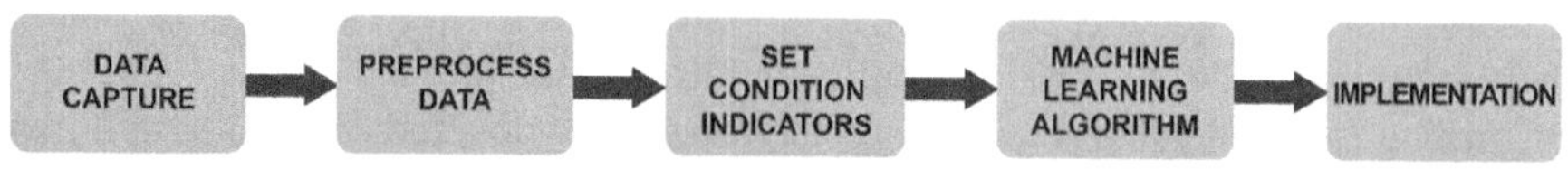

FIGURE 6.8 Typical flowchart for machine learning implementation.

effective maintenance scheduling. The machine's built-in stop detection sensors help to identify the root cause of machine stops and enable operators to take corrective actions to reduce unplanned downtime [13]. Sensor inputs, real-time data processing and adaptive responses converge seamlessly, ensuring a harmonious interaction between operators and the CKM. The implementation of PdM for CKMs is therefore a major asset for the sustained competitiveness of the Mauritian industry.

Machine profiles detailing production settings like machine type and yarn configurations might be included in the database. The database can allow easy management of multiple machine profiles and maintain detailed records of each shift activity. This dataset may serve as a repository for preserving the machine's historical records, facilitating the application of PdM through the categorization and classification of instances of machine stoppages.

6.7 UNLEASHING THE POWER OF "DIGITAL TWINS"

> A digital twin is a set of virtual information constructs that mimics the structure, context, and behaviour of an individual/unique physical asset, is dynamically updated with data from its physical twin throughout its lifecycle, and informs decisions that realize value.
>
> *[14]*

The idea of a "digital twin" (DT) was born at NASA in the 1960s as a "living model" of the Apollo mission. In response to Apollo 13's oxygen tank explosion and subsequent damage to the main engine, NASA employed multiple simulators to evaluate the failure and extended a physical model of the vehicle to include digital components [15].

A DT helps to analyze and accurately plan the implementation of possible changes, address bottlenecks and reduce the environmental impact of operations. While the DT concept predates the COVID pandemic, lockdowns did hasten their introduction in many industries. To create a replica of a manufacturing unit, the tool that is commonly used is a NavVis 360 mapping device, just like Google Car does for Streetview. An operative can walk around scanning the plant area, capturing its layout and key installations. This digital replica of the plant facilitates the study and planning of possible changes from the convenience of a desk.

These virtual replicas open a "digital door" to a wide range of possibilities, as they can be used to simulate and optimize production workflows, reduce costs and identify safety and ergonomic improvements to assembly processes. Digital replicas also enable the input of tracking points. These can be used to easily locate where a part is installed or where a particular piece of equipment is located.

DTs also offer great support to conduct "what-if" scenarios. The impact of changes can be simulated or improvements to production systems safely implemented, with little risk to safety or quality. Digital replicas allow easy access to a detailed 3D survey in which they can apply their modifications, avoiding unnecessary costs, work and time for the development of new products. This helps to make informed decisions about the actual physical implementations.

6.8 CONCLUSION

Post-pandemic, the industry focus on AI tools to build tangible value, slow the corrosive effect of high inflation on production costs and re-capture market share is gaining momentum. It is becoming increasingly clear that the merger of the human imagination, ML algorithms and AI can deliver more efficiency and enhanced value propositions than the human intellect alone. PV is well established, as till recently most machines were designed and built around it. The focus has been on automation and robotization to reduce labor input and associated costs. The transition to PdM is happening slowly but surely. Future inroads of PdM in the industry will depend on how much and what type of risk is acceptable to build AI-driven maintenance protocols into machines. Threats may include ransomware attacks paralyzing production and endangering the safety of the production plant's critical infrastructure.

REFERENCES

1. M. Ben-Daya, S. O. Duffuaa, "Maintenance and quality: The missing link", *Journal of Quality in Maintenance Engineering*, 1(1), 1995, pp. 20–26. https://doi.org/10.1108/13552519510083110
2. M. Faccio, A. Persona, F. Sgarbossa, G. Zanin, "Industrial maintenance policy development: A quantitative framework", *International Journal of Production Economics*, 147, Part A, 2014, pp. 85–93.
3. S. K. Pinjala, L. Pintelon, A. Vereecke, "An empirical investigation on the relationship between business and maintenance strategies", *International Journal of Production Economics*, 104(1), 2006, pp. 214–229.
4. M. Imai, *The Key to Japan's Competitive Success*, 1986, McGraw-Hill.
5. V. Janjić, M. Todorović, D. Jovanović, "Key success factors and benefits of kaizen implementation", *Engineering Management Journal*, 32(2), 2020, pp. 98–106. https://doi.org/10.1080/10429247.2019.1664274
6. F. Almada-Lobo, "The industry 4.0 revolution and the future of manufacturing execution systems (MES)", *Journal of Innovation Management*, 3(4), 2016, pp. 16–21. https://doi.org/10.24840/2183-0606_003.004_0003
7. C. Scheffer, P. Girdhar, *Practical Machinery Vibration Analysis and Predictive Maintenance*, 2004, Newnes Publications, pp. 1–4.
8. R. K. Mobley, *An Introduction to Predictive Maintenance*, Second Edition (Plant Engineering), 2002, Elsevier Science, pp. 2–4.
9. Y. Ran, X. Zhou, P. Lin, Y. Wen, R. Deng, "A survey of predictive maintenance: Systems, purposes and approaches", *IEEE Communications Surveys & Tutorials*, 20, 2019, pp. 1–36.
10. T. Zonta, C. André da Costa, R. da Rosa Righi, Miromar J. de Lima, E. Silveira da Trindade, G. Pyng Li, "Predictive maintenance in the industry 4.0: A systematic literature review", *Computers & Industrial Engineering*, 150, 2020, pp. 106889.
11. M. Achouch, M. Dimitrova, K. Ziane, S. Sattarpanah Karganroudi, R. Dhouib, H. Ibrahim, M. Adda, "On predictive maintenance in industry 4.0: Overview, models, and challenges", *Applied Sciences*, 12, 2022, pp. 8081. https://doi.org/10.3390/app12168081
12. DIFT. Data-driven resource efficiency on circular knitting machines. No Date. https://www.ditf.de/en/index/more-information/data-driven-resource-efficiency-on-circular-knitting-machines.html [Accessed 05 June 2024].

13. A. Elkateb, A. Métwalli, A. Shendy, A. E. B. Abu-Alanine, "Machine learning and IoT – Based predictive maintenance approach for industrial applications", *Alexandria Engineering Journal*, 88, 2024, pp. 298–309.
14. AIAA Institute Position Paper, 2020. https://www.aiaa.org/docs/default-source/uploadedfiles/issues-and-advocacy/policy-papers/digital-twin-institute-position-paper-(december-2020).pdf
15. E. H. Glaessgen, D. S. Stargel, "The digital twin paradigm for future NASA and U.S. airforce vehicles", In *53rd AIAA/ASME/ASCE/AHS/ASC Structures, Structural Dynamics & Materials Conference*, 23–26 April 2012, Honolulu, Hawaii.

7 Enabling Technologies with Industry 4.0 for Renewable Energy in Africa

Robert T. F. Ah King and Bhimsen Rajkumarsingh

7.1 INTRODUCTION

Klaus Schwab, the World Economic Forum's founder and executive chairman, created the term "fourth industrial revolution," which describes a world in which people travel between digital domains and offline realities using connected technology to empower and control their lives [1]. The Industrial Revolution 4.0 shows challenges for the business world, including a lack of sufficient Human Resource (HR) skills, a security issue with communication technology, the reliability of the stability of production machines, stakeholders' inability to change, and the number of job losses due to automation [2].

Africa can improve energy access, increase energy security, and speed the transition to a sustainable and resilient energy system by incorporating Industry 4.0 technologies into its renewable energy industry. However, successful integration necessitates investments in digital infrastructure, skill development, and regulatory frameworks that encourage innovation and technological adoption in the renewable energy sector.

Industry 5.0, also known as the human-centric industrial revolution, emphasizes the integration of human intelligence with sophisticated technology such as artificial intelligence (AI), robotics, and the Internet of Things (IoT). While its benefits are worldwide, Africa's particular socioeconomic situation promises to benefit considerably from Industry 5.0 in various ways.

7.2 INDUSTRY 4.0 TECHNOLOGIES FOR RENEWABLE ENERGY

Industry 4.0 technologies like IoT, robotics, AI, big data and analytics, blockchain, cloud, augmented reality (AR), simulation, and cyber security make low-cost development possible by improving energy and resource utilization [3,4] (Figure 7.1, adapted from Ref. [5]). Industry 4.0 promotes sustainable production by lowering waste output through a streamlined manufacturing process and a heavy emphasis on recycling and remanufacturing.

Green processes are critical for implementing green technology in manufacturing to generate positive sustainability outcomes in the Industry 4.0 era. Vrchota et al. [6]

DOI: 10.1201/9781003511298-7

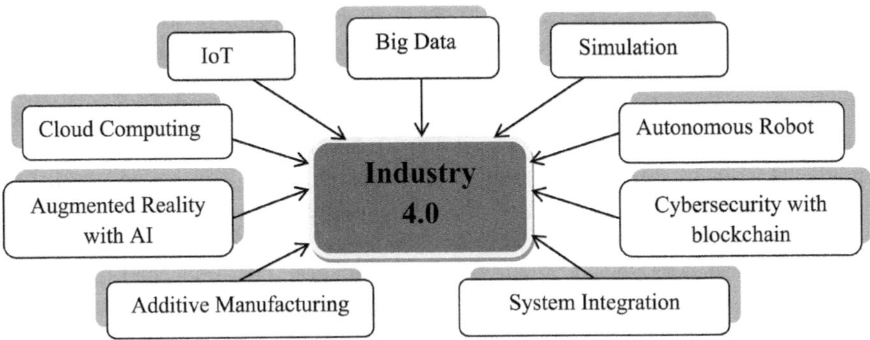

FIGURE 7.1 Technologies of Industry 4.0 [5].

showed that the most prevalent sustainability outcomes include energy savings, emission reduction, resource optimization, cost reduction, productivity and efficiency, improved economic performance, HR development, social welfare, and workplace safety. Leong et al. [7] presented a study on an improved adaptive model for implementing the lean and green strategy in processing sectors to address dynamic industry concerns related to Industry 4.0. This upgraded adaptive model integrates experts' experience and operational data as input when dealing with real-world industry applications. Shahin et al. [8] gave a complete assessment and report on the links between lean tools and Industry 4.0 technology, as well as how implementing these two paradigms concurrently affects factory operating performance.

7.2.1 IoT for Renewable Energy

The IoT is helping to turn traditional factories into smart factories in Industry 4.0 by monitoring and optimizing industrial processes through a network of networked devices, sensors, and software [9]. IoT in smart factories has recently been developed to support the digital revolution in manufacturing and enable firms to operate more efficiently, cost-effectively, and sustainably [10]. Smart factories are becoming increasingly popular, and IoT is a key technology enabling them [11]. The IoT has the potential to provide several benefits to Africa in a variety of industries, including agriculture, healthcare, transportation, energy, and urban development.

A study in Kenya developed a LoRaWAN-based IoT platform for monitoring distribution networks, enabling quick reparative action in Nakuru County [12]. The system triggers a network-monitoring center within 100 ms of fault occurrence. This example illustrates how IoT can be leveraged to enhance renewable energy applications in Africa, addressing energy access challenges while promoting sustainable development.

7.2.2 Autonomous Robots for Renewable Energy

Robotics is a crucial Industry 4.0 technology that provides several industrial applications. This technology has improved automated systems by doing repeated activities more precisely and cost-effectively [13]. The primary goal of robotics and Industry 4.0 is to increase productivity, produce high-quality products at cheap costs, and exceed customer expectations [14]. Robotic process automation (RPA) is a relatively recent office automation technique. RPA is not a real robot; rather, it is software that behaves like a human when conversing with a computer. Activities such as reading emails, viewing attachments, entering data, and preparing reports are accomplished quickly, precisely, and consistently. RPA is classified into three types: attended mode, unattended mode, and hybrid mode, and it may be used for a variety of operations in many sectors, including purchasing, production, HR, sales, and marketing. RPA provides benefits to businesses, customers, and employees [15]. Automation and robots have the ability to uncover new prospects for economic growth, sustainable development, and human progress in Africa. Automation and robotics can be beneficial to Africa through increased productivity in manufacturing, agricultural automation, infrastructure development, healthcare and medical robotics, logistics and supply chain management, mining and resource extraction, education and skills development, and environmental monitoring and conservation.

With the local Moroccan weather in mind, a new self-guiding cleaning robot system has been developed to maximize the cleaning schedule [16]. This study looked at how dust collection affected the glass samples' transmittance and the PV module's overall electrical efficiency for various cleaning frequencies. This example illustrates how autonomous robotics can contribute to the efficient operation and maintenance of renewable energy infrastructure in Africa, supporting the continent's transition toward sustainable energy solutions.

7.2.3 AI for Renewable Energy

AI is an important component of industrial transformation because it enables intelligent equipment to conduct autonomous functions including self-monitoring, interpretation, diagnosis, and analysis [17]. Recent developments in industrial AI have shown the technology's capacity to help manufacturers address the challenges related to the digital evolution of cyber-physical systems, thanks to its data-driven predictive analytics and ability to aid decision-making in highly complex, non-linear, and frequently multiple environments [18]. AI has the ability to alter African industries, stimulate innovation, and accelerate economic growth in the framework of Industry 4.0. Some ways AI can help Industry 4.0 in Africa are in predictive maintenance, smart manufacturing, supply chain optimization, energy efficiency, customized production, quality control and inspection, remote monitoring and control, skill development and workforce training, market intelligence, and competitive analysis.

A Kenyan startup called M-KOPA installed an AI-powered pay-as-you-go solar system to give off-grid households access to reasonably priced electricity [19]. The system analyzes usage trends using machine learning, allowing for customizable payment schedules based on user energy requirements.

7.2.4 Big Data Analytics for Renewable Energy

Two primary goals for Industry 4.0 applications are to ensure maximum uptime throughout the manufacturing chain and to boost productivity while lowering production costs. As the data-driven economy advances, businesses have begun to use big data strategies to achieve their goals. Big data and IoT technologies are critical in developing data-driven applications like predictive maintenance [20]. Big data analytics may provide African industries with important insights and actionable intelligence, boosting innovation, efficiency, and competitiveness in the era of Industry 4.0. Big data analytics can help Africa in Industry 4.0 in predictive maintenance, optimized production processes, supply chain optimization, quality control and defect detection, energy management and efficiency, personalized customer experiences, market intelligence and competitive analysis, risk management and fraud detection, and workforce optimization.

An IoT-enabled smart microgrid system is proposed for sub-Saharan Africa, focusing on Nigeria's energy sector [21]. It uses IBM Watson analytics software to analyze energy-driven economic growth, providing insights for sustainable smart energy systems. This collaboration allows stakeholders to meet increasing energy production demands in the region.

7.2.5 Simulation for Renewable Energy

Simulation is an important tool for developing planning and exploratory models to improve decision-making, as well as for designing and managing sophisticated and intelligent production systems. Simulation can help businesses understand the risks and expenses associated with the adoption, deployment, and operational performance difficulties of Industry 4.0 technology. The bulk of process-related sectors use simulation extensively to monitor real-time data and match it to the physical environment using a virtual model. Simulation is widely used using a virtual model to track real-time data while also monitoring and analyzing relevant physical processes [22]. Simulation can provide African firms and governments with significant analytical capabilities and decision-making tools, allowing them to adapt to changing market conditions, negotiate difficult obstacles, and capitalize on possibilities for growth and development in the context of Industry 4.0. Simulation can help Africa in Industry 4.0 in process optimization, product design and development, training and skills development, predictive analytics and decision support, supply chain resilience and risk management, energy efficiency and sustainability, cybersecurity and threat assessment, smart cities and urban planning, policy modeling, and impact assessment.

A Kenyan renewable energy company is planning to develop a wind farm using advanced simulation software [23]. The company conducts a thorough wind resource assessment, identifying optimal turbine locations based on wind speed data and local topography. Simulation models are used to optimize turbine layouts, maximizing energy yield and minimizing environmental impact, while ensuring project financing and compliance with regulatory requirements.

7.2.6 Blockchain for Renewable Energy

Industry 4.0 involves advances based on emerging digital technology, like blockchain. Blockchain technology can be used to increase security, privacy, and data transparency in both small and large organizations. Blockchain is a well-recognized technology that has the potential to improve the industrial and supply chain environment.

Javaid et al. [24] and Aoun et al. [25] presented a survey of research publications on blockchain and Industry 4.0. Blockchain technology has the ability to accelerate digital transformation, economic empowerment, and sustainable development in Africa by providing new trust, transparency, and decentralization solutions within the context of Industry 4.0. Blockchain can help Africa in Industry 4.0 in supply chain traceability, smart contracts and automation, secure data sharing and collaboration, digital identity and authentication, financial inclusion and access to capital, intellectual property protection, decentralized energy grids, supply chain financing, and trade finance.

A blockchain platform in Ghana allows households with solar panels to register as energy producers, monitor their production and consumption, and trade excess energy with nearby consumers [26]. This decentralized and sustainable energy ecosystem empowers local communities, promoting renewable energy adoption and fostering a resilient and sustainable energy infrastructure for the future.

7.2.7 Cloud Computing for Renewable Energy

Cloud computing refers to the availability of hardware and software resources, specifically data storage (cloud storage) and computational power, without the user's direct active control. Cloud computing allows users to obtain technological services whenever they need them through a "pay-as-you-go system," saving them from the headache of buying and operating data centers and servers [27]. There are four primary subtypes of cloud computing: private, public, hybrid, and multi-cloud. Cloud computing services are classified into four categories: Infrastructure-as-a-Service, Platforms-as-a-Service, Functions as a Service, and Software-as-a-Service [28]. Ivanov et al. [29] proposed a cloud supply chain with a business model that uses cloud-enabled networking of third-party physical and digital assets to construct and operate a supply chain network. Cloud supply chain merges Industry 4.0 principles and technology with digital platforms emerging from the "supply chain-as-a-service" paradigm. Cloud computing provides African firms with a strategic edge by enabling access to scalable, cost-effective, and creative IT solutions that enable digital transformation, agility, and competitiveness in the context of Industry 4.0. Cloud computing can help Africa in Industry 4.0 in scalable infrastructure; cost efficiency; remote access and collaboration; data storage and management; big data analytics and AI, IoT, and edge computing; security and compliance; disaster recovery; and business continuity.

A cloud-based platform in Rwanda manages solar microgrids, utilizing IoT sensors to transmit real-time energy data [30]. Machine learning algorithms optimize energy distribution based on local demand patterns. This ensures reliable electricity access, efficient management, and customer satisfaction. Cloud computing enhances renewable energy solutions' effectiveness and scalability, contributing to sustainable development and improved energy access in rural areas.

7.2.8 Cyber Security for Renewable Energy

Industry 4.0 is a manufacturing revolution that brings disruptive technologies like the IoT and cloud computing to the core of the industry. The consequent enhanced automation and enhanced synergy between stocks, supply networks, and customer needs, together with the dangers and attacks from the Internet [31]. Corallo et al. [32] discussed why cybersecurity is important for the protection of confidential information, compliance with regulations, protection of reputation, prevention of financial losses, and business interruption. The most common types of cybersecurity threats are Malware, Phishing, Ransomware, Denial of Service Attacks, Insider Threats, and Advanced Persistent Threats. The six types of cybersecurity are Network Security, Application Security, Information Security, Cloud Security, IoT Security, and Identity and Access Management. These methods can help protect companies and individuals from cyber attacks. Furthermore, standard practices like utilizing antivirus software, updating software on a regular basis, using strong passwords and multi-factor authentication, and offering cybersecurity education and awareness can all assist in preventing cyber assaults. Cybersecurity is critical for promoting trust, encouraging innovation, and facilitating the safe and secure deployment of Industry 4.0 technologies in Africa. Cybersecurity can help Africa in Industry 4.0 in the protection of digital assets, prevention of cyber attacks, data privacy and compliance, protection of industrial control systems, secure supply chains and third-party partners, incident detection and response, cybersecurity awareness and training, business continuity, and resilience.

A solar power plant in South Africa's Northern Cape employs a robust cybersecurity framework, including regular risk assessments, network segmentation, and access control mechanisms [33]. Real-time monitoring tools detect potential cyber threats, and an incident response team responds promptly. Employee training and awareness programs ensure personnel understand their roles in maintaining cybersecurity. This proactive approach protects critical infrastructure, supports sustainable energy development, and boosts economic growth.

7.2.9 Additive Manufacturing for Renewable Energy

Additive manufacturing, also known as 3D printing, is a manufacturing process that creates products layer by layer using digital 3D models. Unlike traditional subtractive manufacturing methods, which entail cutting, molding, or drilling materials to form products, additive manufacturing adds material layer by layer, most commonly utilizing plastics, metals, ceramics, or composites [34]. Additive manufacturing has the potential to transform manufacturing and propel economic growth, innovation, and sustainability in Africa. African countries may realize the full potential of additive manufacturing in Industry 4.0 by adopting additive manufacturing technology and investing in the appropriate infrastructure, skills, and regulatory frameworks. Additive manufacturing can help Africa in Industry 4.0 in reduced manufacturing costs, customization and personalization, rapid prototyping and product development, localized production and distributed manufacturing, sustainable manufacturing practices, tooling and jig production, skills development and innovation ecosystems, and medical and healthcare applications.

A local research institution in Egypt is working with renewable energy companies to develop innovative wind turbine blade designs using additive manufacturing [35]. Engineers use CAD software to design and simulate blade configurations, which are then tested in simulated wind conditions. The blade designs are then scaled up for production, supporting Egypt's renewable energy goals and economic development initiatives. This demonstrates the potential of 3D printing technology in Africa.

7.2.10 AR for Renewable Energy

AR is a technology that superimposes digital information, such as photographs, videos, or 3D models, in the real-world environment, which is often seen using a smartphone, tablet, or AR headset [34]. Unlike virtual reality, which immerses users in an entirely virtual environment, AR enriches the real-world experience by including virtual features. AR has the ability to address a variety of socioeconomic difficulties in Africa, promoting good change in sectors such as education, healthcare, tourism, infrastructure, agriculture, and economic development. AR can transform education and training by enabling interactive and immersive learning experiences. In Africa, where traditional educational resources may be inadequate, AR can provide students with high-quality educational content, simulations, and virtual labs, improving learning outcomes and boosting digital literacy.

Photovoltaic pumping systems provide water for populations and agriculture in non-electrified areas but often face maintenance and control issues. An AR technology system for training and assisting in maintenance in industrial contexts, featuring a tablet PC display and a scenario-based concept that guides users through training or maintenance tasks has been proposed in Algeria [36].

7.2.11 System Integration for Renewable Energy

System integration is critical to realizing the full potential of Industry 4.0 technologies in Africa, allowing businesses to digitize, optimize, and transform their operations for long-term growth, competitiveness, and resilience in the global marketplace. By adopting system integration, African industries may expedite their digital transformation journey and reap the benefits of Industry 4.0 in promoting economic development, job creation, and social improvement across the continent.

A rural electrification project in Kenya uses an integrated renewable energy system, including solar PV arrays and wind turbines [37]. The energy is stored in battery storage systems and managed through a smart grid. Smart meters in households monitor energy usage, reducing costs. A centralized control center with IoT sensors and cloud-based monitoring provides real-time data analytics, enhancing reliability and sustainability in rural Kenya.

7.3 RENEWABLE ENERGY SOURCES INTO AFRICA'S INDUSTRY 4.0 PROJECTS

Integrating renewable energy sources into Africa's Industry 4.0 projects has the potential to deliver various benefits, including alignment with the continent's sustainable development goals and economic growth. Here's how Africa might benefit from integrating renewable energy in Industry 4.0 (Figure 7.2):

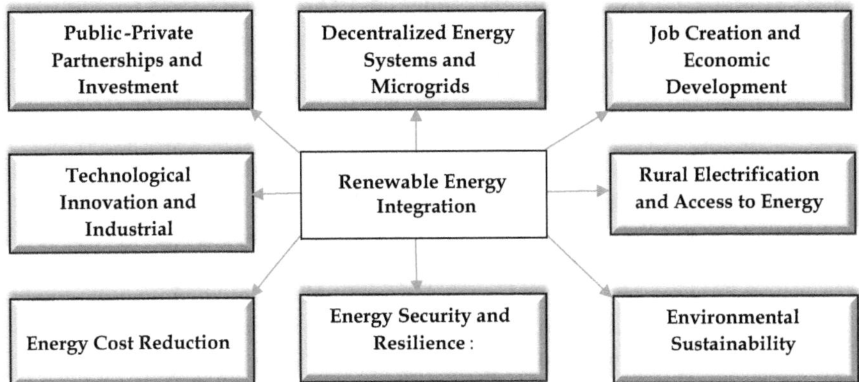

FIGURE 7.2 Renewable energy sources in Africa's Industry 4.0 projects.

1. **Energy Cost Reduction**: Solar, wind, and hydropower are all examples of renewable energy sources that can be less expensive than traditional fossil fuels [38]. African enterprises may minimize their dependency on expensive imported fuels and stabilize energy prices by utilizing renewable energy, thereby increasing their global competitiveness.
2. **Energy Security and Resilience**: Renewable energy sources are numerous and readily available in many African regions. African industries may improve energy security, reduce reliance on volatile fossil fuel markets, and strengthen resilience to energy supply disruptions and price variations by leveraging renewable energy [39].
3. **Environmental Sustainability**: Integrating renewable energy into Industry 4.0 projects can assist African industries in lowering their carbon footprint, mitigating climate change effects, and promoting environmental sustainability. Renewable energy sources emit fewer greenhouse gas emissions and pollutants than fossil fuels, which aids attempts to meet sustainable development goals and protect natural resources [40].
4. **Job Creation and Economic Development**: The deployment of renewable energy infrastructure in Africa has the potential to generate local jobs, improve skills, and diversify the economy [38]. Renewable energy projects necessitate a variety of specialized and semi-skilled labor, with job possibilities available in the building, installation, maintenance, and operation of renewable energy systems.
5. **Rural Electrification and Access to Energy**: Integrating renewable energy into Industry 4.0 efforts can assist in alleviating energy poverty and increasing access to electricity in Africa's rural and neglected communities. Off-grid renewable energy solutions, such as solar mini-grids and decentralized energy systems, can bring reliable and inexpensive electricity to remote locations, opening up new prospects for economic development, education, and healthcare [41].

6. **Technological Innovation and Industrial Competitiveness**: Integrating renewable energy into Industry 4.0 projects promotes technological innovation, research, and development in renewable energy technology, smart grid solutions, and energy management systems [42]. African industries may use renewable energy breakthroughs to boost productivity, maximize resource usage, and gain a competitive advantage in global markets.
7. **Decentralized Energy Systems and Microgrids**: Renewable energy sources enable the creation of decentralized energy systems and microgrids, which promote energy self-sufficiency, community empowerment, and local economic growth. Microgrids fueled by renewable energy can deliver consistent electricity to industrial facilities, commercial buildings, and isolated villages, reducing reliance on centralized grid infrastructure and increasing energy resilience [43].
8. **Public–Private Partnerships and Investment Opportunities**: The integration of renewable energy into Industry 4.0 efforts necessitates cooperation among governments, private sector entities, and international partners. Public–private partnerships can help mobilize investment resources, experience, and technology transfer to assist the deployment of renewable energy projects, infrastructure development, and legislative reforms that promote renewable energy integration across Africa [44].

7.4 CHALLENGES FOR RENEWABLE ENERGY IN AFRICA

Africa holds immense potential for renewable energy development, but several challenges hinder its widespread adoption across the continent. Some of the key challenges are given in Table 7.1.

TABLE 7.1
Challenges for Renewable Energy in Africa

Challenges	Description	References
Lack of infrastructure	Many African locations lack the infrastructure required for large-scale renewable energy projects, such as transmission lines, grid interconnection, and storage facilities. Without proper infrastructure, it is difficult to effectively transmit and use renewable energy.	[45–48]
Limited access to financing	Financing renewable energy projects can be challenging due to limited capital, high upfront costs, and perceived financial risks. Many African countries struggle to attract domestic and foreign investment for renewable energy development.	[49–51]
Policy and regulatory barriers	Inconsistent or obsolete rules, regulations, and bureaucratic roadblocks can stymie renewable energy implementation. Uncertain regulatory environments and a lack of supportive policies frequently discourage investment and slow project development.	[52–54]

(Continued)

TABLE 7.1 (*Continued*)
Challenges for Renewable Energy in Africa

Challenges	Description	References
Intermittency and reliability	Certain renewable energy sources, such as sun and wind, have intermittent and variable output. While advances in energy storage technology can help address these challenges, guaranteeing a consistent and secure energy supply remains difficult, especially in rural or off-grid places.	[55,56]
Skills and capacity building	Many African countries face a dearth of skilled professionals and technicians with knowledge of renewable energy technologies. Building local capacity through education, training, and knowledge transfer is critical to the successful deployment and maintenance of renewable energy systems.	[57]
Land use and environmental concerns	Large-scale renewable energy projects, such as solar and wind farms, may necessitate extensive land acquisition, potentially leading to conflicts over land usage and environmental effects. Balancing renewable energy production, environmental conservation, and community interests is critical.	[58–60]
Energy poverty and accessibility	Millions of people in Africa do not have dependable access to power, particularly in rural and neglected areas. Addressing energy poverty and increasing energy access through renewable energy solutions necessitates focused policies, investments, and novel techniques.	[61–63]
Technological challenges	While renewable energy technologies have made considerable advances in recent years, issues such as device reliability, maintenance, and compatibility with existing infrastructure persist. Continued research and development efforts are required to overcome these technological hurdles.	[64]
Political stability and governance	Political instability, corruption, and governance concerns in several African countries can impede the implementation of renewable energy projects and erode investor trust. Establishing transparent and accountable governance institutions is critical for generating a favorable atmosphere for renewable energy growth.	[65–67]

7.5 FUTURE TRENDS IN THE DIGITAL REVOLUTION: INDUSTRY 5.0

Industry 4.0, often known as the fourth industrial revolution, refers to increased automation for organizational productivity and efficiency gained by combining an industry's virtual and physical worlds. With Industry 4.0 unable to meet the increased need for customization, the term Industry 5.0 was coined to address personalized manufacturing and empower humans in industrial processes. The introduction of the term Industry 5.0 has resulted in a variety of views of what it means to reconcile people and robots [68]. Industry 5.0 presents immense promise for Africa by harnessing technology to promote human-centric, inclusive, and sustainable development. African countries that embrace Industry 5.0 ideas and practices can provide new prospects for economic growth, social improvement, and environmental stewardship in the digital age. Industry 5.0 can help Africa in:

1. **Job Creation and Skills Development**: Human–machine collaboration, in which humans work alongside intelligent machines to improve efficiency, creativity, and invention, is central to Industry 5.0 [69]. Industry 5.0 in Africa has the potential to produce jobs, skills development, and economic empowerment by harnessing human capabilities, creativity, and problem-solving abilities in conjunction with new technologies.
2. **Inclusive Growth and Social Development**: Industry 5.0 seeks to address societal concerns and promote inclusive growth by leveraging technology for social good [70]. In Africa, Industry 5.0 can propel social development projects such as healthcare, education, and agriculture by harnessing sophisticated technologies to increase access, quality, and affordability of key services for underprivileged communities.
3. **Sustainable Industrialization**: Industry 5.0 advocates sustainable industrialization by incorporating environmental concerns into the design, manufacture, and consumption of goods and services [71]. In Africa, Industry 5.0 can help achieve sustainable development goals by optimizing resource use, decreasing waste and emissions, and promoting circular economy concepts in the manufacturing, energy, and transportation sectors.
4. **Digital Transformation and Innovation Ecosystems**: Industry 5.0 promotes digital transformation and innovation ecosystems to enhance economic growth, competitiveness, and resilience in the digital age [72]. Industry 5.0 in Africa can promote innovation, entrepreneurship, and technology adoption by offering platforms, incentives, and assistance for start-ups, Small and Medium Enterprises (SMEs), and research institutions to create and implement new solutions that solve local concerns as well as global market demands.
5. **Value-Added Manufacturing and Global Competitiveness**: Industry 5.0 offers value-added manufacturing processes that focus on customization, personalization, and customer-centricity [73]. In Africa, Industry 5.0 may boost local industries' competitiveness by utilizing modern technologies like additive manufacturing, digital twins, and smart robots to generate high-quality, customized products and services that match different market demands and global standards.
6. **Digital Infrastructure and Connectivity**: Industry 5.0 necessitates strong digital infrastructure and connectivity to enable data-driven decision-making, real-time collaboration, and seamless system and process integration [74]. In Africa, Industry 5.0 can spur investments in digital infrastructure such as broadband networks, data centers, and IoT platforms, thereby bridging the digital gap, expanding access to digital technologies, and enabling participation in the global digital economy.

7. **Smart Cities and Sustainable Urbanization**: Industry 5.0 encourages smart cities and sustainable urbanization programs that use technology to improve quality of life, public services, and environmental sustainability [75]. In Africa, Industry 5.0 can help to construct smart cities, intelligent infrastructure, and digital services that handle urban concerns including transportation, electricity, water, and waste management, while also improving urban resilience and livability.
8. **Collaborative Governance and Stakeholder Engagement**: Industry 5.0 promotes collaborative governance and stakeholder engagement approaches that encourage co-creation, co-design, and co-innovation across the public, private, and civil society sectors [76]. In Africa, Industry 5.0 may support multi-stakeholder partnerships, information sharing, and capacity-building projects that enable local communities, governments, and enterprises to determine the future of industry in a participatory and inclusive manner.

7.6 CONCLUSION

Industry 4.0 technologies can improve energy access, increase energy security, and speed up the transition to sustainable and resilient energy systems in Africa. This chapter has reviewed the different technologies such as IoT, autonomous robots, AI, big data analytics, simulation, blockchain, cloud computing, cyber security, additive manufacturing, AR, and system integration, which can improve energy and resource utilization.

With Industry 4.0, Africa can benefit from renewable energy integration along with the continent's sustainable development goals and economic growth. Energy cost reduction using renewable energy sources, energy security and resilience, environmental sustainability, job creation and economic development, rural electrification and access to energy, technological innovation and industrial competitiveness, decentralized energy systems and microgrids, public–private partnerships, and investment opportunities are potential avenues where Africa can progress.

However, there are several challenges that hinder renewable energy development and its widespread adoption across the continent. These hurdles include lack of infrastructure, limited access to financing, policy and regulatory barriers, intermittency and reliability of renewable energy sources, skills and capacity building, land use and environmental concerns, energy poverty and accessibility, technological challenges, and political stability and governance.

Hopefully, Industry 5.0 presents immense promise for Africa by harnessing technology to promote human-centric, inclusive, and sustainable development. The next industrial revolution will help in job creation and skills development, promote inclusive growth and social development, advocate sustainable industrialization, promote digital transformation and innovation ecosystems, offer value-added manufacturing and global competitiveness, spur digital infrastructure and connectivity, encourage smart cities and sustainable urbanization, and promote collaborative governance and stakeholder engagement.

REFERENCES

1. D. Miller, *Natural Language: The User Interface for the Fourth Industrial Revolution*, Opus Research Report, September 2016.
2. L. Ellitan, "Competing in the era of industrial revolution 4.0 and society 5.0," *Jurnal Maksipreneur: Manajemen, Koperasi, dan Entrepreneurship*, vol. 10, 2020, 1–12. https://doi.org/10.30588/jmp.v10i1.657.
3. L.S. Dalenogare, G.B. Benitez, N.F. Ayala, and A.G. Frank, "The expected contribution of Industry 4.0 technologies for industrial performance," *International Journal of Production Economics*, vol. 204, 2018, 383–394, ISSN 0925-5273, https://doi.org/10.1016/j.ijpe.2018.08.019
4. C. Bai, S. Kusi-Sarpong, and J. Sarkis, "An implementation path for green information technology systems in the Ghanaian mining industry," *Journal of Cleaner Production*, vol. 164, 2017, 1105–1123, ISSN 0959-6526, https://doi.org/10.1016/j.jclepro.2017.05.151.
5. M. Kerin and D.T. Pham, "A review of emerging industry 4.0 technologies in remanufacturing," *Journal of Cleaner Production*, vol. 237, 2019, 117805, ISSN 0959-6526, https://doi.org/10.1016/j.jclepro.2019.117805.
6. J. Vrchota, M. Pech, L. Rolínek, and J. Bednář, "Sustainability outcomes of green processes in relation to industry 4.0 in manufacturing: Systematic review," *Sustainability*, vol. 12, 2020, 5968, https://doi.org/10.3390/su12155968.
7. W.D. Leong, S.Y. Teng, B.S. How, S.L. Ngan, A.A. Rahman, C.P. Tan, S.G. Ponnambalam, and H.L. Lam, "Enhancing the adaptability: Lean and green strategy towards the industry revolution 4.0," *Journal of Cleaner Production*, vol. 273, 2020, 122870, ISSN 0959-6526, https://doi.org/10.1016/j.jclepro.2020.122870.
8. M. Shahin, F.F. Chen, H. Bouzary, and K. Krishnaiyer, "Integration of Lean practices and Industry 4.0 technologies: Smart manufacturing for next-generation enterprises," *The International Journal of Advanced Manufacturing Technology*, vol. 107, 2020, 2927–2936, https://doi.org/10.1007/s00170-020-05124-0.
9. M. Soori, B. Arezoo, and R. Dastres, "Internet of things for smart factories in Industry 4.0, a review," *Internet of Things and Cyber-Physical Systems*, vol. 3, 2023, 192–204, ISSN 2667-3452, https://doi.org/10.1016/j.iotcps.2023.04.006.
10. S. Wang, J. Wan, D. Li, and C. Zhang, "Implementing smart factory of industrie 4.0: An outlook," *International Journal of Distributed Sensor Networks*, vol. 12, no. 1, 2016, https://doi.org/10.1155/2016/3159805.
11. R.Y. Zhong, X. Xu, E. Klotz, and S.T. Newman, "Intelligent manufacturing in the context of Industry 4.0: A review," *Engineering*, vol. 3, no. 5, 2017, 616–630, ISSN 2095-8099, https://doi.org/10.1016/J.ENG.2017.05.015.
12. G.Y. Odongo, R. Musabe, D. Hanyurwimfura, and A.D. Bakari, "An efficient LoRa-enabled smart fault detection and monitoring platform for the power distribution system using self-powered IoT devices," *IEEE Access*, vol. 10, 2022, 73403–73420, https://doi.org/10.1109/ACCESS.2022.3189002.
13. M. Javaid, A. Haleem, R.P. Singh, and R. Suman, "Substantial capabilities of robotics in enhancing Industry 4.0 implementation," *Cognitive Robotics*, vol. 1, 2021, 58–75, ISSN 2667-2413, https://doi.org/10.1016/j.cogr.2021.06.001.
14. R. Goel and P. Gupta, "Robotics and industry 4.0," In Nayyar, A., Kumar, A. (eds) *A Roadmap to Industry 4.0: Smart Production, Sharp Business and Sustainable Development*. Advances in Science, Technology & Innovation. Springer, Cham, 2020, https://doi.org/10.1007/978-3-030-14544-6_9.
15. B. Axmann and H. Harmoko, "Robotic process automation: An overview and comparison to other technology in industry 4.0," In *2020 10th International Conference on Advanced Computer Information Technologies (ACIT)*, Deggendorf, 2020, 559–562, https://doi.org/10.1109/ACIT49673.2020.9208907.

16. A. Azouzoute, M. El Ydrissi, H. Zitouni, C. Hajjaj, and M. Garoum, "Dust accumulation and photovoltaic performance in semi-arid climate: Experimental investigation and design of cleaning robot," In Motahhir, S., Eltamaly, A. M. (eds) *Advanced Technologies for Solar Photovoltaics Energy Systems. Green Energy and Technology.* Springer, Cham, 2021, https://doi.org/10.1007/978-3-030-64565-6_3.
17. I. Ahmed, G. Jeon, and F. Piccialli, "From artificial intelligence to explainable artificial intelligence in Industry 4.0: A survey on what, how, and where," *IEEE Transactions on Industrial Informatics*, vol. 18, no. 8, 2022, 5031–5042, https://doi.org/10.1109/TII.2022.3146552.
18. R.S. Peres, X. Jia, J. Lee, K. Sun, A.W. Colombo, and J. Barata, "Industrial artificial intelligence in industry 4.0 - Systematic review, challenges and outlook," *IEEE Access*, vol. 8, 2020, 220121–220139, https://doi.org/10.1109/ACCESS.2020.3042874.
19. C.M. Ibegbulam, O.J. Aigbovbiosa, J.A. Olowonubi, and S.A. Fatounde, "Role of artificial intelligence in electrification of Africa," *Engineering Science & Technology Journal*, vol. 4, no. 6, 2023, 456–472, https://doi.org/10.51594/estj.v4i6.667.
20. R. Sahal, J.G. Breslin, and M.I. Ali, "Big data and stream processing platforms for Industry 4.0 requirements mapping for a predictive maintenance use case," *Journal of Manufacturing Systems*, vol. 54, 2020, 138–151, ISSN 0278-6125, https://doi.org/10.1016/j.jmsy.2019.11.004.
21. T.M. John, E.G. Ucheaga, O.O. Olowo, J.A. Badejo and A.A. Atayero, "Towards building smart energy systems in sub-Saharan Africa: A conceptual analytics of electric power consumption," In *2016 Future Technologies Conference (FTC)*, San Francisco, CA, 2016, pp. 796–805, https://doi.org/10.1109/FTC.2016.7821695.
22. W. de Paula Ferreira, F. Armellini, and L.A. De Santa-Eulalia, "Simulation in industry 4.0: A state-of-the-art review," *Computers & Industrial Engineering*, vol. 149, 2020, 106868, ISSN 0360-8352, https://doi.org/10.1016/j.cie.2020.106868.
23. L.L. Titus, *Renewable energy projects and community livelihoods in Marsabit County: A case of Lake Turkana wind power project in Marsabit County, Kenya*, Diss. Africa Nazarene University, 2022.
24. M. Javaid, A. Haleem, R.P. Singh, S. Khan, and R. Suman, "Blockchain technology applications for industry 4.0: A literature-based review," *Blockchain: Research and Applications*, vol. 2, no. 4, 2021, 100027, ISSN 2096-7209, https://doi.org/10.1016/j.bcra.2021.100027.
25. A. Aoun, A. Ilinca, M. Ghandour, and H. Ibrahim, "A review of industry 4.0 characteristics and challenges, with potential improvements using blockchain technology," *Computers & Industrial Engineering*, vol. 162, 2021, 107746, ISSN 0360-8352, https://doi.org/10.1016/j.cie.2021.107746.
26. E. Asante Boakye, H. Zhao, and B.N.K. Ahia, "Blockchain technology prospects in transforming Ghana's economy: A phenomenon-based approach," *Information Technology for Development*, vol. 29, no. 2–3, 2022, 348–377, https://doi.org/10.1080/02681102.2022.2073580.
27. P.P. Ray, "An introduction to dew computing: definition, concept and implications," *IEEE Access*, vol. 6, 2017, 723e737.
28. M. Brown, "The 4 types of cloud computing services," ExitCertified. [online] ExitCertified, 2019. Available at: https://www.exitcertified.com/blog/4-cloud-computing-services (Accessed 26 May 2024) Accessed.
29. D. Ivanov, A. Dolgui, and B. Sokolov, "Cloud supply chain: Integrating industry 4.0 and digital platforms in the "Supply Chain-as-a-Service"," *Transportation Research Part E: Logistics and Transportation Review*, vol. 160, 2022, 102676, ISSN 1366-5545, https://doi.org/10.1016/j.tre.2022.102676.
30. S.B. Miles, J. Kersey, E. Cecchini, and D.M. Kammen, "Productive uses of electricity at the energy-health nexus: Financial, technical and social insights from a containerized power system in Rwanda," *Development Engineering*, vol. 7, 2022, 100101, ISSN 2352-7285, https://doi.org/10.1016/j.deveng.2022.100101.

31. V. Mullet, P. Sondi and E. Ramat, "A review of cybersecurity guidelines for manufacturing factories in industry 4.0," *IEEE Access*, vol. 9, 2021, 23235–23263, https://doi.org/10.1109/ACCESS.2021.3056650.
32. A. Corallo, M. Lazoi, and M. Lezzi, "Cybersecurity in the context of industry 4.0: A structured classification of critical assets and business impacts," *Computers in Industry*, vol. 114, 2020, 103165, ISSN 0166-3615, https://doi.org/10.1016/j.compind.2019.103165.
33. A. Kritzinger, and I. Snyman. *Socio-Economic Impact Assessment Report for the Prieska Power Reserve Solar PV Plant and Wind Energy Facility Phase 1,* Prieska, Northern Cape, CENEC, 2022.
34. R. Kumar, R.K. Singh, and Y.K. Dwivedi, "Application of industry 4.0 technologies in SMEs for ethical and sustainable operations: Analysis of challenges," *Journal of Cleaner Production*, vol. 275, 2020, 124063, ISSN 0959-6526, https://doi.org/10.1016/j.jclepro.2020.124063.
35. M. EL- Shimy, "Optimal site matching of wind turbine generator: Case study of the Gulf of Suez region in Egypt," *Renewable Energy*, vol. 35, no. 8, 2010, 1870–1878, ISSN 0960-1481, https://doi.org/10.1016/j.renene.2009.12.013.
36. S. Benbelkacem, N. Zenati, M. Belhocine, A. Bellarbi, M. Tadjine, and S. Malek, "Augmented reality platform for solar systems maintenance assistance," In *EFEEA'10 International Symposium on Environment Friendly Energies in Electrical Applications*, 2–4 November 2010, Ghardaïa, Algeria.
37. M.R. Elkadeem, A. Younes, S.W. Sharshir, P.E. Campana, and S. Wang, "Sustainable siting and design optimization of hybrid renewable energy system: A geospatial multi-criteria analysis," *Applied Energy*, vol. 295, 2021, 117071, ISSN 0306-2619, https://doi.org/10.1016/j.apenergy.2021.117071.
38. G. Mutezo and J. Mulopo, "A review of Africa's transition from fossil fuels to renewable energy using circular economy principles," *Renewable and Sustainable Energy Reviews*, vol. 137, 2021, 110609, ISSN 1364-0321, https://doi.org/10.1016/j.rser.2020.110609.
39. D.A. Alemzero, H. Sun, M. Mohsin, N. Iqbal, M. Nadeem and X.V. Vo, "Assessing energy security in Africa based on multi-dimensional approach of principal composite analysis," *Environmental Science and Pollution Research*, vol. 28, 2021, 2158–2171, https://doi.org/10.1007/s11356-020-10554-0.
40. N. Djellouli, L. Abdelli, M. Elheddad, R. Ahmed, and H. Mahmood, "The effects of non-renewable energy, renewable energy, economic growth, and foreign direct investment on the sustainability of African countries," *Renewable Energy*, vol. 183, 2022, 676–686, ISSN 0960-1481, https://doi.org/10.1016/j.renene.2021.10.066.
41. O.D.T. Odou, R. Bhandari, and R. Adamou, "Hybrid off-grid renewable power system for sustainable rural electrification in Benin," *Renewable Energy*, vol. 145, 2020, 1266–1279, ISSN 0960-1481, https://doi.org/10.1016/j.renene.2019.06.032.
42. A. Jahanger, S.O. Ogwu, J.C. Onwe, and A. Awan, "The prominence of technological innovation and renewable energy for the ecological sustainability in top SDGs nations: Insights from the load capacity factor," *Gondwana Research*, vol. 129, 2024, 381–397, ISSN 1342-937X, https://doi.org/10.1016/j.gr.2023.05.021.
43. L. Richard, C. Boudinet, S.A. Ranaivoson, J.O. Rabarivao, A.E. Befeno, D. Frey, M.-C. Alvarez-Hérault, B. Raison, and N. Saincy, "Development of a DC microgrid with decentralized production and storage: From the lab to field deployment in rural Africa," *Energies*, vol. 15, 2022, 6727, https://doi.org/10.3390/en15186727.
44. S.A. Awuku, A. Bennadji, F. Muhammad-Sukki, and N. Sellami, "Promoting the solar industry in Ghana through effective public-private partnership (PPP): Some lessons from South Africa and Morocco," *Energies*, vol. 15, 2022, 17, https://doi.org/10.3390/en15010017.

45. I.D. Ibrahim, Y. Hamam, Y. Alayli, T. Jamiru, E.R. Sadiku, W.K. Kupolati, J.M. Ndambuki, and A.A. Eze, "A review on Africa energy supply through renewable energy production: Nigeria, Cameroon, Ghana and South Africa as a case study," *Energy Strategy Reviews*, vol. 38, 2021, 100740, ISSN 2211-467X, https://doi.org/10.1016/j.esr.2021.100740.
46. K. Baumli and T. Jamasb, "Assessing private investment in African renewable energy infrastructure: A multi-criteria decision analysis approach," *Sustainability*, 12, 2020, 9425, https://doi.org/10.3390/su12229425.
47. K. Ukoba, T.J. Kunene, P. Harmse, V.T. Lukong, and T. Chien Jen, "The role of renewable energy sources and industry 4.0 focus for Africa: A review," *Applied Sciences*, vol. 13, 2023, 1074, https://doi.org/10.3390/app13021074.
48. Q. Hassan, P. Viktor, T.J. Al-Musawi, B.M. Ali, S. Algburi, H.M. Alzoubi, A.K. Al-Jiboory, A.Z. Sameen, H.M. Salman, and M. Jaszczur, "The renewable energy role in the global energy transformations," *Renewable Energy Focus*, vol. 48, 2024, 100545, ISSN 1755-0084, https://doi.org/10.1016/j.ref.2024.100545.
49. M. Amir and S.Z. Khan, "Assessment of renewable energy: Status, challenges, COVID-19 impacts, opportunities, and sustainable energy solutions in Africa," *Energy and Built Environment*, vol. 3, no. 3, 2022, 348–362, ISSN 2666-1233, https://doi.org/10.1016/j.enbenv.2021.03.002.
50. S.A. Qadir, H. Al-Motairi, F. Tahir, and L. Al-Fagih, "Incentives and strategies for financing the renewable energy transition: A review," *Energy Reports*, vol. 7, 2021, 3590–3606, ISSN 2352-4847, https://doi.org/10.1016/j.egyr.2021.06.041.
51. S.A. Asongu and N.M. Odhiambo, "Inequality, finance and renewable energy consumption in Sub-Saharan Africa," *Renewable Energy*, vol. 165, Part 1, 2021, 678–688, ISSN 0960-1481, https://doi.org/10.1016/j.renene.2020.11.062.
52. E.M. Mungai, S.W. Ndiritu, and I. Da Silva, "Unlocking climate finance potential and policy barriers—A case of renewable energy and energy efficiency in Sub-Saharan Africa," *Resources, Environment and Sustainability*, vol. 7, 2022, 100043, ISSN 2666-9161, https://doi.org/10.1016/j.resenv.2021.100043.
53. I. Todd and D. McCauley, "Assessing policy barriers to the energy transition in South Africa," *Energy Policy*, vol. 158, 2021, 112529, ISSN 0301-4215, https://doi.org/10.1016/j.enpol.2021.112529.
54. G. Falchetta, A.G. Dagnachew, A.F. Hof, and D.J. Milne, "The role of regulatory, market and governance risk for electricity access investment in sub-Saharan Africa," *Energy for Sustainable Development*, vol. 62, 2021, 136–150, ISSN 0973-0826, https://doi.org/10.1016/j.esd.2021.04.002.
55. S. Mokeke and L.Z. Thamae, "The impact of intermittent renewable energy generators on Lesotho national electricity grid," *Electric Power Systems Research*, vol. 196, 2021, 107196, ISSN 0378-7796, https://doi.org/10.1016/j.epsr.2021.107196.
56. L. Deguenon, D. Yamegueu, S.M. Kadri, and A. Gomna, "Overcoming the challenges of integrating variable renewable energy to the grid: A comprehensive review of electrochemical battery storage systems," *Journal of Power Sources*, vol. 580, 2023, 233343, ISSN 0378-7753, https://doi.org/10.1016/j.jpowsour.2023.233343.
57. E. Sutherland, "The fourth industrial revolution – The case of South Africa," *Politikon*, vol. 47, no. 2, 2020, 233–252, https://doi.org/10.1080/02589346.2019.1696003.
58. A. Raihan and A. Tuspekova, "The nexus between economic growth, renewable energy use, agricultural land expansion, and carbon emissions: New insights from Peru," *Energy Nexus*, vol. 6, 2022, 100067, ISSN 2772-4271, https://doi.org/10.1016/j.nexus.2022.100067.
59. R. Ioannidis and D. Koutsoyiannis, "A review of land use, visibility and public perception of renewable energy in the context of landscape impact," *Applied Energy*, vol. 276, 2020, 115367, ISSN 0306-2619, https://doi.org/10.1016/j.apenergy.2020.115367.

60. A.Q. Al-Shetwi, "Sustainable development of renewable energy integrated power sector: Trends, environmental impacts, and recent challenges," *Science of the Total Environment*, vol. 822, 2022, 153645, ISSN 0048-9697, https://doi.org/10.1016/j.scitotenv.2022.153645.
61. P.K. Adom, F.A.- Mensah, M.P. Agradi, and A. Nsabimana, "Energy poverty, development outcomes, and transition to green energy," *Renewable Energy*, vol. 178, 2021, 1337–1352, ISSN 0960-1481, https://doi.org/10.1016/j.renene.2021.06.120.
62. A.O. Acheampong, M.O. Erdiaw-Kwasie, and M. Abunyewah, "Does energy accessibility improve human development? Evidence from energy-poor regions," *Energy Economics*, vol. 96, 2021, 105165, ISSN 0140-9883, https://doi.org/10.1016/j.eneco.2021.105165.
63. J. Zhao, K. Dong, X. Dong, and M. Shahbaz, "How renewable energy alleviate energy poverty? A global analysis," *Renewable Energy*, vol. 186, 2022, 299–311, ISSN 0960-1481, https://doi.org/10.1016/j.renene.2022.01.005.
64. D. Asante, Z. He, N.O. Adjei, and B. Asante, "Exploring the barriers to renewable energy adoption utilising MULTIMOORA-EDAS method," *Energy Policy*, vol. 142, 2020, 111479, ISSN 0301-4215, https://doi.org/10.1016/j.enpol.2020.111479.
65. O.J. Akintande, O.E. Olubusoye, A.F. Adenikinju, and B.T. Olanrewaju, "Modeling the determinants of renewable energy consumption: Evidence from the five most populous nations in Africa," *Energy*, vol. 206, 2020, 117992, ISSN 0360-5442, https://doi.org/10.1016/j.energy.2020.117992.
66. Y. Khan, H. Oubaih, and F.Z. Elgourrami, "The effect of renewable energy sources on carbon dioxide emissions: Evaluating the role of governance, and ICT in Morocco," *Renewable Energy*, vol. 190, 2022, 752–763, ISSN 0960-1481, https://doi.org/10.1016/j.renene.2022.03.140.
67. S.A. Asongu and N.M. Odhiambo, "Enhancing governance for environmental sustainability in sub-Saharan Africa," *Energy Exploration & Exploitation*, vol. 39, no. 1, 2021, 444–463, https://doi.org/10.1177/0144598719900657.
68. A. Akundi, D. Euresti, S. Luna, W. Ankobiah, A. Lopes, and I. Edinbarough, "State of industry 5.0—Analysis and identification of current research trends," *Applied System In*novations, vol. 5, 2022, 27, https://doi.org/10.3390/asi5010027.
69. J. Pizoń and A. Gola, "Human–machine relationship—Perspective and future roadmap for industry 5.0 solutions," *Machines*, vol. 11, 2023, 203, https://doi.org/10.3390/machines11020203.
70. S. Saniuk, S. Grabowska, and M. Straka, "Identification of social and economic expectations: Contextual reasons for the transformation process of industry 4.0 into the industry 5.0 concept," *Sustainability*, vol. 14, 2022, 1391, https://doi.org/10.3390/su14031391.
71. S. Grabowska, S. Saniuk, and B. Gajdzik, "Industry 5.0: Improving humanization and sustainability of industry 4.0," *Scientometrics*, vol. 127, 2022, 3117–3144. https://doi.org/10.1007/s11192-022-04370-1.
72. E.G. Carayannis and J. Morawska-Jancelewicz, "The futures of Europe: Society 5.0 and industry 5.0 as driving forces of future universities," *Journal of Knowlegde Economy*, vol. 13, 2022, 3445–3471, https://doi.org/10.1007/s13132-021-00854-2.
73. R. Sindhwani, S. Afridi, A. Kumar, A. Banaitis, S. Luthra, and P.L. Singh, "Can industry 5.0 revolutionize the wave of resilience and social value creation? A multi-criteria framework to analyze enablers," *Technology in Society*, vol. 68, 2022, 101887, ISSN 0160-791X, https://doi.org/10.1016/j.techsoc.2022.101887.
74. P. Fraga-Lamas, S.I. Lopes, and T.M. Fernández-Caramés, "Green IoT and edge AI as key technological enablers for a sustainable digital transition towards a smart circular economy: An industry 5.0 use case," *Sensors*, vol. 21, 2021, 5745, https://doi.org/10.3390/s21175745.

75. A. Visvizi, R. Malik, G.M. Guazzo, and V. Çekani, "The Industry 5.0 (I50) paradigm, blockchain-based applications and the smart city," *European Journal of Innovation Management*, vol. ahead-of-print No. ahead-of-print, 2024, https://doi.org/10.1108/EJIM-09-2023-0826.
76. Z.-J. Wang, Z.-S. Chen, L. Xiao, Q. Su, K. Govindan, and M.J. Skibniewski, "Blockchain adoption in sustainable supply chains for industry 5.0: A multistakeholder perspective," *Journal of Innovation & Knowledge*, vol. 8, no. 4, 2023, 100425, ISSN 2444-569X, https://doi.org/10.1016/j.jik.2023.100425.

8 Solid Waste Management and Industry 4.0
Case Study in Mauritius

Geeta D. Somaroo, Doorgha Ragoobur, and Iswaree Aubeeluck-Ragoonauth

8.1 INTRODUCTION

In the heart of the Indian Ocean lies Mauritius (20.3484° S, 57.5522° E), a small island developing state (SIDS) spanning 1,865 km^2 [1]. According to Central Statistics Office (CSO) [1], Mauritius had a population of around 1.2 million with a population density of 650 persons/km^2, and the estimated GDP growth rate in 2022 was 8.9%. The substantial socioeconomic advancements observed in the 21st century have unequivocally enhanced the quality of life for Mauritians. Consequently, shifts in production and consumption patterns have resulted in a notable increase in the amount and the composition of solid waste generated on the island [2,3]. Additionally, Mauritius faces the challenge of having limited available space coupled with insufficient financial and technical resources that set barriers to the establishment of an integrated and sustainable waste management system [4]. Among the various SIDS, Mauritius is one of the few small islands with a sanitary landfill, located at Mare Chicose [4]. With over 10 million tons of waste already buried since its establishment in 1997, the landfill's capacity is being stretched to its limits, which is a pressing issue on the island.

In Mauritius, about 1,488 tons of waste is generated every day [5]. Over 50% of this waste emanates from the domestic and commercial sectors, referred to as "municipal solid waste (MSW)" [6]. Figure 8.1 shows the composition of MSW generated in Mauritius. MSW consists predominantly of food waste (27%) and yard waste (27%), followed by paper waste (14%), plastics waste (14%), textile waste (6%), metal waste (3%), and glass waste (3%), signifying a high amount of organic matter [7]. The amount of food and yard waste generated in Mauritius can be attributed to the prevalence of houses with gardens and the minimal consumption of processed foods in the dietary habits of Mauritians [7]. Mauritians still prioritize cooking fresh food using fresh fruits and vegetables [7]. The fraction of plastics and paper waste can be considered to be fairly high [7]. The proportion of plastics (14%) is high compared with other SIDS where the amount of plastics makes up an average of 8% [8]. According to Mauritius Chamber of Commerce and Industry (MCCI) [8], Mauritius produces about 116,000 tons of plastic waste each year. Only around 3,000 tons (approximately 2.5%) of this plastic waste are recycled annually, with an estimated 71,000 tons (about 61%) being disposed of in landfills at Mare Chicose [8]. However,

DOI: 10.1201/9781003511298-8

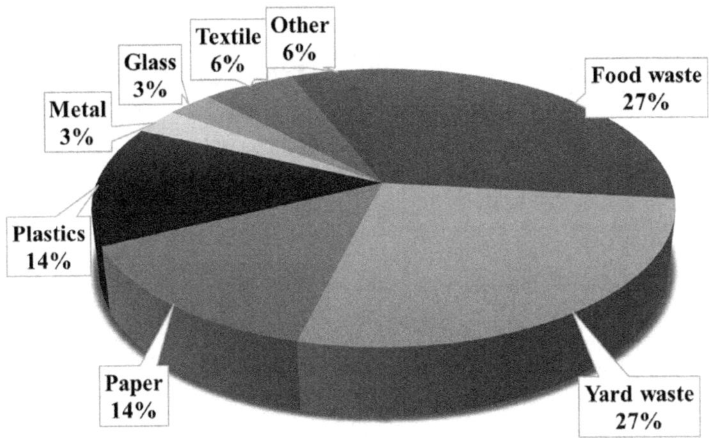

FIGURE 8.1 Composition of municipal solid waste in Mauritius.

Source: P. Kowlesser, Plastics Management in Small Islands Developing States, 2015. http://www.uncrd.or.jp/content/documents/2654Plenary%20Session(3)-Presentation(6)-Prakash%20Kowlesser.pdf

approximately 42,000 tons per year (36.5%) of plastic waste remain unaccounted for [8]. The waste plastic problem in SIDS stems from their heavy reliance on food imports where single-use plastic packaging is preferred [8]. The percentages of metal and glass waste were lower than plastic waste. According to Kowlessur [4], there are markets where these items are recovered for reprocessing, especially metal containers and scrap ferrous materials. In the case of glass, mainly from bottles, the low recycling rate is attributed to people preferring bottles that have a return-deposit system. Currently, there is a deposit-refund scheme for glass bottles, which are collected and recycled by manufacturers. Over the years, there has been a transition from glass to plastic bottles, particularly for soft drinks.

Solid waste management (SWM) encompasses the comprehensive management of solid wastes, involving the guidelines of their generation, storage, collection, transfer, transport, processing, and disposal. This discipline employs best practices that address community health, finances, engineering, preservation, aesthetics, and various ecological aspects. The MSW management system in Mauritius involves the collection of wastes and transfers to the transfer stations and finally to the sole sanitary landfill at Mare Chicose. Figure 8.2 shows the SWM system prevailing in Mauritius.

There is no source separation and sorting of wastes at the transfer stations in Mauritius. There are also very few recycling companies and no waste-to-energy facilities for the treatment of solid wastes. The landfill at Mare Chicose has reached its full capacity, and there is an imperative and pressing need to implement other possible waste treatment opportunities for solid waste.

The Government of Mauritius has devised a number of initiatives to improve resource recovery for recyclers, for instance, the development of a pilot Sorting Unit and a Composting Unit for the separation of dry and wet waste resources, as well as the introduction of a framework to endorse the composting of Green Wastes

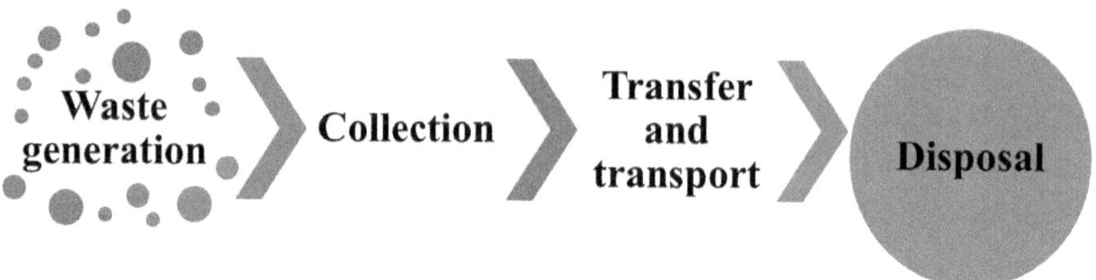

FIGURE 8.2 Solid waste management system in Mauritius.

from Ménages, Marketplaces, Gardens, and Grounds. The Waste Management and Resource Recovery Act 2023 to the Government Gazette of Mauritius No. 32 was promulgated on 20 April 2023. This act aims to establish a regulatory framework for the environmentally safe and effective management of solid and hazardous wastes, promoting a sustainable waste management system through the principles of the circular economy (CE), emphasizing waste reduction, reuse, material recovery, and recycling. Industry 4.0 (I4.0) has the potential to enhance the efficiency and sustainability of SWM practices.

I4.0 can bring an impressive change in the management of solid waste across the world. In this era, digital technologies can immensely enhance the procedure of solid waste recycling [9], substituting the traditional practice of SWM [10]. I4.0 can enhance recycling efficiency and energy efficiency, optimizing manufacturing and operations, and provide solutions to business risks, and recycled products are created with cost efficiency [11]. Digitalization of the SWM deploys advanced technologies like artificial intelligence (AI), Internet of Things (IoT), machine learning (ML), big data analysis, blockchain technologies, image recognition, advanced manufacturing technologies, radiofrequency identification (RFID), and wireless communications [10,12]. A proper implementation of IR 4.0 will enable the waste recycling sector to establish more efficient cooperation between the government and the private sector in terms of job opportunities and also for the promotion of resource conservation. This chapter assesses the possibility of integrating I4.0 technologies and designs in different waste routes, through an analysis of current best practices.

8.2 INTEGRATED SWM

Integrated solid waste management refers to the process of choosing and implementing proper methods, technologies, and management strategies to accomplish defined waste management aims and targets. This waste management hierarchy organizes actions for implementing programs within communities and consists of the following components: reduction of wastes at source, reuse, waste combustion/reprocessing/composting, and landfilling, as shown in Figure 8.3.

For the hierarchical order, recycling should be considered only after all feasible efforts to minimize waste generation at the source have been exhausted. This hierarchy lacks substantial scientific or technical justification. There is no scientific rationale, for instance, that mandates materials recycling as always superior to energy recovery. The hierarchy proves limited when multiple options are combined, as in

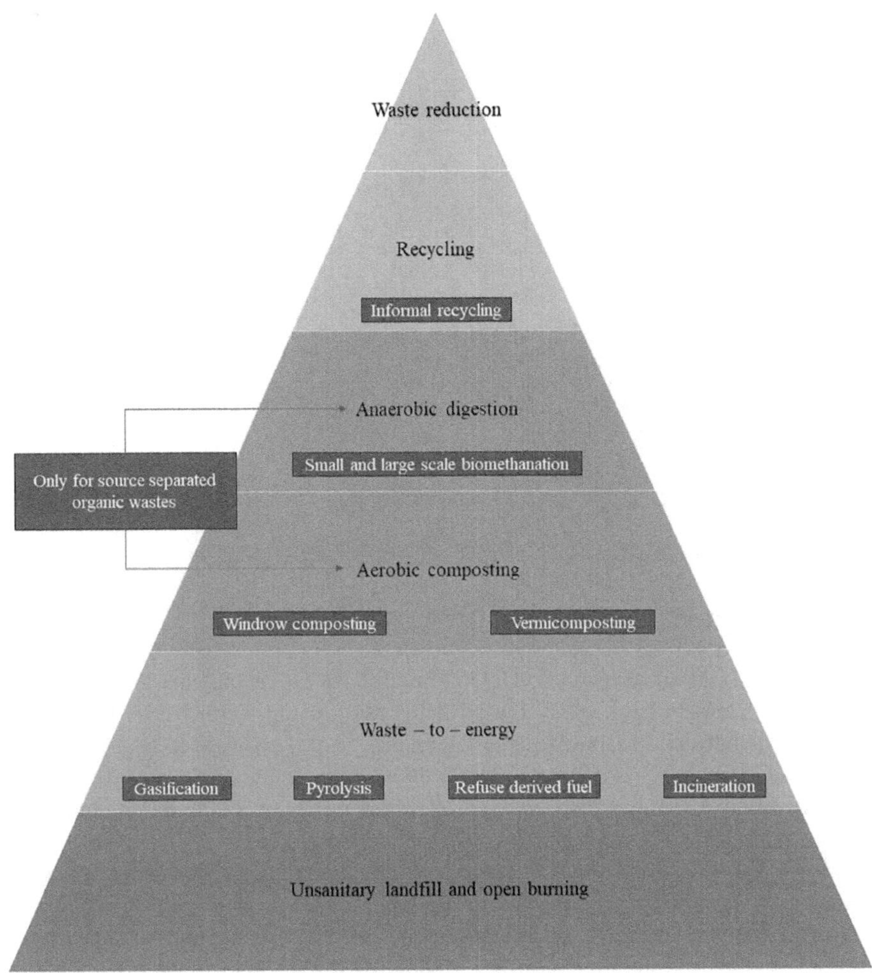

FIGURE 8.3 Hierarchy of integrated solid waste management.

integrated waste management (IWM) systems. Within an IWM approach, the hierarchy fails to predict scenarios where, for example, a combination of biological treatment and thermal treatment of residues might outperform materials recycling paired with landfilling of residues.

What is essential is a comprehensive evaluation of the entire system, a task the hierarchy cannot fulfill. It overlooks cost considerations, thus providing no assistance in assessing the economic viability of waste management systems. Moreover, the hierarchy does not accommodate the diverse array of specific local circumstances where effective waste management systems are essential, such as in small islands, sparsely populated regions, or popular tourist destinations experiencing seasonal population surges. Thus, there exists an all-inclusive tactic for SWM that acknowledges the potential of all options within IWM (Figure 8.4). The question is now how to integrate I4.0 into the holistic approach of SWM.

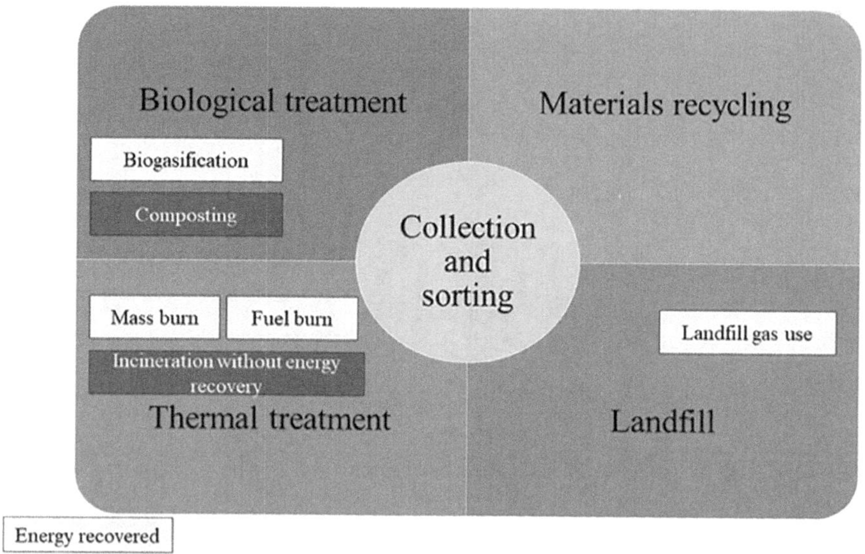

FIGURE 8.4 Holistic approach.

The fourth industrial revolution, also often referred to as Industry 4.0, represents a transition characterized by increased automation using technologies such as AI and ML in industrial manufacturing. Borchard et al. [13] showed that digitalization in waste management is gaining heightened importance. However, this study also identified discrepancies between the implementation of digital technologies and the intended operation reported in 2016/17 compared to the current reality [13]. Moreover, the top three barriers reported to impede the adoption of numerical technologies at the moment include day-to-day operational burdens, significant investment and operational expenses, and the absence of technical standards [13].

Furthermore, developing countries, especially SIDS, can face several challenges in achieving I4.0 in the management of waste. These challenges include financial backing, technical expertise, political climates, institutional capacities, and social considerations [14]. Insufficient local technical expertise among both the populace and government personnel also poses a barrier to the effective practice and effective waste management and treatment [14]. Moreover, most developing countries have been focusing on safe waste disposal without encouraging recycling, whereby the policies fail to add new technology to waste recycling [11]. Sharma et al. [15] argued that the rollout of IoT in developing countries like India would mandatorily require structural framework, standardization, and policy. Khan et al. [16] further highlighted that the main barrier is in the implementation phase where the chosen technologies can be inappropriate and unadaptable. Additionally, digitalization can pose potential risks such as data discretion issues, cybersecurity breaches, legal responsibility concerns, increased conservation costs, and heightened societal discrimination [17].

8.3 CE AND INDUSTRY 4.0

A novel concept that contrasts with the linear economy model is widely recognized today as the CE. The application of CE approaches in waste management by applying upstream and downstream solutions together can lead to significant reductions in greenhouse gas emissions (GHGs) and waste disposal in the next decade and beyond. The CE, as opposed to the linear economy, centers on the values of the 3Rs and the expanded 6Rs (reuse, recycle, redesign, remanufacture, reduce, recover). This approach seeks to optimize the system by minimizing the depletion of natural resources involved in product manufacturing. It accomplishes this through strategies such as reuse, recycling, remanufacturing, redesigning, recovering used products, and waste reduction. These methods not only add value but also prolong the lifespan of products. Figure 8.5 abridges the concept of CE. The conventional linear business model of "take-make-use-dispose" has contributed to a 29% increase in solid waste in Mauritius between 2010 and 2020 and more than doubled over the past two decades [18].

ML, AI, and image recognition are believed to have the potential to automate waste segregation, thereby reducing the menace of exposing employees to hazardous

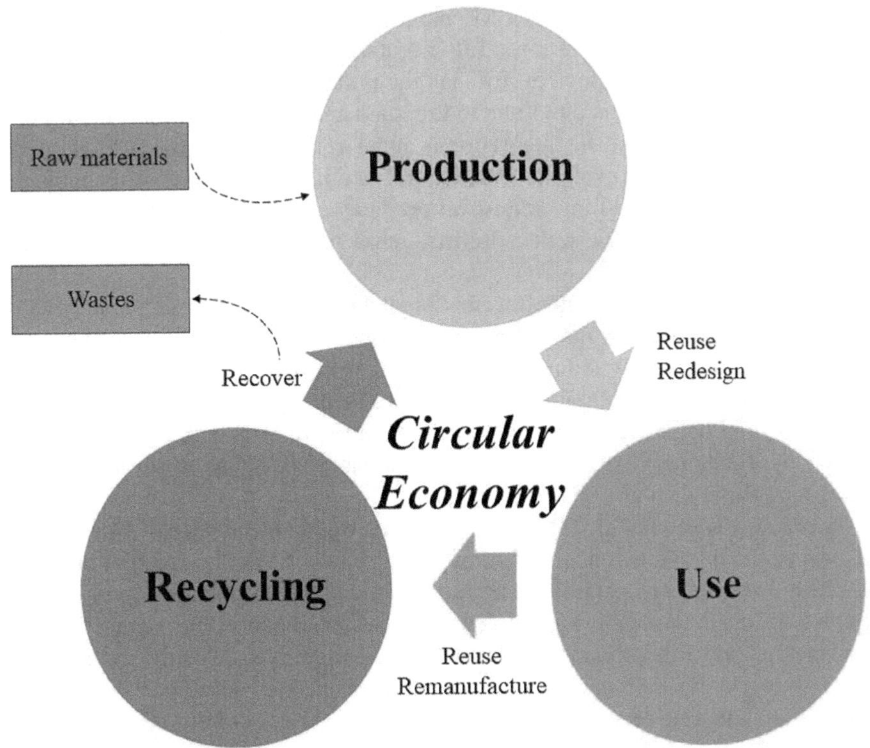

FIGURE 8.5 Overview of circular economy. (Adapted from Ref. [19].)

waste [19]. To explore opportunities in the C for material traceability, RFID and wireless communications can be exploited [19]. With respect to sustainability and a CE, I4.0 can prove to be a fundamental asset to converge the pathway of industries toward more sustainable developments. The main objective of an investigation led by Viles et al. [20] was to identify and redefine the principles of sustainable production within the context of the CE and Industry 4.0 includes the following [20]:

- Principle 1 (Design for circularity)
- Principle 2 (Conserve resources and preserve their value)
- Principle 3 (Manage waste sustainably)
- Principle 4 (Pursue a risk-free environment)
- Principle 5 (Prioritize employees' well-being)
- Principle 6 (Enhance management commitment to sustainability)
- Principle 7 (Make a positive contribution to the community)
- Principle 8 (Promote value chain stakeholder collaboration)
- Principle 9 (Measure and optimize sustainable processes)
- Principle 10 (Boost the use of sustainable technologies)

Evolving digital technologies, such as AI and the IoT, are indispensable enablers of circular business practices [9]. AI enables machines to think, learn, and make decisions autonomously, while the IoT facilitates communication and interaction between computers and devices. IoT has the potential to transform factories into smart towns, villages, and cities and to turn cars and houses into smart homes [21]. For example, in Taiwan, the integration of industrial AI with existing digital technologies has sparked a revolution in industrial intelligent solutions. AI technologies are effectively integrated into industrial products and business services, resulting in faster service delivery, reduced costs, more reliable products, and increased profitability [9].

It is believed that leveraging Information and Communication Technology (ICT) and the IoT presents a new-fangled generation approach to efficiently and competently improve the universal waste management system in SIDS [14]. Fatimah et al. [14] proposed an ICT–IoT integration involving local sensing, data integration, analytics of things, and cognitive actions within the field of waste management. Figure 8.6 shows the fundamental framework of a sustainable and smart waste management system by Fatimah et al. [14].

Moreover, Kolade et al. [22] emphasized that digital technologies in the circular plastic economy can be enhanced through several methods: the creation of mobile applications to engage and link diverse stakeholders, the adoption of digital financial tools to incentivize plastic waste collection, the integration of blockchain solutions for tracking and transparency in circular plastic supply chains, and the application of 3D printing to innovate new products, thereby enhancing consumer value and prolonging the lifecycle of plastic. This approach would empower millions of consumers and households to engage in the circular plastic market, enabling online transactions for recycled products and fostering effective collaboration among waste pickers, transporters, and recyclers [22].

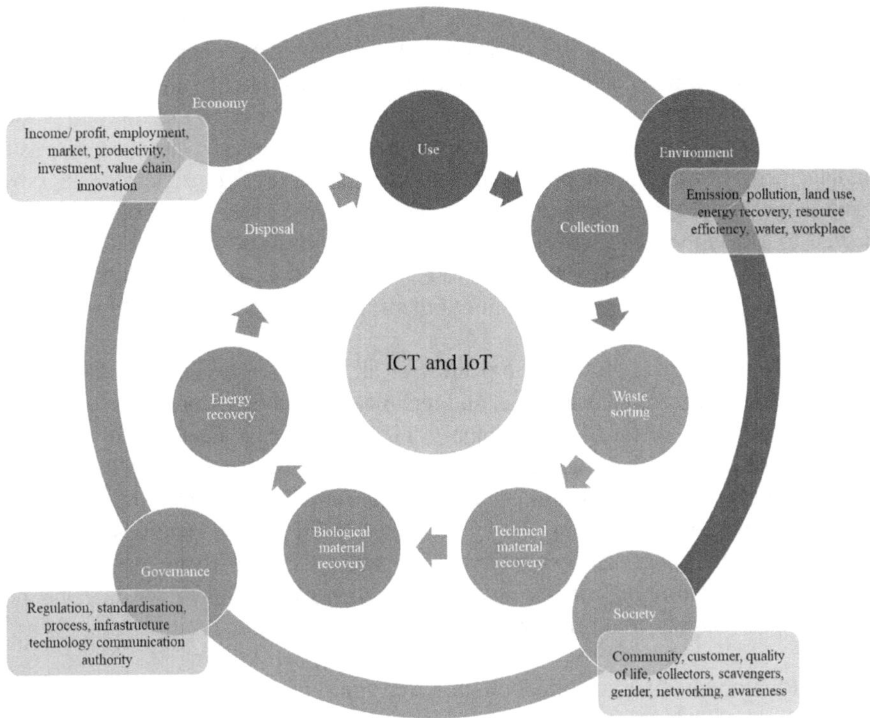

FIGURE 8.6 Sustainable and smart waste management system. (Adapted from Ref. [14].)

8.4 WASTE COLLECTION AND INDUSTRY 4.0

A few researchers examined the possibility of using IoT platforms to enable waste collection in smart cities; for instance, the inclusion of smart sensors in waste bins makes cloud-based waste management systems feasible [23]. The level of recyclable wastes is monitored by the sensors, and then the data recorded are uploaded into a cloud platform, whereby the latter is accessible to the concerned stakeholders, who in turn can optimize and plan the collection of recycled wastes [23]. As is the case for Mauritius and Malaysia, during public events and celebrations, a large volume of waste, including organic and recyclable waste, is usually generated, and this puts a lot of pressure on the authorities for the safe collection and treatment of these wastes [24]. In Mauritius, these wastes are disposed of solely at the sanitary landfill. However, the sole sanitary landfill has reached saturation, and various initiatives are being implemented by the private and public sectors to divert recyclable wastes from the landfill. The introduction of I4.0 in waste management is very limited in Mauritius, and the only smart waste collection that has recently been implemented is at Moka Smart City. Molok bins with IoT systems have been placed in public areas, and, once full, a private recycler collects the recyclable wastes effectively. Through this initiative, about 6,560 tons of recyclable wastes have been diverted from the landfill in 2022 [25].

In Malaysia, the MSW is disposed of in landfills that lack systems for the collection of leachate and gas, which eventually results in the contamination of the terrestrial environment, and large amounts of GHGs are emitted into the atmosphere [24]. Thus, Abdullah et al. [24] reviewed various IoT techniques for improving collection systems in Malaysia, which is considered the most significant aspect of waste management, especially when it comes to recycling waste. The proposed design by Abdullah et al. [24] was the introduction of smart bins. These smart bins are connected to a cloud system, which sends information on the waste type, weight, and volume to a waste management center. When the bins have exceeded 90% full, the system sends information about the bin to the transportation department for the collection of the waste [24].

A study was carried out by Monzambe et al. [26] to statistically examine the demographic, geographic, economic, and technical factors influencing the design of the finest and most sustainable municipal solid waste management system (SWMS) in the era of the Fourth Industrial Revolution. South Africa was deemed as a case study, whereby a conceptual framework for designing SWMS within the context of I4.0 was proposed. This framework, as illustrated in Figure 8.7, identifies three crucial stages where integrating cyber-physical systems (CPS), IoT, and smart factories is essential for optimizing the system's design and operation. Below, we describe the stages, inputs/outputs, and functionalities of these CPS [26]:

- CPS 1 involves waste collection, utilizing smart bins and sensors to inform the waste collection vehicle of the optimal timing and frequency for collection. Data gathered by these sensors is processed in data processing software. CPS 1 is engineered to detect full bins and optimize collection schedules accordingly.
- CPS 2 operates at the initial transportation stage, prior to waste being transported to either a landfill or a waste transfer station (WTS). Inputs for CPS 2 encompass factors such as traffic delays, road closures, and queue times at various WTSs and landfills. This CPS utilizes optimization and simulation models and tools to determine the most efficient destinations and routes.
- CPS 3 integrates information from WTSs and cloud-based systems to determine the optimal final destination and routing for waste. Inputs include

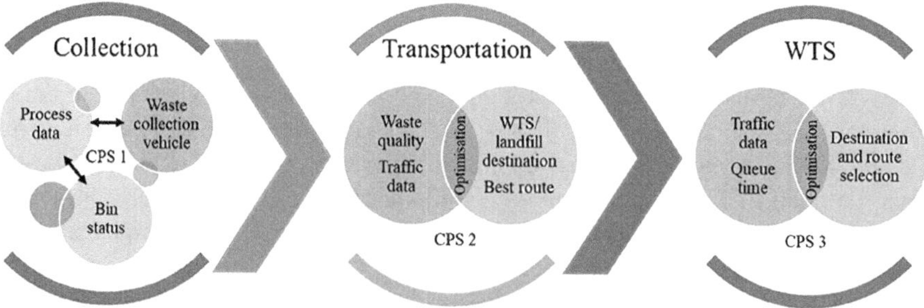

FIGURE 8.7 Conceptual framework for the optimal design of SWMS in the context of Industry 4.0. (Adapted from Ref. [26].)

factors like traffic delays, queue times at various landfills, and quantities of municipal waste. Data processing tools in CPS 3 encompass route optimization, facility selection models, and optimization and simulation software.

Another cyber-physical waste collection model was proposed by Bányai et al. [27] that includes waste collection cloud and its linkages with the waste collection system incorporating cyber-physical technology, involving customers, garbage trucks, waste management sites, and customer support. The monitoring of waste levels in the bin remotely enables the supervision of the entire collection process, as shown in Figure 8.8.

In Mauritius, smart bins can be monitored remotely through either real-time Wi-Fi connection or delayed off-line monitoring. Data gathered from the smart bins can be uploaded to cloud storage, where real-time processing capacities are also available as stated by Bányai et al. [27]. Moreover, the connection between households (waste sources) and waste treatment facilities can be established through garbage trucks, which collect municipal waste and transport it to recycling, composting, or disposal sites. Bányai et al. [27] explored the design and optimization of waste collection routes using optimization algorithms within the waste collection cloud. This optimized collection process not only reduces energy consumption and emissions from garbage trucks but also aligns the collection process with available processing capacities, thereby enhancing environmental awareness [27].

Sharma et al. [15] highlighted that IoT is a pivotal driver in the transformation of smart cities in the advancement of waste management structure and conveyance and to improve better human life in India. Moreover, the usage of IoT in Taiwan has facilitated different cities on the island to communicate effectively about real-time data through the deployment of various RFID sensors [11]. AI images have also been deployed in Taiwan for illegal littering using smart containers and Geographic

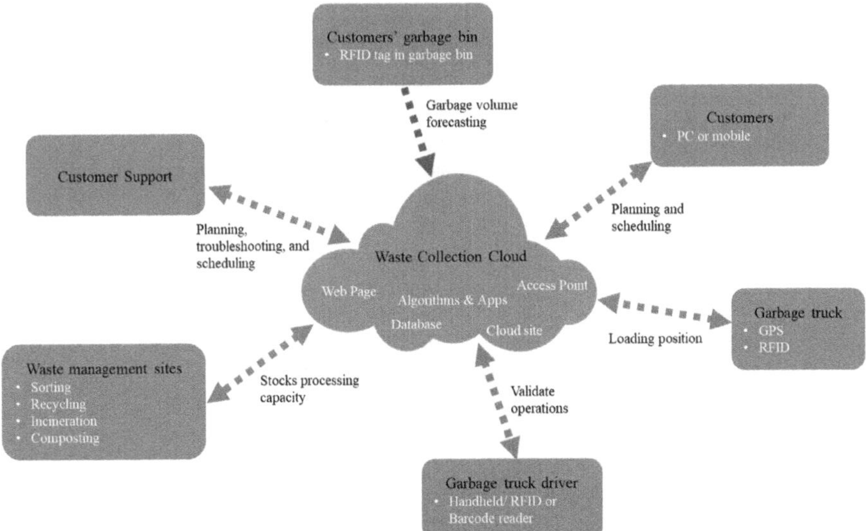

FIGURE 8.8 Model of cyber-physical waste collection system. (Adapted from Ref. [27].)

Information System (GIS) systems, and through these initiatives, Taiwan has reached a recycling rate of 65% of its MSW in the global context [11].

A *toolkit for open and sustainable city planning and analysis* (TOSCA), an open-source GIS software developed by HafenCity University Hamburg, was implemented in Bhubaneswar, India, to serve as a framework for the development of sustainable SWM [28]. TOSCA is coupled with ML algorithms like image recognition of dry waste and gives detailed analytic insight into the physical properties, frequency, object typology, and amount of various forms of waste collected [28]. The data collected directly provide adequate information to the authorities, planners, and governments to improve regular tasks of planning, operating, and managing waste flows in the cities [28]. Table 8.1 gives an overview of various studies, which have been conducted on the integration of Industry 4.0 in SWM in different countries. The table also shows the barriers and recommendations or main conclusions from the studies.

TABLE 8.1
Review of Studies

Country	Waste Route	Digital Tool	Barriers	Main Conclusions/ Recommendations	References
Tunisia	Collection	IoT	Infrastructure limitations; resource constraints	Significant cost reduction	[29]
Indonesia	Collection	IoT	User adoption and engagement; data privacy and security; infrastructure and connectivity; regulatory compliance issues, competition and market dynamics, and socioeconomic considerations.	Technical support and training; behavioral change; infrastructure development	[11]
Taiwan	Sorting	Machine learning; visual identification; AI	High initial cost; technical complexity; scalability; integration with existing infrastructure; public perception and acceptance	The system is still at an early stage and needs to be tested under different conditions to evaluate its performance in diverse operational scenarios	[30]

(*Continued*)

TABLE 8.1 (*Continued*)
Review of Studies

Country	Waste Route	Digital Tool	Barriers	Main Conclusions/ Recommendations	References
Europe	Collection	IoT	Financial constraints; technical capabilities; political limitations; lack of skilled personnel	To explore the potential integration of artificial intelligence and machine learning for data analysis and decision-making in waste management systems	[31]
South Africa	Collection	IoT	Infrastructure limitations;	Collaborative efforts among researchers and nations	[32]
South Africa	Collection	Cyber-physical systems; IoT; Internet of services; smart factories	real-time data; decentralized decision-making	The framework positively recommends I4.0 for optimal waste management system	[26]
Uganda	Collection	IoT	Infrastructure limitations; resource constraints	Job creation for 32 full-time workers and the mobile app has over 1,000 household connections	[22,33]
Germany	Collection	Robotics and sensor technology; online marketplaces; predictive analytics; cloud computing, big data analytics	Data protection and security; too much burden from daily business; high investment and operating costs; technical standards; employee acceptance	Digitalization influences the perception of waste producers and policymakers promoting change in regulations, e.g., EPR	[34]

(*Continued*)

TABLE 8.1 (*Continued*)
Review of Studies

Country	Waste Route	Digital Tool	Barriers	Main Conclusions/ Recommendations	References
India	Collection	IoT	Security and privacy; reliability/lack of mobility; lack of transparency and regulatory norms, policies, and directions; high operational cost and extended payback period; lack of integration among IT network and inadequate internet connectivity; limited skilled workforce and lack of technical knowledge among planners	Amended/ new government policies; IT-enabled infrastructure	[15]
Iraq	Collection	–	Inadequate financing; lack of political will	Model saves electricity, cut down working hours and fuel consumption	[35]

8.5 COMPOSTING AND INDUSTRY 4.0

As mentioned by Abdullah et al. [24], organic waste presents challenges for recycling, yet it can be recovered and repurposed through composting. Composting involves the managed biological breakdown of organic waste with oxygen, resulting in a nutrient-rich product called "compost," which serves as a beneficial soil conditioner. It has been observed that in many households and communities in SIDS, green waste is segregated for composting. However, a significant challenge lies in the limited knowledge about the operational parameters required for effective composting processes [36]. A study carried out by Chen et al. [37] revealed that artificial neural networks, tree-based models, vector machines, and genetic algorithms are the top algorithms in biological treatment. Chen et al. [37] conducted an in-depth investigation of the applications of ML in composting, among other biological treatments

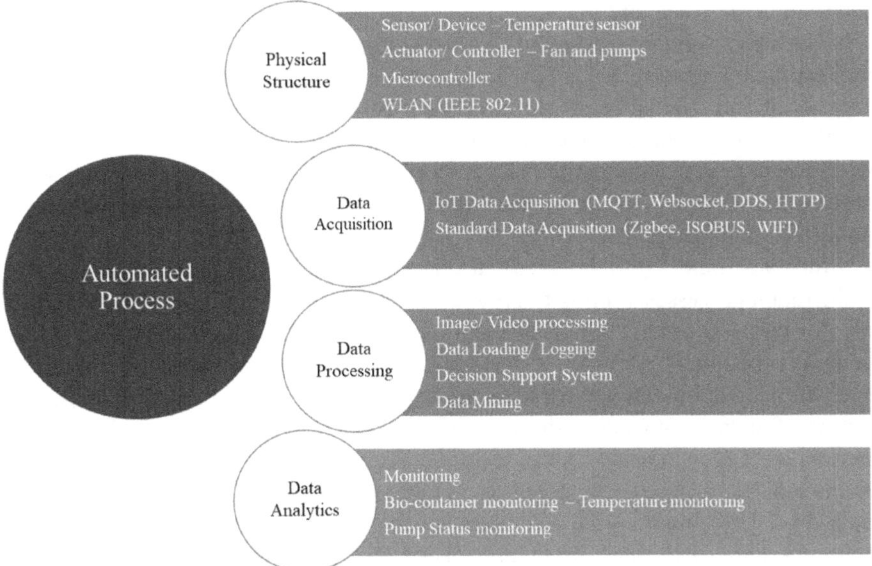

FIGURE 8.9 Major components of IoT-based smart composting facility. (Adapted from Ref. [31].)

to envisage products, improve processes, achieve real-time monitoring, and alleviate pollution emissions.

Olıverı et al. [38] conducted a study on enhancing the composting process using I4.0 technologies, and the authors proposed that I4.0 can improve the monitoring and control of the bio-oxidation phase of composting. This includes equipping the composters with temperature sensors, pumps, and fans, all connected through a local wireless network, as illustrated in Figure 8.9. For example, a fan system activates when the sensor detects an outdoor night temperature of 28°C or lower [19]. Additionally, Cheah et al. [19] also discussed the use of customized mobile enclosures in IoT compost monitoring systems to automate processes for maintaining high-quality compost. This system consists of various monitoring sensors, such as humidity, temperature, and light, and actuators, as well as an aeration system to control the different factors that influence the produce and value of the compost [19].

8.6 WASTE-TO-ENERGY AND INDUSTRY 4.0

Research has shown that the implementation of I4.0 technologies in waste-to-energy processes is relatively rare. However, there are some recent studies that have highlighted the potential benefits of I4.0 technologies. For instance, Luikkonen et al. [39] introduced an adaptive soft sensor that uses real plant data to estimate the nitrogen oxide content in the flue gas of a circulating fluidized bed boiler fueled by demolition wood. The results indicated the necessity of an adaptive approach to understanding the process dynamics, with both linear and nonlinear soft sensors performing well [39]. Additionally, Kabugo et al. [23] explored data-driven soft sensors to forecast syngas heating value and hot flue gas temperature in an Outotec Advanced Staged Gasifier using solid waste fuel. Their study compared several methods including multivariable linear regression, principal component regression, partial least squares regression, and the nonlinear neural network-based NARX model. They found that the NARX model outperformed the other methods in predicting both syngas heating value and flue gas temperature. The authors demonstrated that in situations where process knowledge is limited, data-driven soft sensors serve as valuable tools for predictive analytics [23].

Muri and Hjelme [40] assessed various sensor systems for monitoring waste in incineration processes, aiming to provide parameters for process control and indicators for process alarms. Inductance and hard X-ray sensors were identified as particularly promising for transmissive settings, whereas photonic, inductive, soft X-ray, hard X-ray, and low-frequency radiowave sensors were found effective for reflective settings. Single-point measurements using inductance, radiowave, photonic, or X-ray sensors were recommended for process alarms, while spectral imaging with X-ray or photonic techniques was noted as suitable for both process control and alarm indicators. These findings are pivotal for advancing waste property monitoring in the waste-to-energy industry, improving consistency and energy production efficiency, and ensuring worker safety in waste incineration processes [40].

8.7 CONCLUDING REMARKS

The incorporation of digital tools in SWM has been identified as crucial for fostering a CE and promoting sustainable development. It is believed that the involvement of policymakers and other relevant stakeholders would undeniably promote the inclusion of I4.0 in the management of solid wastes. However, SIDSs are more prone to face various challenges in integrating I4.0 into the management of waste. To this end, the national authority will play a critical role in the implementation of digitization of SWM in developing and emerging developing countries. In Mauritius, smart bins could be developed remotely through either real-time Wi-Fi connection or delayed off-line monitoring. The data gathered from the smart bins could then be uploaded to cloud storage. The other waste treatment technologies are still at an early stage of development for Mauritius. Moreover, education has a major role in driving sustainable SWM at the community level and raising awareness of the environmental, economic, and social advantages of integrating I4.0 in various waste management models. This can also promote consumer engagement and fast adoption of digital tools in waste management.

REFERENCES

1. Statistics Mauritius, "Economic and social indicators," *Digest of Statistics*, 2023. https://statsmauritius.govmu.org/Pages/Statistics/By_Subject/Population/SB_Population.aspx
2. L. I. Fuldauer, M. C. Ives, D. Adshead, S. Thacker and J. W. Hall, "Participatory planning of the future of waste management in small island developing states to deliver on the Sustainable Development Goals," *Journal of Cleaner Production*, vol. 223, pp. 147–162, 2019.

3. W. U. H. Shah, R. Yasmeen, M. Sarfraz and L. Ivascu, "The repercussions of economic growth, industrialization, foreign direct investment, and technology on municipal solid waste: Evidence from OECD economies," *Sustainability*, vol. 15, pp. 1–14, 2023.
4. P. Kowlesser, "Solid waste management in small island developing states, specifically in Mauritius," In Sadhan Kumar Ghosh (Ed.), *Solid Waste Policies and Strategies: Issues, Challenges and Case Studies*, Singapore, Springer, 2020, pp. 165–174.
5. GibsonandHills,*Mauritius–SolidWasteManagementStrategy*,2023.[Online].Available: https://relocationmauritius.wordpress.com/2023/02/23/mauritius-solid-waste-management-strategy/.
6. SWITCH Africa Green programme, *Roadmap and Action Plan for a Circular Economy in the Republic of Mauritius*, 2023. https://circulareconomy.govmu.org/circulareconomy/wp-content/uploads/2023/08/Roadmap-and-Action-Plan-for-a-Circular-Economy-in-the-Republic-of-Mauritius-Final-report.pdf
7. R. Mohee, "Assessing the recovery potential of solid waste in Mauritius," *Resources, Conservation and Recycling*, vol. 36, pp. 33–43, 2002.
8. MCCI, *The MCCI Roadmap for a Waste Plastic Free Mauritius*, 2022. https://www.mcci.org/media/321904/waste-plastic-free-report-web-version-2803.pdf
9. T. A. Kurniawan, A. Maiurova, M. Kustikova, E. Bykovskaia, M. H. D. Othman and H. H. Goh, "Accelerating sustainability transition in St. Petersburg (Russia) through digitalization-based circular economy in waste recycling industry: A strategy to promote carbon neutrality in era of Industry 4.0," *Journal of Cleaner Production*, vol. 363, pp. 1–16, 2022.
10. S. Bag, L. C. Wood, S. K. Mangla and S. Luthra, "Procurement 4.0 and its implications on business process performance in a circular economy," *Resources, Conservation and Recycling*, vol. 152, pp. 1–14, 2020.
11. T. A. Kurniawan, M. H. D. Othman, G. H. Hwang and P. Gikas, "Unlocking digital technologies for waste recycling in Industry 4.0 era: A transformation towards a digitalization-based circular economy in Indonesia," *Journal of Cleaner Production*, vol. 357, pp. 1–16, 2022.
12. S. R. Seyyedi, E. Kowsari, M. Gheibi, A. Chinnappan and S. Ramakrishna, "A comprehensive review integration of digitalization and circular economy in waste management by adopting artificial intelligence approaches: Towards a simulation model," *Journal of Cleaner Production*, vol. 460, pp. 1–14, 2024.
13. R. Borchard, R. Zeiss and J. Recker, "Digitalization of waste management: Insights from German private and public waste management firms," *Waste Management & Research*, vol. 40, pp. 775–792, 2022.
14. Y. A. Fatimah, K. Govindan, R. Murniningsih and A. Setiawan, "Industry 4.0 based sustainable circular economy approach for smart waste management system to achieve sustainable development goals: A case study of Indonesia," *Journal of Cleaner Production*, vol. 269, pp. 1–15, 2020.
15. M. Sharma, S. Joshi, D. Kannan, K. Govindan, R. Singh and H. C. Purohit, "Internet of Things (IoT) adoption barriers of smart cities' waste management: An Indian context," *Journal of Cleaner Production*, vol. 270, pp. 1–21, 2020.
16. S. Khan, R. Anjum., S. T. Raza, N. A. Bazai and M. Ihtisham, "Technologies for municipal solid waste management: Current status, challenges, and future perspectives," *Chemosphere,* vol. 288, pp. 1–12, 2022.
17. T.-D. Bui, J.-W. Tseng, M.-L. Tseng, K.-J. Wu and M. K. Lim, "Municipal solid waste management technological barriers: A hierarchical structure approach in Taiwan," *Resources, Conservation & Recycling*, vol. 190, pp. 1–19, 2023.
18. Business Mauritius & UNDP, *Circular Economy: Optimising Private Sector Investment in Mauritius*, 2021. https://www.undp.org/sites/g/files/zskgke326/files/migration/mu/Circular-Economy--Optimising-private-sector-investment-in-Mauritius.pdf

19. C. G. Cheah, W. Y. Chia, S. F. Lai, K. W. Chew, R. S. Chia and P. L. Show, "Innovation designs of Industry 4.0 based solid waste management: Machinery and digital circular economy," *Environmental Research*, vol. 213, pp. 1–11, 2022.
20. E. Viles, F. Kalemkerian, J. A. Garza-Reyes, J. Antony and J. Santos, "Theorizing the principles of sustainable production in the context of circular economy and Industry 4.0," *Sustainable Production and Consumption*, vol. 33, pp. 1043–1058, 2022.
21. M. Javaid, A. Haleem, R. P. Singh, R. Suman and E. S. Gonzalez, "Understanding the adoption of Industry 4.0 technologies in improving environmental sustainability," *Sustainable Operations and Computers*, vol. 3, pp. 203–217, 2022.
22. O. Kolade, M. Oyinlola, O. Ogunde, C. Ilo and O. Ajala, "Digitally enabled business models for a circular plastic economy in Africa," *Environmental Technology & Innovation*, vol. 35, pp. 1–16, 2024.
23. J. C. Kabugo, S.-. L. Jämsä-Jounela, R. Schiemann and C. Binder, "Industry 4.0 based process data analytics platform: A waste-to-energy plant case study," *Electrical Power and Energy Systems*, vol. 115, pp. 1–18, 2020.
24. N. Abdullah, O. A. Alwesabi and R. Abdullah, "IoT-based smart waste management system in a smart city," *Recent Trends in Data Science and Soft Computing*, vol. 843, pp. 364–371, 2018.
25. ENL, "Waste sorting in Moka," *ENL*, 2023. [Online]. Available: https://www.moka.mu/en/experience-moka/services/smart-waste-sorting-management/.
26. G. M. Monzambe, K. Mpofu and I. A. Daniyan, "Statistical analysis of determinant factors and framework development for the optimal and sustainable design of municipal solid waste management systems in the context of Industry 4.0," In *29th CIRP Design 2019*, Póvoa de Varzim, Portgal, vol. 84, p. 245–250, 2019.
27. T. Bányai, P. Tamás, B. Illés, Ž. Stankevičiūtė and Á. Bányai, "Optimization of municipal waste collection routing: Impact of Industry 4.0 technologies on environmental awareness and sustainability," *International Journal of Environmental Research and Public Health*, vol. 16, pp. 1–26, 2019.
28. J. R. Noennig, M. Dale Molero, A. Mukherjee, Q. Safariallakheili and J. Chang, *Digital Modelling of Solid Waste Management (SWM) for the Indian Urban Development Context*, p. 1, 2022. https://www.hif-hamburg.de/Images/Abschlusskonferenz%202022/pdfs%20Abstracts/Abstract%20J%C3%B6rg%20N%C3%B6nnig.pdf
29. A. Henaien, H. Ben Elhadj and L.C. Fourati, "A sustainable smart IoT-based solid waste management system", *Future Generation Computer Systems*, vol. 157, pp. 587–602, 2024.
30. Y.H. Lin, W.L. Mao and H.I.K. Fathurrahman, "Development of intelligent Municipal Solid waste Sorter for recyclables." *Waste Management*, vol. 174, pp. 597–604, 2024
31. O. Hashemi-Amiri, M. Mohammadi, G. Rahmanifar, M Hajiaghaei-Keshteli, G. Fusco and C. Colombaroni, "An allocation-routing optimization model for integrated solid waste management", *Expert Systems with Applications*, vol. 227, p. 120364, 2023.
32. R. Viswanathan and A. Telukdarie, "The role of 4IR technologies in waste management practices-a bibliographic analysis" Procedia Computer Science, vol. 200, pp. 247–56, 2022.
33. UNDP, "Transforming waste management in Uganda," *Stories*, April 2024 [Accessed on 04 June 2024] Available from: https://stories.undp.org/transforming-waste-management-in-uganda
34. R. Borchard, R. Zeiss, J. Recker, "Digitalization of waste management: Insights from German private and public waste management firms," *Waste Management & Research*, vol 40, issue 6, pp. 775–792, 2022.
35. F. S. Sahib and N.S. Hadi, "Truck route optimization in Karbala city for solid waste collection," *Materials Today: Proceedings*, vol 80, part 3, pp. 2489–2494, 2023.

36. R. Mohee, S. Mauthoor, Z. M. A. Bundhoo, G. Somaroo, N. Soobhany and S. Gunasee, "Current status of solid waste management in small island developing states: A review," *Waste Management*, vol. 43, pp. 539–549, 2015.
37. L. Chen, P. He, H. Zhang, W. Peng, J. Qiu and F. Lü, "Applications of machine learning tools for biological treatment of organic wastes: Perspectives and challenges," *Circular Economy*, vol. 3, in press, pp. 1–37, 2024.
38. L. M. Oliveri, S. Arfò, A. Matarazzo, D. D'Urso and F. Chiacchio, "Improving the composting process of a treatment facility via an Industry 4.0 monitoring and control solution: Performance and economic feasibility assessment," *Journal of Environmental Management*, vol. 345, pp. 1–13, 2023.
39. M. Liukkonen, E. Hälikkä, T. Hiltunen and Y. Hiltunen, "Dynamic soft sensors for NOx emissions in a circulating fluidized bed boiler," *Applied Energy*, vol. 97, pp. 483–490, 2012.
40. H. I. D. I. Muri and D. R. Hjelme, "Sensor technology options for municipal solid waste characterization for optimal operation of waste-to-energy plants," *Energies*, vol. 15, pp. 1–24, 2022.

9 Logistic Management and Industry 4.0

Within the Construction Industry

Asish Seeboo

9.1 INTRODUCTION

Logistics is the lifeblood of any business that relies on moving goods, information, and resources. It is the intricate dance of planning, managing, and controlling the flow of these elements throughout the supply chain. However, how necessary are logistics exactly? The benefits of effective logistics management include: (i) reduced costs: optimised transportation, warehousing, and inventory management lead to significant cost savings; (ii) improved customer satisfaction: timely and accurate deliveries enhance the customer experience; (iii) increased efficiency: streamlined processes ensure smooth flow of goods throughout the supply chain; and (iv) enhanced competitiveness: effective logistics can give businesses a competitive edge in the market. Overall, logistics plays a critical role in business success. By optimising their logistics operations, businesses can achieve significant cost savings, enhance customer satisfaction, gain a competitive advantage, and navigate the complexities of today's dynamic market landscape.

Industry 4.0, Logistics 4.0, and Construction 4.0 are interconnected parts of a more significant trend towards digitalisation and automation. Industry 4.0 is considered the foundation, which refers to using advanced technologies like artificial intelligence (AI), big data, and the Internet of Things (IoT) to create a more connected and automated manufacturing environment. This "smart factory" concept emphasises real-time data exchange, machine-to-machine communication, and intelligent decision-making. Logistics 4.0 builds on Industry 4.0 and leverages the technologies of Industry 4.0 to optimise the entire supply chain. Imagine products with embedded sensors that track location and condition or self-driving trucks navigating optimised routes. Logistics 4.0 aims for increased efficiency, flexibility, and visibility throughout the journey of goods, from production to the end customer. Construction 4.0 borrows from Industry 4.0 and Logistics 4.0. Construction 4.0 is still in its early stages but takes inspiration from Industry 4.0 and Logistics 4.0. Technologies like 3D printing for building components, automated construction robots, and drones for surveying and site monitoring are examples of Construction 4.0. The goal is to improve the construction industry's safety, efficiency, and sustainability. The overall impact

DOI: 10.1201/9781003511298-9

of these three concepts is that they create a more interconnected and intelligent ecosystem. Imagine a scenario where factories automatically adjust production based on real-time demand, logistics networks adapt to real-time disruptions, and construction sites use robots for dangerous tasks. Implementing these technologies requires significant investment and infrastructure development. Cybersecurity threats and the need for a skilled workforce to manage these advancements are essential considerations. Overall, Industry 4.0, Logistics 4.0, and Construction 4.0 represent a future where technology drives greater efficiency, flexibility, and innovation across various industries.

9.2 LOGISTIC MANAGEMENT

Logistics management is the strategic planning, implementing, and controlling of the movement and storage of goods, services, and information across the supply chain. Its core objective is to ensure the correct product gets to the right place, at the right time, in the proper condition, and at the lowest cost. The key aspects of logistics management: (i) planning and coordination: this involves activities like forecasting demand, sourcing materials, and designing efficient transportation routes [1]; (ii) transportation: selecting the most suitable mode of transportation (road, sea, air) for cost, speed, and cargo type is crucial [2]; (iii) warehousing: efficient storage and handling of goods in warehouses is essential for smooth order fulfilment [3]; and (iv) inventory management: maintaining optimal inventory levels to avoid stockouts or overstocking is a balancing act in logistics [4]. Here is a breakdown of its significance in today's business landscape. Efficient logistics minimises waste, optimises transportation (through route planning and mode selection), and ensures efficient use of resources like warehousing space. This translates to significant cost savings for businesses [5]. Furthermore, effective logistics management helps maintain optimal inventory levels, preventing stockouts that disrupt production and overstocking, leading to unnecessary storage costs [6]. Logistics ensures on-time deliveries, which is crucial for customer satisfaction in today's fast-paced world [7]. The prevailing processes embedded in logistics minimise errors in picking, packing, and shipping, leading to higher customer satisfaction [8]. Adopting logistics within any business has two significant competitive advantages: (i) faster response times as efficient logistics enables businesses to react quickly to market changes and respond to customer demands faster than competitors with sluggish supply chains [9], and (ii) better cost competitiveness by optimising costs throughout the supply chain so that businesses can offer competitive prices and attract customers [10]. Further benefits of adopting logistics include (i) improved risk management, which through effective logistics can identify potential disruptions (e.g., supplier delays, transportation issues) and implement proactive solutions to minimise their impact [11], and (ii) enhanced collaboration between different departments within a business (e.g., procurement, production, sales) and with external partners (e.g., suppliers, distributors). By understanding and implementing best practices in logistics management, organisations can significantly improve their overall supply chain efficiency and customer satisfaction.

9.3 LOGISTIC MANAGEMENT VERSUS SUPPLY CHAIN MANAGEMENT

Logistics and supply chain management (SCM) are often used interchangeably but represent distinct yet interconnected concepts. Logistics management focuses on an organisation's operational aspects, primarily dealing with the physical movement, storage, and flow of goods and materials. It concerns the "how" of efficiently getting products from point A to point B. The key functions include (i) transportation management (choosing modes and routes) [12], (ii) warehousing and storage management [13], and (iii) inventory control (maintaining optimal stock levels) [14].

SCM, on the contrary, takes a broader perspective, encompassing all the activities involved in transforming raw materials into finished products and delivering them to the end customer. It includes internal operations (logistics) and collaboration with external partners like suppliers and distributors. It is about the "what" and "why" of the entire supply chain network. Its key functions include:

1. Strategic planning and sourcing of materials
2. Supplier relationship management
3. Demand forecasting and planning
4. Logistics management (as a subset)
5. Customer relationship management

Views diverge on the prevailing relationship between logistics and SCM. Traditionalists position SCM as a broader concept encompassing logistics [15], while unionists emphasise the integration and interdependence of logistics and SCM [16].

Logistics management is like a well-oiled engine within the giant supply chain machine. SCM focuses on optimising the entire network, ensuring smooth interaction between all parts. Effective logistics management is crucial for a well-functioning supply chain, while strong SCM leverages efficient logistics to create a competitive advantage.

9.4 TYPES OF LOGISTIC MANAGEMENT

Logistics management can be broadly categorised into four main types, each focusing on a specific stage within the supply chain:

1. **Inbound Logistics**:
    - Deals with moving and storing raw materials, components, and goods from suppliers to a company's production facilities or warehouses.
    - Key activities include managing procurement, customs clearance, transportation (often involving bulk shipments), and inventory control for incoming materials [17].

2. **Outbound Logistics**:
   - Manages the flow of finished products from a company's production facilities or warehouses to the end customer.
   - This involves order fulfilment, packaging, selecting appropriate transportation modes (often considering speed and smaller quantities), and ensuring timely deliveries [18].
3. **Reverse Logistics**:
   - Focuses on the movement of goods back up the supply chain. This includes product returns, recalls, recycling, and end-of-life product management.
   - Reverse logistics is becoming increasingly important due to environmental regulations and growing customer demand for sustainable practices [19].
4. **Third-Party Logistics (3PL)**:
   - Involves outsourcing some or all of a company's logistics functions to a specialised logistics provider (3PL company).
   - 3PL companies can offer a wide range of services, including warehousing, transportation management, freight forwarding, and value-added services like packaging and labelling [20].
5. **Additional Considerations**:
   - **Fourth-Party Logistics (4PL)**: This is an emerging concept where a 4PL provider acts as a lead logistics integrator, managing a network of 3PL providers to design and oversee a company's entire supply chain.
   - **Green Logistics**: This focuses on implementing sustainable practices throughout the logistics process to minimise environmental impact. This can involve using fuel-efficient transportation modes, optimising packaging to reduce waste, and implementing green warehousing practices [21].

By understanding and effectively implementing these different types of logistics management, companies can optimise their supply chains, reduce costs, improve customer satisfaction, and gain a competitive edge in the marketplace.

9.5 LOGISTIC 4.0

Logistics 4.0, or intelligent logistics, represents the next evolutionary leap in SCM. It leverages emerging technologies like AI, the IoT, Big Data, and automation to create a more intelligent, interconnected, and efficient logistics network (Figure 9.1). Below is a breakdown of the critical characteristics of Logistics 4.0.

- **Increased Automation**: Robots, drones, and autonomous vehicles will increasingly automate tasks like warehousing, picking and packing, and transportation [22].
- **Enhanced Visibility and Transparency**: Real-time data from sensors embedded in products and logistics infrastructure will provide complete transparency throughout the supply chain, allowing for better tracking and proactive management [23].

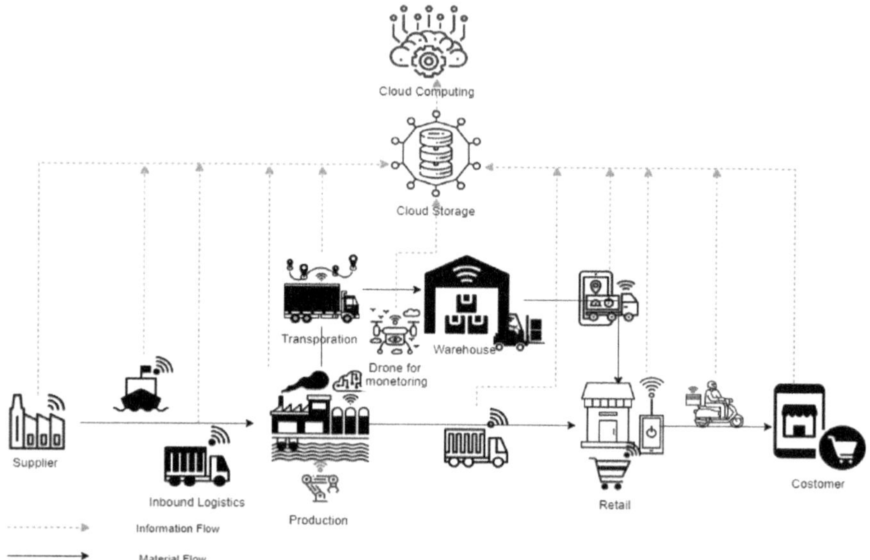

FIGURE 9.1 Illustration of Logistics 4.0.

- **Data-Driven Decision-Making**: Advanced analytics will utilise vast data to optimise transportation routes, inventory levels, and overall supply chain performance [24].
- **Collaboration and Integration**: Logistics 4.0 fosters greater collaboration among all stakeholders in the supply chain, from suppliers to customers, through platforms and information sharing [25].

Benefits of implementing Logistics 4.0 include the following: (i) Increased Efficiency – automation and data-driven insights can streamline processes and reduce waste; (ii) Improved Cost Savings – optimisation of transportation, warehousing, and inventory management leads to cost reductions; (iii) Enhanced Customer Satisfaction – increased visibility and faster deliveries lead to a better customer experience; and (iv) Greater Agility and Resilience – the ability to react quickly to disruptions and changing market demands.

The field of Logistics 4.0 is still evolving, and several researchers are focusing on its concepts and applications. For example, the determinants of Logistics 4.0 adoption modelling examines factors influencing the adoption of Logistics 4.0 practices, particularly in developing countries.

Challenges and considerations faced in the field of logistics include (i) technological integration: implementing and integrating various technologies requires significant investment and expertise; (ii) data security and privacy: ensuring data security and privacy across a complex network is crucial; and (iii) workforce impact: while creating new jobs, automation may also displace some existing jobs, requiring workforce training and reskilling. Despite the challenges, Logistics 4.0 holds immense potential for transforming supply chains and creating a more efficient, sustainable, and customer-centric future.

9.6 INDUSTRY 4.0

Industry 4.0, also known as the Fourth Industrial Revolution, is a paradigm shift in manufacturing characterised by the integration of advanced technologies to create intelligent and connected factories. The core technologies of Industry 4.0 include:

- **Cyber-Physical Systems**: The physical components of a factory are integrated with virtual counterparts, enabling real-time data exchange and autonomous decision-making [26].
- **IoT**: Sensors embedded in machines, products, and infrastructure allow continuous data collection and communication, providing insights into production processes [27].
- **Big Data and Analytics**: The vast amount of data generated by connected devices is analysed to identify patterns, optimise processes, and predict potential issues [28].
- **AI and Machine Learning (ML)**: AI and ML algorithms enable machines to learn from data, make autonomous decisions, and improve process efficiency [29].

Automation, data-driven insights, and real-time optimisation lead to significant efficiency gains, thus increasing productivity. Furthermore, enhanced monitoring and control systems ensure higher quality products and excellent customisation capabilities of the end products. Thus, adapting to changing market demands and production requirements is essential more quickly. Last but not least, the benefits of Industry 4.0 include improved efficiency and optimised processes that lead to cost savings in production and maintenance. It is also important to note that much research explores Industry 4.0's concepts, applications, and potential impacts.

Industry 4.0 faces myriad challenges. Implementing advanced technologies and infrastructure upgrades can be expensive, increasing investment costs. Securing connected systems and protecting sensitive data are crucial challenges, hence the cybersecurity concerns. The shift towards automation requires upskilling and reskilling the workforce to palliate the workforce skills gap. However, Industry 4.0 presents a significant opportunity for increased productivity, efficiency, and innovation in manufacturing. Addressing the challenges and ensuring a smooth transition are crucial for successful implementation.

9.7 THE SYMBIOTIC RELATIONSHIP OF LOGISTIC 4.0 AND INDUSTRY 4.0

Logistics 4.0 and Industry 4.0 are not isolated concepts; they are two sides of the same coin in today's intelligent and connected supply chain landscape. Their integration unlocks a new efficiency, agility, and optimisation level across the entire value chain. To understand the synergy, one must know that Industry 4.0 has laid the foundation for implementing smart factories, which generate vast amounts of real-time data on production, inventory levels, and machine performance (Figure 9.2). On the contrary, Logistics 4.0 leverages the data to optimise transportation routes, automate warehousing processes, and predict potential disruptions.

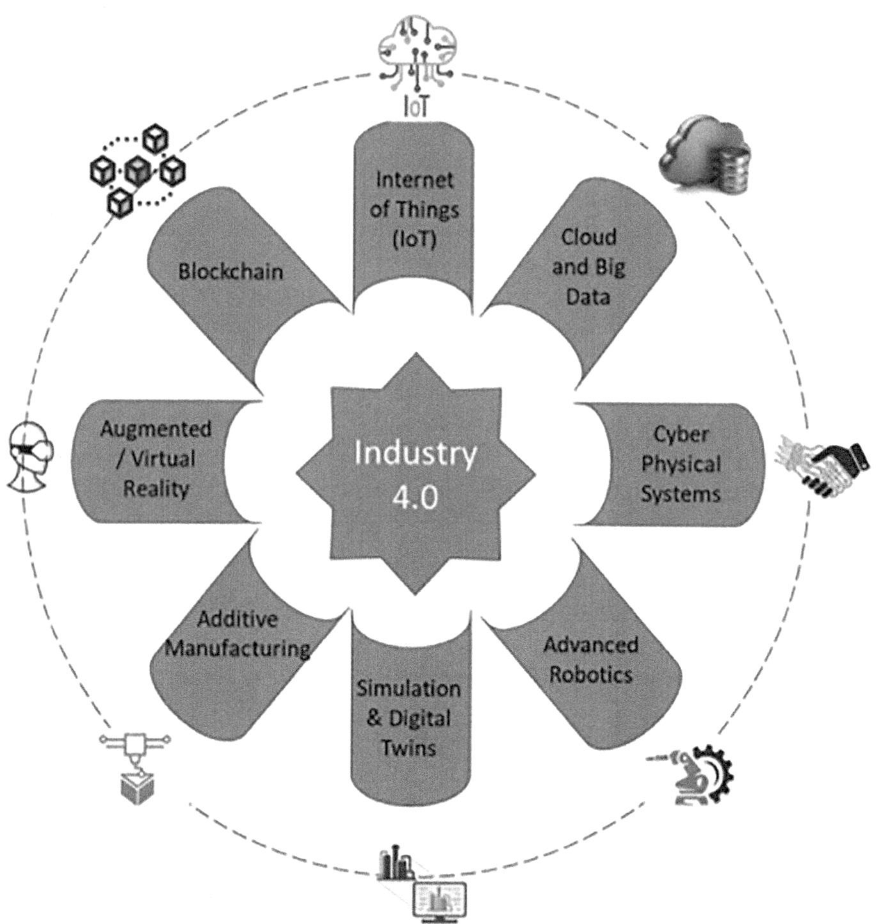

FIGURE 9.2 Logistics 4.0 linkage with Industry 4.0.

9.7.1 Critical Benefits of Integration

- **End-to-End Visibility and Transparency**: Real-time data from both production and logistics provides a holistic view of the supply chain, allowing for proactive management [30].
- **Hyper-Efficient Operations**: Data-driven insights enable optimised inventory management, predictive maintenance, and just-in-time deliveries.
- **Enhanced Customer Experience**: Faster and more reliable deliveries and improved product tracking lead to higher customer satisfaction.
- **Increased Agility and Resilience**: The integrated system can more effectively adapt to disruptions and changing market demands.

9.7.2 Challenges and Considerations

- **Technological Integration Complexity**: Integrating various technologies across different systems requires careful planning and expertise.
- **Data Sharing and Standardisation**: Sharing data securely and establishing data standardisation across the supply chain is crucial.
- **Cybersecurity Threats**: The interconnected network requires robust cybersecurity measures to protect sensitive data.

9.8 THE CONSTRUCTION INDUSTRY: A GLOBAL GIANT SHAPING OUR WORLD

The construction industry is a force to be reckoned with, leaving its mark on every corner of the globe. It is more than just brick and mortar; it is the foundation of our cities, the backbone of our infrastructure, and the architect of our built environment. Let us delve into the world of construction and explore its key features.

9.8.1 Economic Powerhouse

- **Trillion-Dollar Giant**: The industry boasts a massive size, valued at roughly $10 trillion annually. This immense economic footprint highlights its critical role in global GDP.
- **Growth Trajectory**: The future looks bright, with consistent growth projected, exceeding global GDP growth in the coming decade. Several driving forces fuel this positive outlook.
- **Infrastructure Demands**: As populations in developing nations surge and existing infrastructure in developed countries ages, continuous investment is necessary.
- **Residential Boom**: The ever-growing global population constantly demands new housing units, driving residential construction [3].
- **Sustainability Imperative**: The push for renewable energy and improved telecommunication infrastructure opens doors for exciting new construction opportunities.

9.8.2 Challenges and Considerations

- **Productivity Paradox**: Despite its size, the construction industry struggles with lagging productivity compared to other sectors. If addressed effectively, this could translate to a potential $1.6 trillion gain.
- **Sustainability Concerns**: Construction activities are a major consumer of resources and a significant contributor to global emissions. Embracing sustainable practices and materials is crucial [5].

9.8.3 A World of Opportunity

- **Regional Variations**: Growth is expected to be more robust in developing economies due to their rapid urbanisation and infrastructure needs. While

seeing potentially slower growth, developed economies will still require significant investments in infrastructure upgrades.
- **Tech Revolution**: The construction sector is on the cusp of a technological transformation. Digitalisation, automation, and prefabrication hold immense potential for boosting efficiency and productivity.

9.8.4 Building a Sustainable Future

Shifting Gears: The industry is increasingly recognising the importance of sustainability. Environmentally friendly materials, energy-efficient buildings, and waste-reduction practices are becoming the norm.

The construction industry is a complex and dynamic sector that significantly impacts our world. By understanding its current landscape, prospects, and challenges, we can create a more sustainable and prosperous built environment for all.

9.9 SUPPLY CHAIN WITHIN THE CONSTRUCTION SECTOR

The construction industry relies heavily on a complex network of interconnected players and processes – the supply chain. The suggested definitions reflect that SCM is highly complex and diverse [31]. As an evolving field, it encompasses different perspectives and contributions from other research fields [32]. Initially developed along the lines of logistics [32], the incorporation of additional concepts and practices from social and organisational sciences has been observed throughout its evolution [33]. Tiwari et al. [34] have emphasised that researchers are keen on developing and pursuing various decision-making models and exploring areas such as supply chain integration and coordination. This intricate web ensures that the suitable materials, equipment, and services arrive at the job site at the right time and in the right quantities. Here is a closer look at this vital system:

9.9.1 Key Players

- **Manufacturers and Suppliers**: These companies produce and distribute various construction materials, from lumber and concrete to plumbing fixtures and electrical components.
- **Distributors and Wholesalers**: They act as intermediaries, stocking and delivering materials from manufacturers to contractors.
- **Contractors (Prime and Subcontractors)**: They manage the overall construction project and procure materials from suppliers or distributors.
- **Logistics Providers**: These companies handle the transportation and storage of construction materials, ensuring they reach the job site efficiently.

9.9.2 The Flow of Goods and Information

- **Planning and Procurement**: Contractors determine material needs and specifications during the planning phase. Procurement involves sourcing materials from suppliers, negotiating prices, and managing delivery schedules.

- **Just-in-Time Delivery**: Ideally, materials arrive on-site precisely when needed to minimise storage requirements and waste. Effective communication and coordination are paramount.
- **Inventory Management**: Maintaining appropriate stock levels avoids project delays due to material shortages and prevents excessive on-site inventory that can tie up capital.

9.9.3 Challenges in Construction Supply Chains

- **Fragmented Landscape**: The construction industry involves numerous players, making coordination complex.
- **Unpredictable Project Demands**: Construction projects can have fluctuating material needs, making it difficult to maintain consistent inventory levels.
- **External Disruptions**: Global events like material shortages or transportation disruptions can significantly impact supply chains.

9.10 IMPACT OF LOGISTICS ON CONSTRUCTION PROJECTS

Logistics, the planning and execution of efficient movement of materials, equipment, and personnel, plays a critical role in the success of construction projects. Here is a closer look at how effective logistics can significantly impact various aspects of construction:

9.10.1 Positive Impacts

1. **Project Efficiency and Schedule**: Well-planned logistics ensure materials and equipment arrive on-site exactly when needed, minimising delays and disruptions to the construction schedule. This keeps projects on track and avoids costly downtime for workers.
2. **Reduced Costs**: Efficient logistics minimise waste by ensuring that only the necessary materials are delivered at the right time. Optimised transportation routes and storage solutions can also lead to significant cost savings.
3. **Improved Safety**: Proper logistics planning minimises on-site clutter and ensures safe material handling practices, reducing the risk of accidents and injuries for workers.
4. **Enhanced Quality**: When materials are delivered on time and stored correctly, there is less risk of damage or spoilage. This contributes to better overall construction quality.
5. **Increased Worker Productivity**: When materials are readily available and easily accessible, workers can focus on their tasks without wasting time searching or waiting for deliveries. This leads to increased productivity and efficiency.

9.10.2 Negative Impacts of Poor Logistics

1. **Project Delays**: Late deliveries of materials or equipment can cause significant delays, impacting the entire construction schedule. This can lead to cost overruns, penalties for missed deadlines, and disruptions to other trades.
2. **Increased Costs**: Inefficient logistics can result in wasted materials due to overordering, damage from improper storage, or unnecessary transportation costs, which can significantly inflate project costs.
3. **Safety Hazards**: Poorly planned logistics can lead to cluttered worksites, increasing the risk of accidents and injuries. Improper storage of materials can also pose safety hazards.
4. **Reduced Worker Morale**: Delays and disruptions caused by logistical issues can lead to frustration and decreased morale among workers. This can negatively impact productivity and overall project quality.
5. **Reputational Damage**: Repeated delays or safety incidents due to poor logistics can damage a contractor's reputation and make it difficult to secure future projects.
6. Poor logistics can significantly contribute to the failure of construction projects. Here is a breakdown of this critical relationship.

9.10.3 Logistics as a Recipe for Disaster

Imagine a construction project as a complex recipe. For success, each ingredient (material, equipment, personnel) must be added at the right time and in proportion. When logistics falter, it is like using expired ingredients or having them arrive late – the entire recipe gets thrown off.

9.10.4 How Logistics Failures Lead to Project Failure

Inefficient logistics can cause domino effects. Late deliveries of materials or equipment stall construction, leading to delays. This disrupts the entire schedule, incurring costs associated with worker downtime, potential penalties for missed deadlines, and the need for last-minute solutions. Poor logistics planning can lead to cluttered worksites, increasing the risk of accidents and injuries. Improper storage of bulky materials can also pose safety threats. Delays and disruptions caused by logistical issues can lead to frustration and decreased morale among workers. This can negatively impact productivity and overall project quality. Workers may spend time searching for missing materials or dealing with inefficiencies, hindering their ability to focus on tasks. When materials arrive late or are damaged due to improper storage or handling, it can compromise the quality of construction. Workers may be forced to use subpar materials or improvise solutions, impacting the final product.

9.10.5 Examples of Logistic Failures

Ordering too much or too little of a specific material can disrupt the workflow and cause delays. Lack of communication between contractors, suppliers, and logistics providers can lead to missed deliveries or incorrect materials delivered to the site.

Transportation disruptions, lousy weather, or supplier issues can throw a wrench into the best-laid plans if there is no contingency plan. Manual processes for tracking inventory and deliveries can be prone to errors and inefficiencies.

9.10.6 The Importance of Proactive Logistics Management

By prioritising effective logistics throughout the construction process, stakeholders can significantly reduce the risk of project failure. Developing a detailed logistics plan upfront is crucial. Considering potential disruptions, this plan should map out material needs, delivery schedules, storage requirements, and transportation routes. Open communication and collaboration are essential. All project stakeholders must work together seamlessly to anticipate and address logistical challenges. Logistics software for inventory management, real-time tracking, and communication can significantly improve efficiency and transparency. The construction world is full of unforeseen circumstances. A flexible logistics approach can adjust to changes or delays, minimising disruptions and ensuring the project stays on track. By prioritising robust logistics and building a culture of proactive planning, collaboration, and adaptation, construction projects can significantly improve their chances of success and avoid the pitfalls associated with logistical failures.

9.10.7 Best Practices for Effective Construction Logistics

1. **Detailed Planning**: It is crucial to develop a comprehensive logistics plan at the project's outset. This plan should consider material needs, delivery schedules, storage requirements, and transportation routes.
2. **Collaboration**: Open communication and collaboration between all project stakeholders, including contractors, suppliers, and logistics providers, is essential for smooth operation.
3. **Technology Integration**: Utilising logistics software for inventory management, real-time tracking, and communication can significantly improve efficiency.
4. **Flexibility**: Construction projects are inherently dynamic, so it is essential to have a flexible logistics plan to adapt to unforeseen changes or delays.
5. By prioritising effective logistics throughout the construction process, stakeholders can ensure timely project completion, reduce costs, improve safety, and enhance overall project quality.

9.11 CONSTRUCTION 4.0 IS THE FUTURE

Construction 4.0, or Industrial 4.0 for construction, represents the construction industry's digital transformation. According to the European Industry Construction Federation (FIEC), "Construction 4.0" is the counterpart of Industry 4.0 in the Architecture, Engineering & Construction industry, and it refers to the digitalisation of the construction industry [36]. Rastogi [35] stated that the primary goal of Construction 4.0 is to create a digital construction site that monitors progress throughout the life cycle of a project using different technologies. Adopting Construction 4.0 will change the construction process, organisation, and project structures, shifting the fragmented construction industry into an integrated industry [36]. It uses cutting-edge technologies to revolutionise design, build, and manage structures. Here is a breakdown of this exciting movement:

9.11.1 Core Elements

- **Digitalisation**: Embracing digital tools and technologies across all construction project phases, from 3D modelling (building information modelling [BIM]) to data analytics for project management.
- **Automation**: Introducing robots, drones, and other automated equipment to perform repetitive tasks, improve safety, and enhance efficiency.
- **Advanced Materials**: Utilising innovative materials like 3D-printed concrete elements or self-healing materials to streamline construction and improve building performance.
- **Connectivity**: Ensuring seamless communication and data exchange between all project stakeholders through cloud-based platforms and the IoT.

9.11.2 Benefits of Construction 4.0

- **Enhanced Productivity**: Technology streamlines processes, reduces waste, and optimises workflows, leading to faster project completion times.
- **Improved Safety**: Automation can take over hazardous tasks, reducing workplace accidents and injuries.
- **Reduced Costs**: Efficiency gains and minimised rework translate to significant cost savings for construction projects.
- **Increased Sustainability**: Advanced materials and better planning can lead to more energy-efficient buildings and a smaller environmental footprint.
- **Better Decision-Making**: Data analytics provide valuable insights for informed decision-making throughout the construction lifecycle.

9.11.3 Challenges of Construction 4.0

- **Skilled Workforce**: The industry needs to bridge the skills gap by training workers to adapt to new technologies and digital tools.
- **Integration Costs**: Implementing advanced technologies can involve significant upfront costs, requiring investment and planning.
- **Data Security**: With increased reliance on digital information, robust cybersecurity measures are crucial to protect sensitive project data.
- **Standardisation**: The construction industry lacks standardised data exchange and technology implementation approaches, hindering collaboration.

9.12 LOGISTIC SOFTWARE USED IN THE CONSTRUCTION SECTOR

Construction project management software, like Procore or Oracle Primavera Cloud, often has built-in logistics modules. These modules facilitate communication, document sharing, and real-time updates on deliveries, material needs, and inventory levels among all project stakeholders. Delivery management software like Trackforce or Onfleet focuses specifically on managing deliveries. They enable contractors to schedule deliveries, track materials in real time, and ensure transparency throughout the supply chain. Inventory management software tools like eBuilder or Buildertrend help contractors track material inventory on-site and across warehouses. This helps avoid overstocking or running out of critical materials, leading to optimised storage space and reduced waste. Logistics planning and execution, such as transportation management systems, has software like DAT Freight and Analytics or Manhattan Associates Transportation Management, which helps plan and optimise transportation routes, negotiate with carriers, and track deliveries in real time. This ensures efficient movement of materials and equipment to the job site. Warehouse management systems such as Bluejay or Fishbowl improve efficiency in warehouses storing construction materials. They optimise storage layouts, automate picking and packing processes, and ensure materials are readily available for delivery.

For advanced solutions for complex projects, BIM platforms like Revit or AutoCAD can be integrated with logistics software to create a digital twin of the construction project. This allows for simulating material flows, identifying potential bottlenecks, and optimising logistics planning before construction begins. While not strictly software, these emerging technologies are revolutionising construction logistics. Drones can be used for faster and safer site inspections, while 3D printing on-site can reduce the need for specific materials to be transported long distances.

Selecting the most suitable software depends on project size, complexity, and specific needs. Some factors to consider include (i) identification of the areas where logistics pose the most significant challenges (e.g., communication, inventory management, transportation); (ii) construction logistics software solutions range from basic to comprehensive, with varying costs; (iii) choosing software that can grow with your company and adapt to different project requirements; (iv) user-friendly interfaces and intuitive features are essential for maximising software adoption and benefits.

By implementing effective logistics software solutions, construction companies can significantly improve streamlining workflows, real-time communication, and optimised transportation, leading to faster project completion times. Thus, reducing waste, optimising inventory management, and utilising efficient transportation minimise unnecessary expenses, resulting in improved visibility and communication and a safer work environment. Timely deliveries, better quality control, and on-budget completion contribute to happier clients.

9.13 CONCLUSION

Logistics 4.0 and Industry 4.0 integration represent a transformative force in SCM. By embracing these advancements, companies can unlock significant competitive advantages, improve efficiency, and deliver superior customer experiences.

Construction 4.0 is still in its early stages but holds immense promise for the industry's future. Construction can become a more efficient, sustainable, and future-proof sector by embracing innovation, collaboration, and a commitment to upskilling the workforce. The construction sector increasingly relies on specialised logistic software to streamline operations, improve efficiency, and minimise risks.

Investing in robust logistics software is a strategic decision for construction companies of all sizes. It empowers them to navigate the complexities of the construction supply chain, optimise resource allocation, and deliver successful projects.

REFERENCES

1. Jia, H., Li, C., & Zuo, X. "A review of collaborative logistics planning under supply chain disruptions", *International Journal of Production Research*, 57(14), 2019, pp. 4251–4271.
2. Dolce, P., & Nuccia, M. S, "The greening of logistics: Sustainable freight transport and last-mile delivery", *International Journal of Logistics Management*, 30(1), 2019, pp. 148–174.
3. Akyol, E., & Güner, S. "The impact of warehouse layout design on order fulfillment performance in e-commerce", *International Journal of Logistics Management*, 29(2), 2018, pp. 394–418.
4. Sandeep, G., & Ramanathan, R., "A review of inventory management research: Demand forecasting, supply chain coordination, and sustainable practices", *Production and Planning & Control*, 30(10), 2019, pp. 844–866.
5. Sander, T., *Industry 4.0 and Circular Economy: A New Paradigm for Logistics.* We built Worlds in 2021. https://medium.com/life-is-too-short-for-bad-logistics/industry-4-0-and-the-circular-economy-a-new-paradigm-for-logistics-1114be8abb98
6. Stock, J. R., & Lambert, D. M., *Supply Chain Management*, Springer International Publishing, 2018.
7. Christopher, M., & Holweg, M., *Logistics & Supply Chain Management: Creating Value in the New Era*, Pearson Education Limited, 2018.
8. Mentzer, J. T., DeWitt, W., Keebler, J. S., Minnich, S., Nix, N., & Zacharia, Z. G., "Defining supply chain management", *Journal of Supply Chain Management*, 37(4), 2001, pp. 25–39.
9. Gunasekaran, A., Lai, K. H., & Cheng, T. C., "Responsive supply chain design and management: A review of literature and research directions", *International Journal of Production Economics*, 144(1), 2013, pp. 1–24. doi: 10.1016/j.ijpe.2012.08.008.
10. Carter, J. R., Ellram, L. M., & Kaufmann, P., "Integrative sourcing: A theoretical framework and empirical investigation", *Journal of Supply Chain Management*, 39(1), 2003, pp. 8–20.
11. Vrijens, D., & Long, H. D., "A framework for supply chain risk management based on a conceptual model", *International Journal of Production Economics*, 174, 2016, pp. 110–120.
12. Dolce, P., & Nuccia, M. S., "The greening of logistics: Sustainable freight transport and last-mile delivery", *International Journal of Logistics Management*, 30(1), 2019, pp. 148–174.
13. Sengupta, A., "Supply chain inventory control model for Nestle India Ltd". *International Journal of Scientific and Engineering Research*, 8(8), 2017, pp. 523–527. Retrieved from https://www.ijser.org/researchpaper/Supply-Chain-Inventory-Control-Model-for-Nestle-India-Ltd.pdf
14. Akyol, B., & Güner, S., "Warehouse design decisions: A classification review and future research directions", *International Journal of Production Economics*, 221, 2020, pp. 107482.
15. Sandeep, G., & Ramanathan, R., "A review of inventory management research: Demand forecasting, supply chain coordination, and sustainable practices", *Production and Planning & Control*, 30(10), 2019, pp. 844–866.
16. Christopher, M., & Gattorna, J., *Supply Chain Management: A Strategic Perspective*, Pearson Education Limited, 2011.

17. Jüttner, U., Christopher, M., & Godsell, P., "Demand chain management for customer centricity", *European Journal of Marketing*, 37(11/12), 2003, pp. 1613–1634.
18. Kaiplanen, A., & Rajahonka, M., "Inbound logistics and the circular economy: A review of relevant literature", *International Journal of Logistics Management*, 29(2), 2018, pp. 473–497.
19. Dolce, P., & Nuccia, M. S., "The greening of logistics: Sustainable freight transport and last-mile delivery", *International Journal of Logistics Management*, 30(1), 2019, pp. 148–174.
20. Vachon, S., & Klassen, M. J., "The role of reverse logistics in closed-loop supply chains", *Journal of Business Logistics*, 41(1), 2020, pp. 3–17.
21. Akyol, B., & Büyüközkan, G., "Third-party logistics service providers' selection: A fuzzy AHP approach with scenario analysis", *Computers & Industrial Engineering*, 123, 2018, pp. 311–329.
22. Dolce, P., & Nuccia, M. S., "The greening of logistics: Sustainable freight transport and last-mile delivery", *International Journal of Logistics Management*, 30(1), 2019, pp. 148–174.
23. Holmström, J., & Aalto, P., "Automation for logistics 4.0: An empirical exploration of technology and organisational readiness", *International Journal of Logistics Management*, 31(2), 2020, pp. 553–574.
24. Christopher, M., & Holgate, M., *Digital Supply Chain Transformation*, Kogan Page Publishers, 2020.
25. Sodhi, M., & Tang, L., *Integrated Supply Chain Management* (3rd ed.), Pearson Education Limited, 2018.
26. Wang, S., & Hsu, C.-H., "A review of enablers for collaborative logistics 4.0", *Sustainability*, 12(14), 2020, pp. 5542.
27. Lee, J., Bagheri, B., & Kao, H.-A., "A cyber-physical systems architecture for industry 4.0-based manufacturing systems", *Manufacturing Letters*, 3(1), 2015, pp. 18–23.
28. Jazdi, N., "Cyber-physical systems in the context of Industry 4.0", In *2014 IEEE International Conference on Automation, Quality and Testing, Robotics (AQTR)*, Cluj-Napoca, Romania, 2014, pp. 1–4, IEEE.
29. Lee, J., Kao, H.-A., & Yang, S., "Service innovation and smart analytics for industry 4.0 and big data environment", *Procedia CIRP*, 16, 2014, pp. 3–8.
30. Bahrin, M. A. K., et al., "Industry 4.0: A review on industrial automation and robotic", *Jurnal Teknologi*, 78(6–13), 2016, pp. 137–143.
31. Adetunji, I., Price, A. D. F., & Fleming, P. "Achieving sustainability in the construction supply chain", *Proceedings of the Institution of Civil Engineers - Engineering Sustainability*, 161, 2008, pp. 161–172.
32. Croom, S., Romano, P., & Giannakis, M. "Supply chain management: An analytical framework for critical literature review", *European Journal of Purchasing & Supply Management*, 6, 2000, pp. 67–83.
33. Isatto, E. L., Azambuja, M., & Formoso, C. T. "The role of commitments in the management of construction make-to-order supply chains", *Journal of Management in Engineering*, 31, 2015, pp. 04014053.
34. Tiwari, R., Shepherd, H., & Pandey, R. K. "Supply chain management in construction: A literature survey", *International Journal of Management Research and Business Strategy*, 3, 2014, pp. 7–28.
35. Rastogi, D. S. "Construction 4.0: The 4th generation revolution", In *The Indian Lean Construction Conference 2017*, IIT Madras, 2017, p. 12.
36. García de Soto B., Agustí-Juan I., Joss S., & Hunhevicz, J. "Implications of construction 4.0 to the workforce and organisational structures", *The International Journal of Construction Management*, 22, 2019, pp. 1–13.

10 Drone Technology in Industry 4.0

Challenges and Obstacles in Urban Environments

Jovana Kuljanin, Tanja Zivojinović, Natasa Bojkovic, and Cristina Barrado

10.1 INTRODUCTION

Drones, also known as unmanned aerial vehicles (UAVs), are unpiloted aircraft remotely controlled by human operators or autonomously programmed for specific functions. They come in various sizes and configurations, ranging from small consumer models to large military-grade aerial vehicles. With the capability to conduct an impressive range of tasks, drone operations play a significant role in driving innovation across various industries.

Drone technology is a powerful tool for data collection, processing, and real-time information delivering, which is a foundational element of Industry 4.0 and beyond. Using digital technology for real-time tracking and monitoring, inspections, site surveying, and other information enables the optimization of operations and value creation across the entire manufacturing, industrial, and supply chain ecosystem. With the integration of two rapidly evolving technologies, machine learning and artificial intelligence, drones will become fully autonomous and capable of data analysis and data-driven decision-making. Being able to perform tasks with minimum human inputs, drones perfectly align with the goals and requirements of Industry 4.0, in which increasing automation is the core concept.

In urban settings where the development of infrastructure generally cannot keep up with rapid urbanization, drones can offer tremendous benefits. Drones might improve the quality of urban life and address various challenges in many ways: from improving transportation and delivery services to supporting emergency response and urban planning initiatives. Potential areas of applications include monitoring of traffic, crowd or public events, environmental monitoring, agricultural monitoring, infrastructure, land use or other visual inspections, parcel delivery, emergency response (like firefighting, rescue operations, ambulance, medical transportation), and media and entertainment industry (Table 10.1).

DOI: 10.1201/9781003511298-10

TABLE 10.1
Drones Areas of Applications

Sector	Mobility and Transport	Energy	Public Safety and Security	Delivery and E-Commerce	Agriculture	Mining and Construction	Telecom	Insurance	Others
Mission	Railway inspection	Infrastructure sites	Police and fire	Parcel delivery	Crop and livestock monitoring	Bridge, crane and buildings	Cell tower inspection	Roof and site inspections	Real estate, private security, media
	Delivering goods	Pipeline and power lines	In-vehicle units	Medical and supply delivery	Crop spraying	Site surveying	Connectivity provision	Disaster impact	University and research
		Tethered wind energy production	Disasters		Large land monitoring				
			Wildlife						
			Border control						

Source: From SESAR Joint Undertaking [1].

The integration of advanced technologies like the Internet of Things (IoT) or cloud computing will streamline drone operations and enhance their capability to perform planning, execution, and data processing with high accuracy and reliability. Specifically, the integration of IoT involves equipping drones with sensors to collect data about surroundings, communication modules to transmit data in real time to ground stations, and onboard computers or microcontrollers for analysis of data collected from sensors in real time. With cloud computing, large volumes of data generated from drone sensors, cameras, and other onboard devices can be securely uploaded and stored in the cloud and further processed using the powerful capabilities of cloud computing platforms. Further on, cloud-based control systems enable centralized management of drone fleets, mission planning, and coordination of multiple drones for collaborative tasks. The delivery of top-notch cloud computing services for drones is made possible by 5G technology.

Yet, the deployment of drones in cities is still limited and many applications are waiting to be opened up. Some of the main challenges are associated with integrating drones into air traffic management systems, restriction of drone flights in densely populated areas, signal interference, privacy concerns, safety and cyber security risks, and noise pollution. Obviously, to ensure the safe and efficient integration of drones into the urban environment, technological advancements together with regulatory framework, community engagement, and education are needed.

Projections for the European drone market indicate continued growth, with estimates varying depending on factors such as technological advancements, regulatory developments, and market demand. A significant contributor to growth is also a decrease in manufacturing expenses [2]. According to the European Commission [3], the compound annual growth rate (CAGR) is expected to range between 8.5% and 16.2% for the period 2020–2030. In the most optimistic scenario, the estimated value of the drone market will reach $20.4 billion by 2030. Some projections indicate CAGR growth exceeding 18% in the coming years, leading to the market's value of a hundred billion dollars by the end of the 2020s [4]. Furthermore, the total number of direct employees in the drone industry in Europe is expected to reach around 145,000 by 2030, with a CAGR of 12.3%. Asia, North America, and Europe are the biggest markets in the drone industry, and, according to forecasts by 2025, these parts of the world will remain frontrunners [5]. While Europe increasing its funds in new products and services, the USA and China lead in global investment. However, it is essential to note that these projections are subject to change based on evolving market dynamics and unforeseen factors.

The content of the text is organized into four sections. Section 10.2 delves into the concept of operations (ConOps), examining current frameworks in the USA and Europe and discussing implementation strategies and airspace management. Section 10.3 explores three main types of missions: parcel delivery, medical delivery, and loitering missions, providing detailed case studies and comparisons of key characteristics and technologies. Section 10.4 addresses the challenges and obstacles faced in drone operations, including technological, societal, environmental, regulatory, privacy, and safety issues. Finally, the chapter ends with a comprehensive summary of the discussed topics and insights into future potential and necessary advancements for the widespread use of drone delivery systems.

10.2 CONCEPT OF OPERATIONS

The procedures, services, and requirements governing the deployment and utilization of drones while adhering to safety regulations and protocols are outlined in the ConOps. In general, as pointed out in Maly et al. [6], the ConOps document outlines and defines a system within its intended context of use. It forms the basis for all subsequent development activities, making accuracy and detail crucial. The document provides details of the operational environment and safety aspects. It can also assist in the certification process and in presenting findings to stakeholders. In summary, the ConOps provides a thorough overview of how the system will operate.

10.2.1 Current Outlooks of ConOps – USA and Europe

ConOps varies by region depending on many issues like regulatory environments, societal norms, or airspace management systems. In the following text, the key points of current outlooks of the ConOps in the USA and Europe are described in brief.

10.2.1.1 USA

In the USA, the Federal Aviation Administration (FAA) NextGen Office released an initial ConOps (v1.0) for unmanned aircraft systems (UAS) Traffic Management (UTM) in 2018 [7]. This baseline document was followed by v2.0 published in 2020 [8]. FAA UTM ConOps aims to identify operational requirements and technical specifications to enable UAS operations at low altitudes. In 2023, the FAA submitted the Implementation Plan, which addresses current work on UTM implementation [9]. The plan also outlines policy gaps affecting UTM implementation that must be resolved to achieve the full operational capability of operations beyond visual line of sight (BVLOS).

The FAA UTM (ConOps v2.0) covers a range of topics focusing on the integration of drones into existing controlled and uncontrolled airspace. It specifies participants and their roles, services and supporting infrastructure, operations, allocation of responsibilities, remote identification, and airspace management. The document also addresses several operational scenarios like interactions between UAS and low-altitude manned aircraft or complex operations of visual line of sight (VLOS) and BVLOS in uncontrolled and controlled airspace.

Some of the inherent components of the FAA UTM ecosystem are noted below.

- Access to the airspace
  The operators should apply for performance authorization granted by the FAA, which evaluates if the performance standards are met. The performance authorization is valid within the specific geographical area with discretely defined boundaries, the so-called authorized area of operation. To grant access to operate in controlled airspace (Class B/C/D/E), the FAA issues airspace authorization for a limited period of time, where after the operator gets approval from Air Traffic Control (ATC). Operations below 400 feet above ground level are not subject to authorization from ATC.

- UAS Service Suppliers (USSs)
  USS is a third-party entity that assists operators in maintaining compliance with regulations and safety standards and offer weather and environmental data services that may affect flight safety and performance. USSs coordinate the movements of multiple drones and ensure safe separation from manned aircraft and other obstacles without straight FAA involvement. Through the USS network, all information concerning situational awareness is exchanged on behalf of operators.
- Operation Plan
  The operation plan is shared with other operators prior to operation conduction. It indicates operation volumes, which are designated areas of airspace typically defined by spatial and temporal elements. The spatial component includes specific dimensions in three-dimensional space (latitude, longitude, and altitude) while the temporal specifies entry and exit times for the operator's UA. Sharing intent data prior to and during the operation enables sharing separation responsibility among operators, as well as de-confliction of overlapping airspace.
- Data sources and communication
  The main data sources within the UTM ecosystem are Supplemental Data Service Providers (SDSPs), Flight Information Management System (FIMS), and National Airspace System (NAS) Data Sources. UAS SDSPs complement the basic functions provided by USS and enhance situational awareness. They may provide information on weather and environment, flight restrictions, surveillance and geospatial data, and mapping services. The information is delivered to the USS network or operators. Information exchange between FAA and UTM participants is being implemented through FIMS, which is a cloud-based component managed by FAA. For the FAA, FIMS is an access point for information about UA operations and their possible impact on NAS. FIMS is also a communication bridge between NAS data sources and the UTM ecosystem.

10.2.1.2 Europe

The work on UTM vision in Europe was the joint effort of several bodies and organizations, the European Commission (EC), the European Aviation Safety Agency (EASA), the SESAR Joint Undertaking (SJU), EUROCONTROL, and Joint Authorities on Rulemaking for Un-Manned Systems. Drone technology is seen as a part of the European strategy for green and digital transition, which was announced in the European Strategy for Sustainable and Smart Mobility adopted in 2020. This document has foreseen the adoption of A Drone Strategy 2.0 for a Smart and Sustainable Unmanned Aircraft Eco-system in Europe [3], which should contribute to the further development of drone technology and its regulatory and commercial environment.

In 2016, the EC launched the U-space initiative as part of its efforts to develop a regulatory framework for drones in Europe. The U-space concept aimed to enable safe, efficient, and secure drone operations in urban environments and other populated areas.

Within the SESAR project CORUS (2017–2019), three editions of the U-space ConOps were released aiming to define rules, standards, and procedures. The latest fourth edition, produced in 2023, is the updated version of the third edition which specifies a set of services, technologies, and procedures designed to support safe, efficient, and secure integration of UAS into airspace. Among distinctive components of U-space, the following can be distinguished [10]:

- Operational categories
  All drone operations must be categorized as defined by the EASA. This framework divides drone operations into three categories: Open, Specific, and Certified. Each category has its own set of rules and requirements. The "Open" category covers drone operations that present a low risk to people on the ground and other airspace users. Operation with increased risk and therefore not appropriate for this category may be permitted under the "Specific" or "Certified" category. For operations that fall under the "Specific" category, operators should perform Specific Operations Risk Assessment (SORA) methodology and flights should be approved by the National Aviation Authority. For high-risk operations, operators need to obtain certification or authorization from the relevant authority, similar to manned aviation.
- Airspace types and access conditions
  According to the service provided, U-space defines three types of volume in the very-low-level (VLL) airspace, which are X, Y, and Z.
  Type X: No conflict resolution offered
  Type Y: Pre-flight conflict resolution offered
  Type Z: Pre-flight conflict resolution and in-flight separation are offered
  Accordingly, requirements for access differ. An approved plan for operations managed by U-space (called "operation plan" or "U-plan") is required in both Y and Z volumes, while there is no such obligation in X volume.
- U-space services
  According to the airspace types and U-space levels, different services are envisioned. Some of them are mandatory, while others are optional or recommended. The comprehensive description of the services is given in the latest, fourth edition of ConOps. Services are classified into several categories: Identification and Tracking, Airspace Management, Mission Management, Conflict Management, Emergency Management, Monitoring, Environmental Data, and Interface with ATC. The associate services are outlined in detail and also compared to previous ConOps releases and related to EU regulations.
- Social aspects
  Recommendations for strategic and operational decisions that improve social acceptance of UAS missions are discussed. These are related to minimum altitudes, hovering, speed limits, routing, etc. A series of surveys have been conducted to reveal the attitude of EU citizens toward active

involvement in drone operations in terms of being active users of delivery services. Safety, nuisance, and privacy concerns are found to be the issues of the most concern.

The comparison of FAA UTM and U-space ConOps is performed in detail by Lieb and Volkert [11]. Airspace classes, participants and their roles, services and supporting infrastructure, operations types and categories, air traffic management, as well as foreseen implementation are compared. The authors point to the main differences between the two ConOps. These are for example security aspects which are more deeply and, in more detail, specified in FAA UTM or EASA's drone operation categories for drone and U-space operation standards in contrast to performance authorization required by FAA UTM. The detected similarities and differences may serve as a valuable ground for learning from each other and further improvement of both concepts.

10.2.2 Implementation of ConOps

The EASA collaborated with industry stakeholders to develop the U-space Blueprint in 2017 [12], a roadmap which outlined the key principles, services, and technologies required for U-space progressive implementation. The document specifies four levels of U-space expected evolution ranging from U1 to U4.

U1 – Foundation Services: Lays the groundwork for more advanced U-space services by establishing essential infrastructure and regulatory frameworks; key functionalities include registration and identification of drones, geofencing to prevent flights in restricted areas, and basic communication and surveillance capabilities.

U2 – Initial Services: Introduces initial services for dynamic space management like real-time tracking of drone flights, including position, altitude, and velocity information; planning and approval of flights, designation of temporary no-fly zones and the coordination of drone operations in shared airspace.

U3 – Advanced Services: Incorporates automated conflict resolution algorithms to detect and avoid potential collisions between drones and other airspace users; optimize flight paths and airspace usage to minimize congestion and enhance efficiency.

U4 – Full Services: Supports advanced services such as cooperative traffic management, which involves collaborative decision-making between drones, manned aircraft, and ground control centers; seamless integration with existing air traffic management systems, enabling interoperability and coordination across different airspace sectors.

The implementation of FAA UTM is envisioned to be conducted progressively, in the so-called "spiral development," starting from low-complexity and low-environmentally risk operations. The development stages are established on three risk metrics:

- the number of people and amount of property on the ground,
- the number of manned aircraft in close proximity to the UAS operations, and
- the density of UAS operations.

Implementation of more complex operations will be possible thanks to the modular and discrete nature of services. The modular service approach enables technological progress to be accompanied by appropriate measures of supervision and regulation. It also allows for launching less demanding operations using existing communication networks, surveillance systems, data processing and analytics, cybersecurity protocols, etc., without the need for full-scale architecture.

10.2.3 Airspace Structure for Managing Drone Operations

Urban airspaces at VLLs are expected to host numerous drone flights operating in close proximity. These airspaces will also need to support a diverse array of missions that vary significantly in terms of range, altitude, and levels of autonomy. A well-designed airspace structure can cater to these diverse needs, allowing for a wide range of applications. Structured airspace ensures that drones can communicate effectively with ATC and other drones, enabling coordinated operations and reducing the risk of conflicts. Cities have numerous obstacles such as buildings, towers, and other infrastructure. Structured airspace helps drones navigate these complex environments safely, preventing accidents, and ensuring efficient routing, noise management, and efficient use of airspace. Drone operation within structured airspace can help gain public trust and acceptance, which is crucial for the widespread adoption of drone technology.

There are various existing and emerging airspace topologies for managing drone operations, which conceptually differ in addressing urban challenges. Studies on airspace structuring define two categories: structured and unstructured. However, since even unstructured concepts have some forms of sectionalization (e.g., layers), it is probably more precise to say that there are more- or less-structured designs. As explained in the research of the DACUS project [13], while the unstructured concept tends to foster the market by increasing the demand, a more structured airspace could reduce it. The opposite applies to the capacity because structured airspace is easier to manage and separate from manned aircraft.

In the unstructured design of airspace, drones are free to use the airspace except for the physical constraints, allowing for direct routing in their operations and hence maximum flexibility. To ensure safety, however, an airborne separation assurance algorithm is required so that drones can avoid collisions while following their optimal paths. Unstructured airspace design can be further classified into those controlled by centralized systems and those operated in a distributed manner (for a detailed analysis see Lee et al. [14]). Further smart city development will enable the use of cloud devices and services for drone operations. In that sense, a promising way to manage drones in less-structured airspace is using the IoT in combination with smart onboard devices [15].

As regards the structured airspace concept, various configurations have been proposed so far, two of which are singled out and described below.

10.2.3.1 Matrix Configuration

The framework is proposed by Low [16] and Mohamed Salleh et al. [17] as the way to discretize the airspace by forming uniform airspace units, that is standardized air blocks, cells, or grid-based cubes (Figure 10.1a). The drones follow trajectories

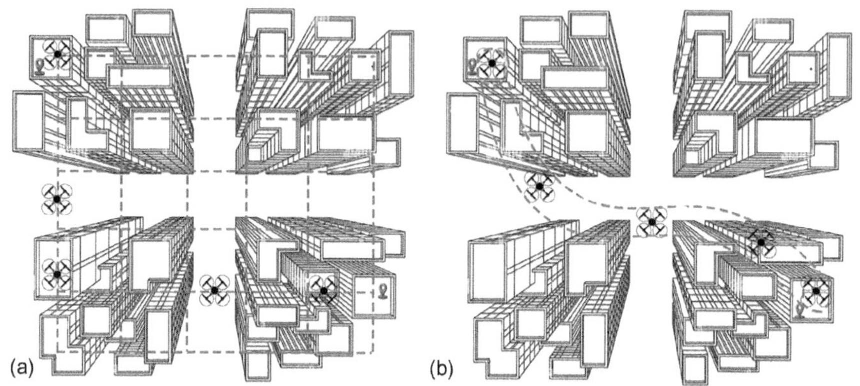

FIGURE 10.1 Matrix (a) and corridors (b) configurations schemes (compiled by authors).

through designated waypoints with the freedom to change movement direction. A distinct and unique address is allocated to each air block so that flight operations can be managed by processing these spatial data and checking for flight plan overlapping and possible conflicts. The air blocks are organized into altitude-based layers. The higher the layer, the higher the drone speed, and, accordingly, the size of the blocks can be bigger. On low-level layers, operations are performed at low altitudes, and because of the need to navigate around obstacles in the city, the blocks should be smaller in size to enable greater flight flexibility and frequent landing and take-off. One of the challenges of the air-matrix structure is the scalability. As noted in Lee et al. [14], supporting a growing number of drones within the grid could lead to congestion and require a large amount of data to be stored and managed. Another issue that needs to be addressed is related to the routes that must follow straight lines at right angles which reduces the efficiency of airspace utilization. For more detailed analysis of the matrix airspace concept see Pang et al. [18] and Tran and Nguyen [19].

10.2.3.2 Corridors

In this concept of airspace structure, the idea is to establish drone-dedicated routes/lanes. This also includes lanes directly connecting origins and destinations, without transit nodes, which is suitable for linking high traffic-density nodes. Such form is known as air tubes and can be conceived as flight highways or expressways (Figure 10.1b). They can be implemented in the air-matrix structure to handle high-speed traffic. To support a spatially greater range of flight plans, a network of lanes or tunnels in urban areas could be developed. According to Pathiyil et al. [20], there are several possibilities for that purpose. One is to follow routes over waterways, coastlines, or other less-dwelt urban areas. Another possibility is to utilize urban transportation infrastructure, including metro rail networks and highways. They present an opportunity to establish UAV lanes and tunnels either above or alongside existing land transportation routes while maintaining a safe distance from vehicles and buildings. A dynamic geofencing, adjusting continuously over time to maintain the required separation distance should be set around the projected drone positions according to the flight plan.

10.3 MISSIONS

10.3.1 Parcel Delivery

Similar to commercial aviation, the application of drones was initially restricted to military/police service where they assist in a variety of missions. On the one hand, they act as military-industrial couriers and enable delivering supplies and equipment [21], while, on the other hand, they provide enhanced aerial monitoring and collecting information to the soldiers in the surrounding environment. Since then, levering the development in technologies, the drone found its broad applications in other labor-intensive and complex tasks across different industry sectors. They are essentially practical for tasks that can be considered dull, hazardous, or dirty [22]. These applications encompass a broad spectrum of uses, from detecting defects in oil and gas pipelines to monitoring crop health, identifying hotspots in fire situations, conducting surveillance for mining and construction activities, cinematography, and delivering packages, among others [23].

Among many successful trials in each of the mentioned cases, UAS are apparently becoming an appealing means of transportation in modern logistics operations. Leveraging on operational advantages such as shorter delivery time, higher accuracy of the service, and environment-friendly operations, the drone has already found its prominent role in last-mile delivery. The types of parcels that are typically transported include a large set of commodities: from light deliveries of food, groceries, and homecare products to very sophisticated medical laboratory samples. The major players in the retail and logistics industries like Amazon, Google, the United Parcel Service (UPS), DHL, and Alibaba have seized the momentum of this promising concept. They have made substantial investments in validating the proof of concept through large-scale experiments in delivery drone systems (Table 10.2). In addition, the industries have been constantly supported by drone delivery service providers,

TABLE 10.2
Delivery Companies and Some Features on Their Service (Compiled by the Authors)

Provider	Country of Operations	Drone Type	Range	Payload
Amazon	USA (Texas, California), United Kingdom, Italy	MK27-2	Up to 14km (9 miles)	Up to 2.3kg (5 lbs)
UPS	USA (North Carolina, Florida, Ohio)	Matternet M2	Up to 20km (12.4 miles)	Up to 2kg (4.4 lbs)
Alphabet (Google)	USA (Dallas-Fort Worth)	Wing	Up to almost 10km (6 miles)	Up to 1.36kg (3 lbs)
DHL	Germany (Bavaria)	Paketcopter/ Parcelcopter	Up to 8km (4.97 miles)	Not available

among which most of them are also technology developers and manufacturers who sell full-stack drone delivery hardware and software systems (e.g., Zipline, Wing, Matternet, Swoop Aero, Manna).

In this context, the idea of parcel delivery, which was born more than a decade ago, eventually reached a required level of technological maturation. This milestone was marked by the approval of the first commercial drone delivery flight by the FAA on July 17, 2015 [24]. This flight was conducted by Flirtney company, transporting the medicine from Lonesome Pine Airport to the Remote Area Medical Clinic in the Virginia-Kentucky District Fairgrounds in the USA. This event opened the door for the integration of drones into mainstream logistics operations, paving the way for innovative and efficient parcel delivery solutions. Shortly after this event, in 2016, Amazon's drone delivery program, Prime Air, completed its first customer drone delivery by using an automated track from the Cambridge fulfillment area, which is the base of drones to the launch area [25]. In 2020, Amazon received FAA Part 135 approval to "safely and efficiently deliver package to customers" through its fleet of Prime Air delivery drones. This certification enabled Amazon to operate an autonomous drone in BVLOS conditions in areas with low population density and packages weighing 5 lbs or less. One year prior to Amazon, the same certification was granted to the subsidiaries of two large package delivery companies, UPS Flight Forward and Google's Wing, permitting these two companies to fly an unlimited number of drones. This also gave momentum to technology developers to pursue the production certificate. In this regard, the US Matternet, a leading developer of commercial drone delivery systems for urban and suburban environments, became the first drone delivery system to achieve Type Certification by the FAA in 2022 [26]. This goal was reached after the company completed the 4-year safety and reliability assessment for its Matternet M2 drone putting the company in a strong competitive advantage in the drone delivery market. The company operates its technology directly for customers or in partnership with logistics organizations, such as UPS.

Apart from the USA, which paved the way in this lucrative market segment, drone delivery attracted a lot of attention in other countries and world regions. In Europe, the largest logistics company DHL was the first one that launched commercial delivery via drone. The project aimed at delivering medicine and other critical goods from a pharmacy located 12 km from Germany's island of Juist in the North Sea [27]. In 2017, Siroop, the online sales platform belonging to the retail distributor Coop and the mobile company Swisscom, as well as the Californian Matternet which provided its drones, performed a series of demonstration flights enabling online shopping delivery in Zurich [28]. The project was performed in collaboration with the German automobile manufacturer Mercedes Benz who provided their electrical cars serving as landing platforms for drones. In this light, the recently closed project called the "mobil-e-Hub" led by the University of Mannheim, proposed a novel concept which aimed to integrate the parcel logistics of autonomous delivery drones into existing public transport [29]. To achieve this goal, the project uses AI to accurately predict the routes of public transport vehicles on which roof the drone will drop the parcel to its final destination.

Moreover, the EASA recently provided a list of the cities across the European region which are currently involved in large demonstration activities to integrate drone delivery into urban and suburban environments [30]. As seen in Table 10.3,

TABLE 10.3
Application of Drone Delivery in the European Region

European Cities	Stakeholder Category/Name	Business Use Case	Benefits for the Society and Environment
Dublin, Ireland	UAS operator/ Wing	• Good delivery • Drone demonstration	With 7–9% of retail transactions by 2033, the benefits would be: • 480 million vehicle kilometers less of traffic congestion, • 60 fewer accidents each year thanks to the reduced traffic on roads • 28,000 t of less CO_2 emissions, equivalent to the carbon storage of close to 830,000 trees.
Espoo, Uusimaa, Finland	UAS operator/ Wing	• Good delivery • Foods delivery • Drone demonstration	With 6% or more of consumer deliveries and pick-ups in Helsinki by 2030, the benefits would be: • Million fewer vehicle delivery kilometers reducing congestion, • 38 avoided road accidents yearly taking deliveries off the road and putting them in the sky, • Up to 2,000 t CO_2-emissions reductions yearly, the equivalent of planting 250 ha of additional forest.
Torsby, Värmland, Sweden	UAS operator/ Aviant AS	• Medicine delivery • Foods delivery • Goods delivery	• Access to home delivery services, which was previously only available near city centers, • Less pollution due to reduced road traffic, • Instant access to urgent goods when suddenly needed.
Trondheim, Trøndelag, Norway	UAS operator/ Aviant AS	• Medicine delivery • Foods delivery • Goods delivery	• Access to home delivery services, which was previously only available near city centers, • Less pollution due to reduced road traffic, • Instant access to urgent goods when suddenly needed.

Source: Form EASA [30].

drone operations are foreseen in the areas which are difficult to access by other means of transportation offering unique advantages in accessing remote or rugged terrain. For similar reasons, drone delivery has found its broad application in the region of Oceania, mainly Australia and Vanuatu, due to its large territory and dispersion of the people. The Australian Civil Aviation Safety Authority already approved the operation of two drone delivery services, Wing Aviation and Swoop Aero, providing supplies in remote communities [31]. In a similar vein, the testing flights were conducted by Japan Post in delivering parcels up to 2.5 kg to private properties in mountainous areas, which would be dangerous to access by road transportation [32].

However, the application of the drone is expected to provide benefits other than those related to easier access to remote and geographically inaccessible areas. As seen in Table 10.3, most of the benefits derived from drone operations refer to the reduction in traffic congestion and environmental impact and an increase in safety achieved by reduced traffic on roads. Indeed, drone-based delivery systems, which are envisioned to have a considerably lower carbon footprint, have the potential to replace the current system relying on vans/motorcycles for short distances. The magnitude of the particular benefits will largely depend on the ability of drone operators to divert the portion of the parcel delivery from road transportation. This will, in turn, be driven by the cost of the service which is currently estimated to be $484 (£395) for Amazon's parcel deliveries [33]. However, projections indicate a substantial reduction in costs to $63 (£51) by 2025, making drone delivery a more financially viable option. The study by Aurambout et al. [34] estimated that up to 7% of EU citizens could benefit from last-mile-drone delivery services under the scenario considered as the most technologically realistic.

To facilitate the seamless integration of drone delivery on a large scale, the reduction in costs must be complemented by the careful design of airspace structures. This includes not only rural areas but also urban centers, where over half of the world's population resides. The design of airspace infrastructure in urban environments remains a topic of ongoing debate among both academics and industry practitioners as it has to bridge a gap between some contradictory requirements including safety, efficiency, and regulatory compliance. Some feasible solutions were previously presented in Section 10.2, although not all of them are suitable for large-scale operations. For instance, the non-structured airspace allows direct routing in drone operations, although a prescribed airborne separation needs to be large enough to avoid other drones while flying their optimal route, directly affecting the system capacity. On the contrary, a street (grid)-based network will be able to accommodate a larger number of drones, which in turn, requires the careful design of corridors and flight rules with the highest level of traffic coordination. Consequently, the airspace structure design will have a direct impact on the logistic chains encountering the transportation of parcel delivery from logistic (distribution) centers to final customers. On the one hand, it will define the route operated by the given drone, and on the other, it will directly influence the battery consumption and thus, energy efficiency. In this way, the inherently costly drone-based parcel delivery will face the "last-mile" problem which has been already acknowledged as a challenging one by many logistic operators. This final delivery step is deemed as most complex one due to the fact that economies of scale cannot be further exploited and delivery costs are no longer shared with other packages. This issue is particularly exacerbated in rural areas, where the demand for parcel delivery is more dispersed across the territory, leading to increased kilometers travelled in a fragmented manner.

The current academic literature on parcel delivery logistics systems draws a clear distinction between truck-based and drone-based delivery systems, each characterized by its own set of constraints [35]. Moreover, the drone delivery system is observed to function either independently as a stand-alone drone delivery or in a hybrid mode when integrated with truck-based operations. The stand-alone concept of drone parcel delivery can be generally summarized by the following steps: Once a user places an order at an e-commerce store, the delivery operator initiates the process of assigning a drone to pick up the ordered package and deliver it. This assignment process involves selecting the most suitable drone from the fleet in real time, considering factors such as the destination(s) of the delivery and the current battery status of the drones. The drones may be in various states, including inactive, in-service, or at a charging center for battery swapping. Once assigned, the selected drone proceeds directly to the pickup location (typically called distribution centers [DCs] or warehouses) and then to the delivery destination, bypassing any constraints typically encountered by ground-based delivery routes. After completing the trip, the drone assesses its battery range and decides whether to proceed to the recharging station for battery swapping, move on to the next requested destination(s), or remain in its current location and await the next delivery request. The successful implementation of this concept will necessitate a robust network of stakeholders, each playing a crucial role at various stages of operations to ensure the smooth and efficient delivery of service. These stakeholders will include:

- **E-Commerce Platforms:** Responsible for receiving orders from customers and initiating the delivery process.
- **Delivery Service Providers/Drone Owning Company:** Facilitate the assignment of drones to pick up and deliver the ordered packages, coordinating the entire delivery process.
- **Drone Manufacturers:** Develop and maintain the fleet of autonomous drones, ensuring they are equipped with the necessary technology for safe and efficient operations.
- **Charging Station Operators:** Manage the charging stations where drones can swap batteries as needed to extend their range and minimize downtime.
- **Ground Support Team:** Manage drone vertiports (launch pads) including loading and unloading of packages;
- **UAS Traffic Management:** Ensure safe and efficient UAS traffic flow management, flight operation management, and meteorological information;
- **Regulatory Bodies:** Establish and enforce regulations governing drone operations, ensuring safety and compliance with airspace regulations.
- **Local Authorities:** Provide support and approval for the establishment of drone delivery routes and infrastructure within their jurisdictions.
- **Customers:** Receive timely and efficient delivery of their orders, providing feedback to improve the service.

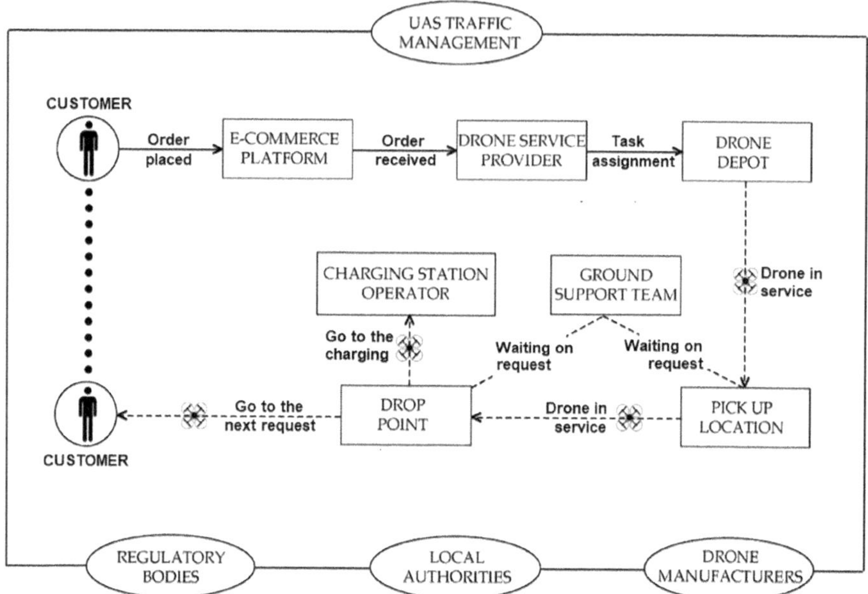

FIGURE 10.2 Process workflow for efficient and timely drone delivery systems (compiled by authors).

This process is illustrated in Figure 10.2 whose final objective is to ensure efficient and timely delivery while optimizing the utilization of the drone fleet. However, with the current technology settings, battery lifetime and range limitations are the main obstacles for large-scale applications in stand-alone delivery systems.

The introduction of drones in the "last-mile" operations opens a "pandora box" among academics for revisiting/reconsidering the existing transportation problems and approaches pertinent to efficient transportation. Therefore, it is not surprising that the problem of last-mile delivery is now bringing together the existing transportation approaches, combining them with recent developments in Information and Communication Technologies, Intelligent Transport Systems, Industry 4.0, and new transport vehicles [36]. In this regard, research Li et al. [37] provided an extensive literature review on urban drone delivery and classified the articles into five categories, depending on the group of the problem addressed, namely, the traveling salesman problem (TSP), vehicle routing problem (VRP), drone delivery scheduling problems, drone optimization problems, and urban last-mile problems. In this respect, the traditionally applied methodologies need to encounter certain constraints emerging from the distinctive drone features. For instance, drones have a limited range due to the limited energy capacity of their batteries (the current flight time is within the range of 40 minutes), which also affects the area covered [38]. To increase the flight range, the drone will need to use charging stations. These facilities resemble landing platforms that will support prompt battery recharge or swapping. Some of the innovative solutions to overcome this problem suggest that charging stations can be located at major traffic circles within the maximum range of the drone, facilitating the simultaneous accommodation of multiple drones [35]. The drones are still vulnerable to difficult methodological conditions (e.g., wind speeds, rains, snow) that may affect the drone's maneuverability and performance or even lead to damage to the drone structure and electronics. Despite ongoing technological advancements, drones still require significant development to achieve efficiency and robustness. This necessitates the employment of highly skilled personnel and operators, resulting in increased production and operational costs. All these results in contradictory objectives between customers and providers. While customers are keen to deliver a product in a short time and at the lowest possible cost, the provider's primary concern is to reduce the total costs (operational, maintenance, and recharging costs) and the fleet size, thereby increasing the number of customers served and the revenue earned. Moreover, from an academic perspective drone delivery systems are characterized by stochastic demand, battery range limitations, unpredictable trip durations, and the stochastic nature of trips themselves. Consequently, modeling, optimizing, evaluating performance, and analyzing these systems pose significant challenges. One of the extensively researched and most pertinent issues is the TSP and VRP, which, strictly speaking, represents a subset of thereof. TSP is a combinatorial optimization problem that determines, for a list of cities, the shortest possible route that goes through each city once and returns to the origin city. Developments in drone technology are fostering the emergence of novel variants of the TSP (typically referred to as TSP-D). The publication addressing the TSP-D problem assumes that the last-mile parcel delivery will be executed through hybrid delivery systems (a combination of drones and vehicles), making a distinction between collaboration and parallel approach (or scenario). Khoufi et al. [39] performed a comprehensive literature review on drone path optimization problems acknowledging the implication of adequate selection of trajectories on total costs. The authors also provided a detailed taxonomy of routing problems (i.e., TSP and VRP) in the context of drone operations based on the number of drones and number of ground vehicles (truck, van) employed. Authors of the paper Murray and Chu [40] were the first one who provided two mixed integer linear programming formulations on the TSP-D problem with the objective of minimizing the time required to serve all customers and return both vehicles to the depot. The authors addressed a scenario in which a drone works in collaboration with a delivery truck to distribute parcels to final customers. The first variant is known as the Flying Sidekick TSP (FSTSP). It aims at optimizing customer assignments for a UAV operating in conjunction with a delivery truck. In the FSTSP framework, both the drone and truck originate from a shared DC and must conclude their tour back at the same DC. During the journey, the drone may be deployed from the truck at any customer location to commence a delivery service. Following this, it meets with the vehicle at a subsequent customer location. The second optimization problem proposed by Murray and Chu [40] is called the Parallel Drone Scheduling TSP (PDSTSP) and aims to minimize the latest time that a vehicle returns to the depot. In the Parcel Delivery Sidekick TSP (PDSTSP), there is a single depot from which a single delivery truck and a fleet of one or more identical drones depart and return. The truck follows a TSP route to serve customers, while the UAVs provide direct service to customers from the DC. Unlike the FSTSP, this approach does not require time coordination between a drone and a truck. The problem addressed by Murray and Chu [40] can be efficiently solved by exact methods only if it is small in size. As the problems become larger in size and prove to be NP (nondeterministic polynomial time) hard, their efficient solution requires the application of some of the heuristics methods which typically guarantee a near-optimal solution if constructed properly. Research Agatz et al. [41] examined the possibility of improving truck and drone schedules for parcel delivery by applying both novel integer programming and two heuristics (the exact partitioning and the greedy heuristics). While the exact algorithms ensure the optimal solution for the instances of reasonable size, the heuristics efficiently dealt with large instances of 100 nodes in only a few seconds. In particular, the study highlighted the advantages of integrating drones into delivery services when paired with trucks, as evidenced by significant time savings averaging between a factor of 1.4 and 2. The paper by Marinelli et al. [42] extended the work of Agatz et al. [41] by modifying the objective function to maximize drone coverage and usage in parcel delivery. The formulation enables a truck can deliver and pick a drone up not only at a node, that is, the depot or at a customer location but also along a routing arc (en route). By employing the Greedy Randomized Adaptive Search Procedure (GRASP) including en route drone operations, the authors concluded that this approach can derive benefits only in the case when truck and drone have the same speed with an average increase in battery savings of 10%. Relying on the formulation of the TSP-D problem proposed by Murray and Chu [40], research Ha et al. [43] considered a new variant of TSP-D in which the objective is to minimize operational costs including total transportation costs and one created by wasting time a vehicle has to wait for the other as well as to minimize the delivery competition time. To efficiently solve this problem, the authors used GRASP and the local search method. Study Kitjacharoenchai et al. [44] addressed the problem of TSP when there is more than one drone vehicle and truck with the aim of minimizing the delivery time. By using the adaptive insertion heuristic, the obtained results indicated that the total delivery time can be reduced by 39% when employing one truck and multiple drones compared to the situation when the truck is the only means of transportation. However, the results also pointed to one important aspect for logistic operators; i.e., the effect of drones gradually diminished once the number of trucks increased.

10.3.2 Medical Delivery

Despite sharing many commonalities with the delivery of goods, medical parcel delivery has distinctive characteristics, primarily due to the urgency required for these transportation services. As highlighted in the previous subsection, the application of drones in the medical sector is one of the most promising uses of this new technology, likely to gain higher public confidence and acceptance. In the medical transportation logistics chain, drones offer significant benefits by avoiding traffic congestion and reaching remote or poorly accessible areas that lack adequate road infrastructure.

The application of drone technology typically covers two distinctive missions based on the criticality of the transportation service and the nature of the goods transported. The first mission refers to routine healthcare deliveries which assume the transportation of medical supplies such as blood and plasma [45], vaccines [46], medication, and other essential medical equipment including defibrillators, ventilators, and other portable devices. The transportation of these supplies typically takes place between clinics and laboratories to speed up the diagnosis process. The second mission is related to the application of drones as a reliable and efficient means of transportation in emergency responses. For example, in situations such as disaster relief or accident scenes, the traditional transportation is compromised and the drone can promptly deliver first-aid kits and emergency supplies. Due to its ability to access areas that are not well-served by traditional transportation networks, it is not surprising that several countries in Africa have already launched the medical drone delivery service as an efficient way to distribute life-saving medical products. Moreover, in countries such as Rwanda where 83% of the population lives in rural areas, remote hospitals typically face the issue of blood overstocking derived from the strategy to collect blood and its products well in advance as a way to prevent the shortage due to initial long delivery time [47]. This implies that a large amount of these valuable resources may be wasted in the case when there is no sufficient demand. Therefore, in 2016, Rwanda's government engaged a USA-based automated logistic company, Zipline, to optimize blood deliveries by carrying the blood from a distribution hub to the health care facility. Analyzing the 13,000 drone orders that took place between 2016 and 2018, the study performed by Nisingizwe et al. [48] revealed that the average time for half of the orders accounted for 41 minutes in the case of drones compared to 2 hours by road transportation. Following a similar path, in 2019, the Government of Ghana established a drone network to facilitate the prompt immunization campaign [49]. Similar to Rwanda's case, the aim is to deliver medical stuff from the DCs to health clinics managed through Zipline's hardware and software systems.

Similar to Zipline, Matternet also took a prominent role in the drone delivery sector by performing test flights to deliver medicine and supplies to remote areas in Lesotho, Bhutan, and the Dominican Republic [50]. In addition to the important undertaking of delivering medicine to the landside of Haiti after the earthquake in 2010 [51], the company primarily focused on supporting deliveries from aid organization to remote areas of the country through a network of drone stations situated 6 miles apart from each other. Drawing the conclusion from the test flights in Lesotho, the drone delivery is aimed to be cost efficient as the company estimated that the cost incurred to operate 50 base stations and 150 drones (around $900,000) throughout the capital city of Maseru would be lower compared to building 2-km one-lane road (around $1 million). In 2022, Matternet announced the test flights for the on-demand delivery of diagnostic samples between the Triemli and Waid Hospitals on a 5-km BVLOS route over the city of Zurich in Switzerland [52]. Leveraging on the experience gained in Zurich, the company initiated its drone delivery service in the city of Berlin ensuring the transportation of critical medical samples between Europe's largest hospital laboratory, Labor Berlin, and 13 affiliated hospitals. The company received authorization from Germany's Federal Aviation Office to operate its M2 drone delivery system under BVLOS conditions. The introduction of drones into delivery systems aims to enhance supply chain efficiencies while reducing carbon emissions from the nearly 250,000 miles per day that cars travel within the city [53]. In addition to Matternet, RigiTech is a Swiss-based logistic company and drone manufacturer that efficiently demonstrated its compliance with BVLOS SORA regulations by successfully flying over Lake Geneva to a laboratory in Coppet and back [54]. The company is expanding its geographical scope with a new partnership in Uruguay [55], Netherlands [56], Maldives [57], and many other regions to fully integrate its UAVs into routine operation by medical transport and response units.

Although both companies (i.e., Matternet and RigiTech) aimed to provide cost-effective and reliable delivery service, their main difference resides in the drone design and configurations. Namely, Matternet's M2 drone belongs to the group of rotary-wing aircraft making it suitable for drone delivery of medicines in urban environments with a payload capacity of around 4.5 lbs and a maximum range of over 12 miles (see Table 10.2). On the contrary, RigiTech's Eiger incorporates tilting wing technology by combining the characteristics of both helicopter and plane technology, i.e., it possesses the capability to take off and land vertically as a helicopter but then cruise like an airplane to reach longer distances. Such configuration is typically more efficient as it allows connecting facilities 80 km apart in less than 40 minutes, which makes its application plausible in last-mile inter-city logistics. Nevertheless, both types of drones meet the high safety standards. For instance, Matternet uses geofencing technology to track the aircraft route and Automatic Dependent Surveillance-Broadcast (ADS-B) sensors to detect air traffic along the route. The M2 is able to take off and land within a space of 3 feet by 3 feet, allowing it to integrate into tight urban environments like apartment buildings and hospitals. This enables great flexibility as most drones nowadays are not able to navigate without specially designed infrastructure, like landing pads. In a similar vein, RigiTech assesses air risks in real-time using integrated data including live traffic information, airspace data, and alerts for conflict zones allowing the operator to be aware of potential hazards. In addition to safety, which is a prime objective in aviation, drone transportation also needs to ensure that the quality of medicine is not jeopardized, as some medicaments could be potentially sensitive to the effect of temperature and vibrations [58]. To overcome this issue, RigiTech designed a specialized medical box with temperature control which ensures the integrity of sensitive samples throughout the journey [59].

10.3.3 Loitering Mission

Loitering mission is today the most currently seen use case in urban environments. Examples of loitering operations entail surveillance flights, mostly in the hands of police, law enforcement bodies, and security agencies. Other examples of loitering missions are the streaming of local sportive events, the inspection of urban infrastructures, and the aerial photography works used, for instance, in real estate sales.

According to the Oxford dictionary, loitering means to stand or wait around without apparent purpose. When applied to drones it refers to those flights where the drone is mainly stationary or flying in a holding pattern for an extended period. Loitering could probably be better referred to instead of hovering, that is, remaining in one place in the air, because it adds a vertical dimension. But the definition of loitering has a noticeable aspect when mentioning the (lack of an) apparent purpose. Any flight has always a purpose, even if this is just entertainment. The actual purpose of such loitering flights is the focus of this section.

We could envision the following loitering drone missions that exist today, other than entertainment:

- police surveillance,
- transportation network monitoring,
- infrastructure inspection,
- real state,
- cinema or photography,
- commercials,
- environmental measures,
- emergency response,
- fauna and flora care, and
- construction and building maintenance.

In the following paragraphs we will briefly introduce each one of these loitering use cases, ending with more detail for the last one, the use of drones in the construction arena.

Police surveillance is one of the most frequent loitering missions that occur today in cities in developed countries. A good list of police surveillance use cases, extended with emergency response, can be found on the Drone Respondents website [60]. Police have the benefit of the legislation to get faster permission to fly due to their non-profit, public safety, and nature. During the pandemic, when police had to watch for curfew observance, the use of drones grew very fast [61–64]. Drones provided very good results with low cost and low risk. Two risks were minored: the aeronautical ground risk of the flights given that the streets were mainly empty, and the virus spread, given that drones could address the citizens without physical contact. After such a period the drone had become part of the police crew's usual set of tools and many other tasks were envisioned: land use update, capacity limits assurance, thief watching, road traffic status capturing, fast accident reconstruction, etc.

Capturing the status of road traffic in real time is an important part of monitoring the overall transportation network, which is becoming increasingly multi-modal and inter-urban [65–67]. Road traffic is just one component of a complex system that includes various modes of transport such as trains, trams, metro, bus and HOV (high-occupancy vehicle) lanes, cable cars, aircraft, and even future technologies like taxi drones. The smooth connectivity of the full transportation system needs a well-designed interconnectivity and the infrastructures that support it are in good health. Slippery roads, hurt pavement, unstable bridges, broken traffic lights, dirty railways, etc. affect the capacity of the transportation means as much as the instant vehicle flow. The capability of monitoring this infrastructure has been provided by connected road cameras, by adding sensors under the IoT paradigm. Although this type of equipment gets cheaper as years go by, they cannot reach absolutely all points of the extensive multi-modal infrastructure, especially in remote areas. Moreover, this equipment becomes part of the infrastructure to monitor and maintain, raising the costs, while not avoiding having blind spots. Conventional aerial means, such as helicopters, have been used to reach points out of the surveillance equipment. But due to their high cost they are reserved only for special occasions. Drones reduce the cost of such missions, multiplying the efficiency of the inspections and, in the long term, avoiding recurrent costs of the deployed equipment.

Drones are used for infrastructure inspection, such as bridges, wind mills, communication antennas, and runways, due to their capability of reaching access to areas that are difficult or dangerous for humans, providing high safety, high efficiency, and low cost [68–75]. Given that drones are equipped with high-resolution cameras they can capture detailed images and videos of tall infrastructures and buildings and help inspectors examine wide surfaces, identifying potential issues like cracks, corrosion, or damage, with no danger to them. The use of thermal cameras helps to detect temperature variations on energy infrastructures such as high-power lines, which typically indicate structural problems or areas that need further inspection. Radar, sonar, and LiDAR are also common payloads used in inspection tasks, since they can obtain very detailed three-dimensional data that are perfect for this type of use cases. By repeating measurements at regular intervals over time and comparing them to baseline data, drones can detect deformation and movement for instance in bridges and ancient buildings. Drone inspections can be conducted more quickly and cost-effectively than traditional methods, reducing the need for lane closures and traffic disruptions. The data collected from drone inspections are analyzed offline by qualified engineers to ensure accurate assessment of the infrastructure's condition, although new approaches are being used to apply artificial intelligence in the detection of damages [76–79].

Drones are increasingly being utilized in the modern real estate industry for purposes such as property inventory, aerial photography for listings, and surveying neighborhoods [80–82]. Real estate companies use drones mainly to capture stunning aerial footage of properties, showcase the surrounding area, and provide potential buyers with a new aerial perspective. Additionally, drones can be employed for fast property appraisals, to check property status (i.e. by the inspection of the roof), and, when construction to be sold is still in work, to monitor its progress. The use of drones in real estate offers a cost-effective and efficient way-to-market of properties and gathers essential data for decision-making of potential new buyers.

Drones have been used in various ways in photography and cinema to expand the visual field and perception of reality [83–85]. Some examples include documentary films like *National Bird* and *The Human Flow*, in which drones capture aerial shots that go beyond the confines of the human gaze. Artists have employed drone cinematography to stimulate the viewer's embodied perception and expand their knowledge of the world.

Another artistic use of drones is the fireworks produced with drones [86,87]. Swarms of 49 drones in the first drone light show performance took place at a local music festival, SPAXELS, on the famous Danube River in Austria in 2012. Currently they have expanded to multiple events and cities over the world with amazing drone performance by BotLab Dynamics in India, using 5,500 drones in a single drone light show on April 6, 2024.

The versatility and cost-effectiveness of drones have made them an increasingly popular tool for advertising films, allowing brands to create visually stunning and engaging content that captures the attention of audiences [88,89]. Drones have been used to film commercials for consumer products like smartphones and cameras.

The aerial footage captured by drones can highlight the capabilities of these devices, such as their camera quality and video stabilization features. In the tourism industry, drones have been leveraged to film commercials showcasing scenic destinations from unique aerial angles. This helps attract potential visitors by providing a compelling visual representation of the location's beauty and attractions. Auto companies such as Tesla, Corvette, Audi, or Toyota have long used aerial footage to create car commercials [90]. In the past, this involved the use of a helicopter or airplane and was extremely expensive. Nowadays, great aerial footage for car commercials can be had much more cheaply using drones. Drones allow for dynamic aerial shots of vehicles driving through scenic landscapes that would be difficult or impossible to capture with traditional camera setups.

The atmosphere pollution and the greenhouse gas emissions are the origin of the rapid global climate change. The impact of urban air quality on human health issues, such as respiratory conditions, asthma, heart diseases, and even low birth weight in newborns, is well known. The use of smart environmental drones equipped with advanced particulate matter sensors is being proposed to monitor urban air quality effectively [91–93]. For instance, Durgun and Durgun [93] introduce the use of smart environmental drones as a promising approach for dynamic air quality monitoring in Turhal, a district of Tokat province, Turkey. Drones equipped with low-cost particulate matter sensors are capable of accurately measuring $PM_{2.5}$ and PM_{10} concentrations, flying autonomously at 10 m/s, cover a range of 5 km, mapping pollution levels across urban areas, and obtain a 95% accuracy rate in sensor readings, lowering time and costs. Similar approaches are being tested in Chennai city, India [94], Region Lagunera and other parts of Mexico [95], or Dhaka, Bangladesh [96]. A comprehensive survey can be found in Burgués and Marco [97].

The United Nations Secretary General Strategy on New Technologies [98] supports the use of AI and robotics to accelerate the achievement of the 2030 Sustainable Development Agenda. There are several examples of governmental, non-governmental, and intergovernmental organizations using drones in disaster response, mainly for search-and-rescue operations (in floods, wildfires, earthquakes, etc.) and also for situational awareness for the authorities. The emergency response also includes safe and regular migration processes. In this spirit, projects such as Freeda and Frontex propose the use of drones to enable the coverage of a broader area, speed up search processes, and increase the efficiency and effectiveness of migrants' rescue operations at sea. Most of these missions have a first phase where the drone has to fly away to reach the disaster area, and then a second phase where a loitering path is executed. Different strategies have been proposed for loitering, basically following flight regular patterns that ensure the complete coverage of the area [99] but also random search following some heuristic [100].

Precision agriculture is a concept in which technology is intensively used to improve the harvest production of farms. This concept has been supported by drones in a large number of papers as shown in this survey [101]. In the same topic, drones have been operated in polar areas to assess glacier health and to protect mammals and other fauna from extinction danger [102]. For instance, Hodgson [103] proposes the use of drones as high-resolution, remotely sensed data sources to inform species conservation, improve ecosystem management, and assess mitigation strategies for biodiversity loss using colonial birds as a study group in Australia.

The above classification of types of loitering missions is not a strong cluster and has many overlaps. For instance, the Castelldefels local police drone unit used a drone to observe the road traffic patterns around an area with a high ratio of accidents and proposed a global re-urbanization plan for the area. In this police surveillance mission, we observe also infrastructure inspection and transportation network monitoring mission types.

For the rest of the section we will discuss deep into the drone loitering missions that Industry 4.0 can most benefit from, which is the construction industry.

Construction of buildings is a business that represents between 5% and 10% of the national gross product of most developed countries. New methods are being imposed in construction to modernize the logistics chain that includes the off-site partial prefabricating of full parts of the building, the extensive use of simulators with high-definition rendering, or the intensive use of internet connections with providers' storage systems. The addition of drones for construction is well covered in Mahajan [104] including discussion about the challenges, opportunities, limitations, and strategies for the adoption of drones in construction.

Drones can be used in construction from the early stage of land purchase to the post-construction stage of the project. They are used during the pre-construction phase, the construction phase, and the post-construction phase in supplementing the conventional methods of contour mapping, site progress monitoring, construction quality management, etc. The utilization of drones in the worldwide construction industry has become one of the most compelling trends and experienced a 239% growth year-over-year, greater than any other commercial area [105]. For the tedious and time-consuming task of monitoring the progress of a construction project with respect to the planning and scheduling, the site engineers have to keep a real-time check over every space of the construction. Contractors rely on the engineers and their studies for taking early decisions when planning seems not being followed, to minimize the consequent associated expenses. The use of drones gives rapid information about the real-time progress on all parts of the site, supporting contractors in reducing the costs and at the same time drones permit the scale of the businesses as they are able to manage a large number of sites.

According to Tatum and Liu [106] the construction companies manifested in a survey that the most popular use of drones is by far the capture of the progress of the construction work. The other highly popular use cases are the creation of promotional videos, inspections, and site management. On the contrary, the use of drones for security is the less popular use case, which is conducted by fixed-cameras surveillance systems instead. The same work explains that fixed-wing drones are more suitable for large-area construction works, while rotary-wing drones are used in confined areas.

Construction involves several consecutive phases: design, construction, maintenance, and, eventually, demolition.

- The design phase starts with the decision about the most suitable site. Drones play a significant role in this process by providing valuable insights through aerial imagery collection with high resolution [107,108]. Drone imagery provides a global view of the site, with it, the project architects and engineers assess its characteristics. In consequence, the project managers can evaluate the expenses more accurately. For instance, they can decide if the site meets the expected logistical requirements by identifying any limitation or challenge related to site access. Access to the site has a high impact on construction activities because it has economic effects on personnel costs and the transportation of materials and equipment. In addition to visual imagery, several works [109–112] report the use of drones equipped with LiDAR and/or thermal sensors for the measurement and obtaining the topography of the area, recognizing the existing structures, and determining the boundaries. Topography data, obtained from these sensors or from the overlapping of the visual images and applying photogrammetry techniques, include the precise measurements of distances and terrain elevations, contributing to the creation of detailed 3D models, topographic maps, digital elevation models, orthomosaics, and inventory maps. On top of such maps, architects can establish survey control points for precise geo-referencing, ensuring the accuracy of the necessary mapping activities [113]. In addition, the environment is better incorporated into the design processes [114]. Other works [115–117] focus on explaining how to organize the preparation of the mapping process using drones. Steps include the pre-flight planning, where the survey area is defined, flight paths are determined and necessary permits and safety measures are put in place; the flight itself, in which the images are captured; and the post-processing step, which relays in specialized software such as geographical information systems and others [118].
- For the construction phase the most common use case reported is the use of drones in search-and-rescue operations (more than 30%), followed by monitoring (17.5%) and grading measurements (15.7%) [119]. Effective communication and quality assurance are the less-reported use cases, although both are very promising for the future. For instance, earthwork and grading monitoring have been shown to improve efficiency [120–122], accuracy [123–126], and safety [109,127,128]. For instance, Park et al. [125] explain the use of drones for the construction of a 771-household capacity building complex in Seoul, Republic of Korea. Drones were used to document the project from the beginning to the end, to calculate cut and fill volume data, and to monitor the changing of slopes. Tasks such as volume calculation, cut and fill analysis, and slope monitoring showed many advantages in quality, time, and cost. As explained in the previous section, the same drones being used for delivering can be used in construction for the transport of materials needed by the construction industry. Timely and efficiently, drones help reduce the time and cost associated with traditional delivery methods and are especially useful in construction sites with limited access or where heavy machinery cannot be used. Drones have also become perfect assets for improving the safety of the workers and the security of the site during the construction phase. Drones and

their payload help in mitigating hazards by potentially identifying dangerous debris accumulations, movements in structures, malfunctioning equipment, or the unuse of protection equipment that poses a threat to personnel on site. The drone perspective facilitates the assessment of safety across the whole site thanks to the identification of any potential risk and the application of preventive measures. In the case study of the safety in skyscraper construction in Santiago de Chile [129], the authors were able to document dangerous situations, such as workers without safety ropes or the lack of guardrails. In addition, drones can help in managing a large-scale emergency by providing real-time situational awareness in such cases. Drones can assist emergency responders in assessing the actual situation, in identifying damaged points and, thus, planning the rescue operations with higher guarantees for the people. Thermal cameras have proved to be very helpful in locating missing persons and improving the emergency response time. Live surveillance footage taken by drones may be later used for investigation purposes, which finally is valuable for protecting the site and securing all construction equipment. But the major role is in quality control. Drones can run inspections and defect detection on a frequency basis. Using high-definition imagery, the identification of minor defects, including corrosion, cracks, and even surface imperfections is now possible with much more quality than ever. The systematic data captured by the drones is used to create 3D models, either virtual or physical, using 3D printers or digital twins [130], which helps in identifying minor deviations in the construction process. The early detection of errors enables seamless rectification that supports the desired quality standards [131,132]. Drone data is also used to complement the construction documentation, providing suitable appendixes to the company for their clients. Moreover, they enable the company managers to assess the status of different construction activities and the identification of delays in comparison with the construction planning. Resource allocation adjustments can be taken early to keep the project on schedule and avoid penalization. Additionally, drones facilitate communication among the involved team, enabling a common vision and the same understanding of the progress of the construction work. Thanks to a holistic visualization of the construction site and its evolution, which provides a shared understanding of the project's status among all stakeholders, the likelihood of any misunderstanding in coordination is largely reduced [133]. By leveraging the use of drones for the control of the quality of the project, the accuracy and frequency of inspections are enhanced and the costs are reduced. Real-time drone monitoring outcomes contribute to better construction quality thanks to improved coordination, seamless communication, and an enhanced agreement between all the stakeholders involved in construction.

- In the maintenance phase the drones are mostly used for inspection and maintenance of large infrastructure as explained above: bridges, wind mills, power lines, etc. In addition, in the real estate business inspections are increasingly using drones. But both types of inspections are done by third parties not involved in the construction. The cases where the construction company is interested in post-construction recognition are rare and mainly futuristic. For instance, Li et al. [134] propose the use of drones to clean the outside face of the windows of skyscrapers. Also, a study by Reckling et al. [135] proposes them for underground inspection of the sewage system. Elevator maintenance, TV or satellite antennas, solar panels, or other roof machinery can at some point become the target of post-construction maintenance, especially for very high constructions.

Summarizing, one can conclude that drones, equipped with cameras and sensors, allow for detailed topographic maps, 3D models, and digital twins to be created and thus are increasingly being used in construction for tasks like initial site selection, construction progress monitoring, and quality inspection. They can quickly and accurately map construction sites, monitor progress, and identify potential issues. However, there are still some limitations and challenges to using drones in construction [136]. Below a brief list is given:

- Regulation around drones, such as registration, flight permission, coordination, and certification, varies worldwide. Indeed, they need to be followed, but the huge casuistic is limiting the scale of the business potential.
- In addition to the regulation that ensures safety for mitigating air and ground risks, drones still require enhanced safety features for on-site personnel. Training is necessary for those workers who share the time and space with drones, to avoid dangerous interactions between them.
- Companies that use drones on a day-by-day basis will need to expand their professional manpower with drone pilots and software experts to extract data from the drone images to use them effectively or add a new subcontractor for those works. If drones become part of the construction company then it will require additional personnel for drone management and legal maintenance.
- The sporadic use of drones in the construction dynamics may be an obstacle if construction work has to stop to let drones do their job. The systematic integration of drones in the logistic chain of construction companies is a must to obtain the economic benefits expected.

To address these challenges, improvements are needed in terms of technology, procedures, systems, and regulation. Even more, drones will end up working in combination with other emerging technologies, like computer vision and artificial intelligence. This will enhance drones' capabilities in construction, at the price of more expertise of personnel and further legal requirements.

10.4 CHALLENGES AND OBSTACLES

10.4.1 Technological Aspect

The range of drones is constrained by their battery life and flight time. As Kirschstein [137] explains, a key challenge is to reduce power consumption and mitigate the effects of wind direction, travel speed, and customer concentration, which collectively impact the efficiency of drone delivery systems. Furthermore, targeting areas with higher customer concentrations for delivery requires further exploration, as it may offer cost advantages over deploying drones in rural areas. This involves addressing the number of customers, traffic conditions, and even battery impairment. The related issue of determining the optimal size of necessary facilities based on battery duration is discussed in Aiello et al. [138]. Additionally, Torabbeigi et al. [139] highlight the need to address the effect of payload on battery consumption rate and flight time.

Expanding coverage for last-mile delivery can be achieved by installing multiple charging stations, which warrants a detailed impact analysis. The location of these charging stations is also important, as noted by Pinto and Lagorio [140], who suggested developing a model incorporating strategically located charging stations along their routes. This approach, which is already technically feasible, allows drones to recharge during their journeys to final destinations, thus overcoming the limitations imposed by battery life and enhancing the overall efficiency and coverage of drone delivery systems. Furthermore, drone charging time must be considered as it affects time constraints in mathematical models. To better adapt the drone charging process to real-world scenarios, factors such as partial charging and charging speed should be considered, as recommended by Daknama and Kraus [141]. Instead of charging, batteries can be replaced with new charged batteries to reduce delivery time. This process is most often associated with a combined drone truck operation system where trucks serve as carrier platforms (thus saving energy) and drones act as the primary means of transportation [142].

10.4.2 Societal Impacts

According to research by Kellermann et al. [143], the initial perspective on these impacts is quite mixed. While societal benefits represent over a third of the anticipated advantages, the predominant concern lies in the negative societal implications of using drones, which account for almost two-thirds of the potential problems. Society's main concerns regarding drone operations are related to the potential decline of traditional retail, job losses, heightened stress levels, diminished social interactions, and the establishment of an elite mobility regime [143]. In addition to these challenges, some authors point out that the continuous presence of drones could weaken the current perception of privacy regarding the collection, storage, and use of private data [144–147]. Moreover, certain researchers caution that urban drone delivery could alter people consumption and mobility patterns [148]. Also, a study by Çetin et al. [149] indicates that the greater use of drones generates psychological fear of the unknown in certain segments of society.

10.4.3 Environmental Issues

Environmental considerations greatly impact users' perceptions of drone operations. Due to its fully electric propulsion, drones are seen as a more environmentally friendly solution for passenger and goods transportation [149]. Research indicates that the five primary environmental concerns of drones are noise, visual impact, energy consumption, CO_2 emissions, and land impact [150]. An extensive study by Schäffer et al. [151] revealed that the level of noise emissions largely depends on the type of drone, payload, and its operational status or flight maneuvers. Although the literature suggests that drone noise is significantly more annoying than road traffic or aircraft noise at the same level, studies on the effects of drone noise on humans are scarce.

The subject of many studies is also visual pollution, which represents a challenge for drone operators who must improve route planning [37]. The choice of different energy sources will greatly influence the drone's rate of energy consumption, affecting its performance efficiency, design, noise output, emissions, and user acceptance [152]. Although electric power is considered a zero-emissions technology, the charging of batteries may still rely on energy from non-renewable sources. Additionally, producing, servicing, and disposing of batteries consumes energy and valuable materials, which may not be sustainable. The requirement for dedicated land infrastructure, including charging stations, control stations, and warehouses, can result in substantial resource and land consumption [153]. The other environmental concerns up for discussion are risks to wildlife and the energy efficiency of drone operations [143].

10.4.4 Regulation

Drone regulations are continually evolving to keep pace with technological advancements and increasing drone use across various sectors. Countries across the world are currently facing a huge challenge to incorporate drones into the aviation regulatory framework. While some countries are proactive and innovative in this regard, others rather act as observers waiting for the results of other nations before proceeding to regulation design.

Study Jones [154] provided a comprehensive outlook on the regulatory approaches taken by different countries across the world to examine how the regulation may drive further drone applications. Based on the extensive regulation review, the authors identified six broad regulation approaches and their variants taking into account four elements, i.e., pilot's license, aircraft registration, restricted zones, and insurance. On one side of the spectrum are countries that do not allow drones at all for commercial use ("Outright ban") or have a formal process for commercial drone licensing, but requirements are either impossible to meet or licenses do not appear to have been approved ("Effective ban"). The moderate approach is taken by the countries in which drone operation is permitted within the pilot's VLOS, thus limiting potential range ("VLOS" required). There is a group of countries that allows experimental uses BVLOS under certain restrictions and pilot ratings ("experimental BVLOS"). Moreover, certain countries have implemented relatively liberal legislation regarding commercial drone use ("permissive") with a notable amount of regulation that may give operational guidelines or require licensing. Finally, the last group of countries adopt a "wait-and-see" strategy, choosing to observe the progress of other nations' policies before taking action. Among these six approaches, it is evident that only BVLOS operations will enable more complex and longer-range missions, but at the same time it will impose a significant advancement in the capabilities of drones.

As a rooftop organization in civil aviation, the International Civil Aviation Organization (ICAO), has been actively working to develop a comprehensive framework for the regulation of UASs. This involves creating international standards and recommended practices that ensure safety, security, and efficiency in drone operations worldwide. In its ICAO Circular 328 AN/190, titled "Unmanned Aircraft Systems (UAS)," it provides guidance on the regulatory framework for UASs. It covers various aspects including definitions and classifications, regulatory framework, safety and security, operational guidelines, airworthiness and certification, air traffic management, privacy and data protection, and BVLOS operations [155]. In particular, ICAO is developing guidelines for the implementation of UTM systems, which are crucial for integrating drones into low-altitude airspace. UTM systems aim to manage drone traffic and ensure safe operations, particularly in urban areas and other complex environments. In addition to ICAO, there are several notable countries that have applied or are actively working toward enabling BVLOS drone operations such as the United States, Canada, United Kingdom, Australia, Singapore, China, and New Zealand. All these countries already established a framework for BVLOS operations allowing this type of operations under certain conditions.

In Europe, tremendous progress in tailoring regulation is made by EASA, although specific countries such as France and Germany also have a prominent role. To ease the application of a regulatory framework for the U-space across different stakeholders, EASA recently published the easy access rules for U-space (Regulation (EU) 2021/664) [156].

10.4.5 Privacy and Safety

While drones offer significant benefits in areas like delivery, emergency response, and environmental monitoring, these benefits must be balanced against the potential risks to privacy and safety. Drones are typically equipped with high-resolution cameras and other sensors that can inadvertently capture images and data of private property and individuals without their consent. Moreover, there is a risk of drones being used for intentional surveillance, potentially infringing on individuals' privacy by monitoring their activities without their knowledge or approval. In many regions, existing privacy laws and regulations have not kept pace with the rapid development and deployment of drone technology. This regulatory lag creates gaps in protecting individuals' privacy. The study by Lidynia et al. [157] underlined the need for a more transparent use of drones to satisfy the exposed people who are concerned about unpermitted recordings rather than injuries caused by crashing aerial vehicles.

Another issue structured around the privacy issue is associated with data storage and usage. Namely, the vast amount of data collected by drones needs to be stored securely to prevent unauthorized access and breaches. There are concerns about how the collected data is used, shared, and potentially sold. Misuse of such data can lead to privacy violations [158]. Another important concern is safety, as similar to commercial aviation, drones can collide with each other especially in dense environments when facing the problem of loss of communications and software or hardware failures [159]. Consequently, drones may cause substantial damage to property on the ground or may even lead to loss or severe injuries to humans. Overcoming privacy and safety challenges in drone applications requires a multi-faceted approach. One of the promising solutions acknowledged by numerous bodies in this field is the implementation of no-fly zones over sensitive locations (such as residential areas or government buildings). In addition, no-fly zones can be dynamically adjusted during events, emergencies, or at the request of property owners to provide additional protection when necessary. In a similar vein, restrictions on the use of cameras in drones can help protect privacy. This can include disabling cameras in certain zones or during specific types of operations. The risk of invasive surveillance can be substantially reduced by setting minimum altitude limits and hover restrictions, as claimed by Çetin et al. [149].

10.5 CONCLUSION

Drones have emerged as a transformative technology with the potential to revolutionize various industries by providing innovative solutions for data collection, processing, and real-time information delivery. The integration of advanced technologies such as machine learning, artificial intelligence, IoT, and cloud computing has enhanced the capabilities of drones, aligning them perfectly with the goals of Industry 4.0. The projected growth of the drone market, particularly in Europe, underscores the need for continued technological advancements, regulatory developments, and community engagement to fully realize the potential of drones. As the market evolves, it is essential to adapt to changing dynamics and unforeseen factors to harness the full benefits of drone technology.

The ConOps is vital for the deployment and utilization of drones, outlining procedures, services, and requirements while adhering to safety regulations and protocols. The ConOps document serves as a foundational blueprint, detailing operational environments and safety aspects, and aiding in the certification process and stakeholder communication. In the USA, the FAA's UTM ConOps versions and the 2023 Implementation Plan highlight efforts to integrate drones into existing airspace, emphasizing performance authorization, USSs, and comprehensive data communication systems. Europe's U-space initiative, driven by various regulatory bodies, focuses on creating a smart and sustainable unmanned aircraft ecosystem. Both regions are progressively implementing their ConOps, with Europe detailing a roadmap from foundational services to full integration, and the USA adopting a modular approach based on risk metrics. Urban airspace management, crucial for the widespread adoption of drone technology, varies from unstructured designs to structured concepts like matrix configurations and dedicated corridors, addressing urban challenges and ensuring safe and efficient operations. The ongoing evolution of ConOps will be key to the successful integration of drones into airspace, supporting innovation and economic growth.

The evolution of drone technology from military and law enforcement applications to commercial and logistical uses has profoundly impacted parcel delivery systems. Leveraging advancements in drone technology, companies have successfully demonstrated the viability of using UAS for last-mile delivery, offering advantages such as reduced delivery times, improved accuracy, and environmentally friendly operations. The commercial drone delivery milestone was notably marked by the

FAA's approval of the first commercial drone flight in 2015, followed by Amazon's successful implementation of its Prime Air program. These developments have spurred significant investments and technological innovations in the industry, fostering collaborations between drone manufacturers, logistics companies, and regulatory bodies.

Globally, various countries and regions have embraced drone delivery, addressing unique logistical challenges and enhancing accessibility to remote and difficult-to-reach areas. The integration of drones into urban and suburban environments is underway, with numerous demonstration projects highlighting the potential societal and environmental benefits, such as reduced traffic congestion and lower CO_2 emissions. Despite these advancements, several challenges remain. The high cost of drone delivery services, limited battery life, range constraints, and the need for sophisticated airspace infrastructure and regulatory frameworks pose significant hurdles. Moreover, the complexities of last-mile delivery, particularly in rural areas, necessitate innovative solutions to optimize operational efficiency and cost-effectiveness.

The problem of last-mile delivery has revived interest in traditional transportation and optimization problems, such as the TSP and VRP. New variants like TSP-D have emerged, exploring hybrid delivery systems combining drones and ground vehicles. These models aim to minimize delivery times and operational costs, although efficient solutions require advanced heuristics due to the inherent complexity of these problems.

The utilization of drones in medical parcel delivery presents a scenario where urgency and accessibility are paramount. Unlike conventional goods delivery, medical deliveries demand a higher level of reliability and speed, characteristics that drones inherently possess. The technology's ability to bypass traffic congestion and reach remote or poorly accessible areas underlines its transformative potential.

The two primary missions of drone application in the medical field – routine healthcare deliveries and emergency response – highlight the versatility and critical impact of this technology. Routine deliveries, such as those of blood, vaccines, and essential medical equipment, are streamlined, ensuring quicker diagnostic processes and better resource management. Emergency responses benefit significantly from the drones' ability to swiftly deliver vital supplies to disaster-stricken or isolated areas, showcasing their life-saving capabilities.

The successful implementation of medical drone delivery services in various countries, such as Rwanda and Ghana, underscores the practical benefits and growing acceptance of this technology. Companies like Zipline, Matternet, and RigiTech have demonstrated the feasibility and efficiency of drone logistics, with significant reductions in delivery times and costs. These companies' innovations in drone design and operational strategies further enhance the reliability, safety, and effectiveness of medical deliveries.

Moreover, the integration of advanced safety features and temperature-controlled medical boxes ensures that the delivery of sensitive medical supplies remains secure and intact. As the technology evolves and becomes more integrated into healthcare systems worldwide, the future of medical parcel delivery via drones looks promising, offering a reliable, efficient, and potentially life-saving solution to global healthcare challenges.

Loitering missions have become the most prevalent use case for drones in urban environments. These missions include activities such as police surveillance, transportation network monitoring, infrastructure inspection, and aerial photography for real estate. Additional applications span diverse areas like environmental monitoring and emergency response. Each of these use cases demonstrates the versatility and critical importance of drones in modern urban operations.

Except for the aforementioned, drones have become an invaluable asset in the construction industry, offering benefits from the pre-construction phase through to post-construction. Their ability to rapidly provide real-time data significantly enhances site progress monitoring, construction quality management, and decision-making processes. Despite their advantages, challenges such as regulatory compliance, safety concerns, and the need for skilled personnel persist. Addressing these issues through technological advancements and improved integration with other emerging technologies will be crucial. Ultimately, the successful incorporation of drones promises to revolutionize construction efficiency and project management.

In addition to the undisputable advantages of using drones in inner-city areas for different tasks, several challenges must be addressed to ensure their safe and efficient integration into urban environments. Addressing the technological aspects, drones face limitations in battery life and flight time, with challenges including power consumption, wind direction, and customer concentration. Solutions involve optimizing battery usage, strategically placing charging stations, and considering partial charging or battery swapping. The societal impacts of drone use are mixed, with concerns over privacy, job losses, and changes in social interactions. Environmentally, drones are seen as a greener option, but issues like noise, visual impact, and energy consumption need attention. Regulatory frameworks are evolving, with different countries adopting various approaches to drone integration. Privacy and safety concerns require measures like no-fly zones, camera restrictions, and altitude limits to protect individuals and property.

Achieving widespread adoption of drones will depend on overcoming current technological and regulatory challenges. Continuous research and development, coupled with strategic collaborations among stakeholders, will be crucial in realizing the full potential of drone systems.

REFERENCES

1. SESAR Joint Undertaking, European drones outlook study. Unlocking the value for Europe, 2016 (available online: https://op.europa.eu/en/publication-detail/-/publication/93d90664-28b3-11e7-ab65-01aa75ed71a1/language-en).
2. L. Kapustina, N. Izakova, E. Makovkina, M. Khmelkov, The global drone market: main development trends. *SHS Web of Conferences*, vol. 129, p. 11004, EDP Sciences, December 2021.
3. European Commission, *A Drone Strategy 2.0 for a Smart and Sustainable Unmanned Aircraft Eco-System in Europe*. Brussels, Belgium: EC, 2022 (available online: https://eur-lex.europa.eu/legal-content/EN/TXT/?uri=CELEX%3A52022DC0652).
4. Presedence Research, *Unmanned Aerial Vehicle (UAV) Drones Market Size and Growth*, May 2024 (available online: https://www.precedenceresearch.com/unmanned-aerial-vehicle-drones-market).

5. Drone Industry Insights, *Drone Market Size 2020–2025*, June 2020 (available online: https://droneii.com/?srsltid=AfmBOopxW4utjTVSu7W4zi7OoFcuQNGNQJgQ9lb3hnqhJ7zBzgeCFEWx).
6. M. Maly, J. Hoffelner, C. Krammer, M. Wechner, Concept of operations in aviation industry–A procedure based approach for developing novel aerial systems. In *VFS 79th Annual Forum*, West Palm Beach, FL, May 2023.
7. N. NextGen, *Concept of Operations v1.0 Unmanned Aircraft System (UAS) Traffic Management (UTM). Foundational Principles, Roles and Responsibilities, Use Cases and Operational Threads.* Washington, DC, May 2018.
8. N. NextGen, *Concept of Operations v2.0 Unmanned Aircraft System (UAS) Traffic Management (UTM). Foundational Principles, Roles and Responsibilities, Use Cases and Operational Threads.* Washington, DC, March 2020.
9. Federal Aviation Administration, *Unmanned Aircraft Systems (UAS) Traffic Management (UTM) Implementation Plan.* Washington, DC, July 2023.
10. EUROCONTROL, *U-Space ConOps and Architecture*, CORUS-XUAM Project Deliverable, July 2023 (available online: https://www.sesarju.eu/sites/default/files/documents/reports/U-space%20CONOPS%204th%20edition.pdf).
11. J. Lieb, A. Volkert, Unmanned aircraft systems traffic management: A comparsion on the FAA UTM and the European CORUS ConOps based on U-space. In *2020 AIAA/IEEE 39th Digital Avionics Systems Conference (DASC)*, San Antonio, TX, pp. 1–6, IEEE, October 2020.
12. SESAR Joint Undertaking, *U-space – Blueprint.* Brussels, Belgium: SESAR, 2017 (available online).
13. Y. Seprey, D. El Malem, H. Eduardo, M. Büddefeld, R. Kumar, R. Moldes Teijeiro, E. Garcia Collado, A. Hately, H. P. Jonsson, *Structures and Rules in Capacity Constrained (Urban) Environments, Exploratory Research*, DACUS Project Exploratory Research, March 2021 (available online: https://dacus-research.eu/wp-content/uploads/2021/05/DACUS_D5.1_Structures_and_Rules_in_Capacity_Constrained_Environments_00.02.00.pdf).
14. U.J. Lee, S.-J. Ahn, D.-Y. Choi, S.-M. Chin, D.-S. Jang, Airspace designs and operations for UAS traffic management at low altitude. *Aerospace*, vol. 10, no. 9, 737, August 2023.
15. D.D. Nguyen, Cloud-based drone management system in smart cities. In *Development and Future of Internet of Drones (IoD): Insights, Trends and Road Ahead*, edited by R. Krishnamurthi, A. Nayyar, A.E. Hassanien. Cham: Springer International Publishing, pp. 211–230, February 2021.
16. K.H. Low, *Framework for Urban Traffic Management of Unmanned Aircraft System (UTM-UAS).* Singapore: ICAO, 2017 (available online: https://www.icao.int/Meetings/UAS2017/Documents/Kim%20Huat%20Lo_Singapore_UTM_%20Day%201.pdf).
17. M.F.B. Mohamed Salleh, C. Wanchao, Z. Wang, S. Huang, D.Y. Tan, T. Huang, K.H. Low, Preliminary concept of adaptive urban airspace management for unmanned aircraft operations. In *2018 AIAA Information Systems-AIAA Infotech@ Aerospace*, Kissimmee, FL, p. 2260, January 2018.
18. B. Pang, W. Dai, T. Ra, K.H. Low, A concept of airspace configuration and operational rules for UAS in current airspace. In *2020 AIAA/IEEE 39th Digital Avionics Systems Conference (DASC)*, San Antonio, TX, pp. 1–9, IEEE, October 2020.
19. T.H. Tran, D.D. Nguyen, Management and regulation of drone operation in urban environment: A case study. *Social Sciences*, vol. 11, no. 10, p. 474, October 2022.
20. L. Pathiyil, K.H. Low, B.H. Soon, S. Mao, Enabling safe operations of unmanned aircraft systems in an urban environment: A preliminary study. In *The International Symposium on Enhanced Solutions for Aircraft and Vehicle Surveillance Applications (ESAVS 2016)*, Berlin, Germany, p. 10, German Institute of Navigation and the German Aerospace Center (DLR), April 2016.
21. ZenaDrone, *Overview*, June 2024 (available online: https://www.zenadrone.com/).

22. W.C. Chiang, Y. Li, J. Shang, T.L. Urban, Impact of drone delivery on sustainability and cost: Realizing the UAV potential through vehicle routing optimization. *Applied Energy*, vol. 242, pp. 1164–1175, May 2019.
23. Wipro, 2021. *The Future of Delivery with Drones: Contactless, Accurate and High-Speed*, August 2021 (available online: https://www.wipro.com/business-process/the-future-of-delivery-with-drones-contactless-accurate-and-high-speed/).
24. FAA, *Timeline of Drone Integration*, June 2024 (available online: https://www.faa.gov/uas/resources/timeline).
25. BBC, *Amazon Makes First Drone Delivery*, December 2016 (available online: https://www.bbc.com/news/technology-38320067).
26. Matternet, *Matternet Receives FAA Production Certificate for Its M2 Drone Delivery System*, Press Release November 2022 (available online: https://www.matternet.com/milestones).
27. M. Heutger, M. Kückelhaus, *Unmanned Aerial Vehicle in Logistics a DHL Perspective on Implications and Use Cases for the Logistics Industry.* Bonn, Germany: DHL Customer Solutions & Innovation, 2014 (available online: https://www.dhl.com/discover/content/dam/dhl/downloads/interim/full/dhl-trend-report-uav.pdf).
28. M. Farine, *Du Café Livré Par Drone à Zurich*, Le Temps, November 2017 (available online: https://www.letemps.ch/cyber/cafe-livre-drone-zurich?srsltid=AfmBOooy0tBNOV2drMYeH0nXTgTOCHYZugywf0XIYwo0zXhuj0czoVx0).
29. University of Mannheim, *Drones Delivering Lab Samples Offer Much Potential for City Logistics*, July 2023 (available online: https://www.uni-mannheim.de/en/news/drones-delivering-lab-samples-offer-much-potential-for-city-logistics/).
30. EASA, *Drones & Air Mobility Landscape Understanding Drones and Innovative Air Mobility – Role of Cities*, June 2024 (available online: https://www.easa.europa.eu/en/domains/drones-air-mobility/drones-air-mobility-landscape/roles-of-cities).
31. Australian Government, *Drones: Goods Delivery*, June 2024 (available online: https://www.casa.gov.au/drones/industry-initiatives/drone-delivery-services#Approveddeliverylocations).
32. Unmanned Airspace, 2020. *Japan Post Tests BVLOS Parcel Delivery with View to Full Operations by 2025*, May 2020 (available online: https://www.unmannedairspace.info/latest-news-and-information/japan-posts-tests-bvlos-parcel-delivery-with-view-to-full-operations-by-2025/).
33. The Conversation, *Amazon Delivery Drones: How the Sky Could Be the Limit for Market Dominance*, October 2023 (available online: https://theconversation.com/amazon-delivery-drones-how-the-sky-could-be-the-limit-for-market-dominance-216047).
34. J.P. Aurambout, K. Gkoumas, B. Ciuffo, Last mile delivery by drones: An estimation of viable market potential and access to citizens across European cities. *European Transport Research Review*, vol. 11, no 1, pp. 1–21, Jun 2019.
35. P.R. Gabani, U.B. Gala, V.S. Narwane, R.D. Raut, U.H. Govindarajan, B.E. Narkhede, A viability study using conceptual models for last mile drone logistics operations in populated urban cities of India. *IET Collaborative Intelligent Manufacturing*, vol. 3, no. 3, pp. 262–272, September 2021.
36. L. Ranieri, S. Digiesi, B. Silvestri, M. Roccotelli, A review of last mile logistics innovations in an externalities cost reduction vision. *Sustainability*, vol. 10, no. 3, pp. 1–18, March 2018.
37. X. Li, J. Tupayachi, A. Sharmin, M. Ferguson, Drone-aided delivery methods, challenge, and the future: A methodological review. *Drones*, vol. 7, no. 3, pp. 1–26, March 2023.
38. K.A.O. Suzuki, F.P. Kemper, J.R. Morrison, Automatic battery replacement system for UAVs: Analysis and design. *Journal of Intelligent & Robotic Systems*, vol. 65, pp. 563–586, January 2012.

39. I. Khoufi, A. Laouiti, C. Adjih, A survey of recent extended variants of the traveling salesman and vehicle routing problems for unmanned aerial vehicles. *Drones,* vol. 3, no. 3, pp. 66–96, August 2019.
40. C.C. Murray, A.G. Chu, The flying sidekick traveling salesman problem: Optimization of drone-assisted parcel delivery. *Transportation Research Part Emerging Technologies*, vol. 54, pp. 86–109, May 2015.
41. N. Agatz, P. Bouman, M. Schmidt, Optimization approaches for the traveling salesman problem with drone. *Transportation Science*, vol. 52, no. 4, pp. 965–981, April 2018.
42. M. Marinelli, L. Caggiani, M. Ottomanelli, M. Dell'Orco, En-route truck-drone parcel delivery for optimal vehicle routing strategies. *IET Intelligent Transport Systems*, vol. 12, no. 4, pp. 253–261, May 2018.
43. Q.M. Ha, Y. Deville, Q.D. Pham, M.H. Hà, On the min-cost traveling salesman problem with drone. *Transportation Research Part C: Emerging Technologies*, vol. 86, pp. 597–621, January 2018.
44. P. Kitjacharoenchai, M. Ventresca, M. Moshref-Javadi, S. Lee, J.M. Tanchoco, P.A. Brunese, Multiple traveling salesman problem with drones: Mathematical model and heuristic approach. *Computer and Industrial Engineering*, vol. 129, pp. 14–30, March 2019.
45. C.A. Thiels, J.M. Aho, S.P. Zietlow, D.H. Jenkins, Use of unmanned aerial vehicles for medical product transport. *Air Medical Journal*, vol. 34, no. 2, pp. 104–108, April 2015.
46. L.A. Haidari et al., The economic and operational value of using drones to transport vaccines. *Vaccine,* vol. 34, no. 34, pp. 4062–4067, July 2016.
47. Wired, *Drones Have Transformed Blood Delivery in Rwanda*, April 2022 (available online: https://www.wired.com/story/drones-have-transformed-blood-delivery-in-rwanda/).
48. M.P. Nisingizwe et al., Effect of unmanned aerial vehicle (drone) delivery on blood product delivery time and wastage in Rwanda: A retrospective, cross-sectional study and time series analysis. *The Lancet Global Health*, vol. 10, no. 4, pp. 564–569, April 2022.
49. Gavi, *Ghana Launches the World's Largest Vaccine Drone Delivery Network*, April 2019 (available online: https://www.gavi.org/news/media-room/ghana-launches-worlds-largest-vaccine-drone-delivery-network).
50. A. Choi-Fitzpatrick et al., Up in the Air: A Global Estimate of Non-Violent Drone Use 2009–2015. Joan B. Kroc School of Peace Studies—University of San Diego: Degheri Alumni Center, CA, USA, 2016.
51. D. Bamburry, Drones: Designed for product delivery. *Design Management Review,* vol. 26, no. 1, pp. 40–48, July 2015.
52. Businesswire, *Matternet Launches World's Longest Urban Drone Delivery Route Connecting Hospitals and Laboratories in Zurich, Switzerland*, December 2022 (available online: https://www.businesswire.com/news/home/20221212005097/en/Matternet-Launches-World%E2%80%99s-Longest-Urban-Drone-Delivery-Route-Connecting-Hospitals-and-Laboratories-in-Zurich-Switzerland).
53. Drone Life, *Matternet M2 Will Fly BVLOS Medical Drone Delivery in the Heart of Berlin*, December 2023 (available online: https://dronelife.com/2023/12/12/matternet-m2-will-fly-bvlos-medical-drone-delivery-in-the-heart-of-berlin/).
54. RigiTech, *RigiTech Performs BVLOS Flights over Lake Geneva in Switzerland*, October 2022 (available online: https://rigi.tech/rigitech-performs-bvlos-flights-over-lake-geneva-in-switzerland/).
55. RigiTech, *RigiTech's BVLOS Drone Delivery Lands in Uruguay, A First in Latin America*, July 2023 (available online: https://rigi.tech/rigitechs-bvlos-drone-delivery-lands-in-uruguay-a-first-in-latin-america/#:~:text=RigiTech's%20BVLOS%20Drone%20Delivery%20

Lands%20in%20Uruguay%2C%20A%20First%20in%20Latin%20America,-July%2012%2C%202023&text=Tacuaremb%C3%B3%2C%20Uruguay.,BVLOS%20operation%20in%20Latin%20America.).
56. RigiTech, *RigiTech Onboards Medical Drone Service to Enhance Healthcare Access in The Netherlands*, June 2024a (available online: https://rigi.tech/rigi-tech-onboards-medical-drone-service-to-enhance-healthcare-access-in-the-netherlands/).
57. eVTOL Insights, *Drone Services RigiTech "to Deliver Medical Healthcare Across Maldives*, February 2024 (available online: https://evtolinsights.com/2024/02/drone-services-rigitech-to-deliver-medical-healthcare-across-maldives/).
58. M.S.Y. Hii, P. Courtney, P.G. Royall, An evaluation of the delivery of medicines using drones. *Drones*, vol. 3, no. 3, pp. 1–20, June 2019.
59. RigiTech, *Laboratory Logistics*, June 2024 (available online: https://rigi.tech/labs/).
60. Drone Responders, *Preparedness, Response, Resilience*, June 2024 (available online: https://www.droneresponders.org/).
61. K. Gupta, S. Bansal, R. Goel, R., Uses of drones in fighting covid-19 pandemic. In *2021 10th International Conference on System Modeling & Advancement in Research Trends (SMART)*, Moradabad, India, pp. 651–655, December 2021.
62. P.L. Nedelea, T.O. Popa, E. Manolescu, C. Bouros, G. Grigorasi, D. Andritoi, C. Pascale, A. Andrei, D.C. Cimpoesu, Telemedicine system applicability using drones in pandemic emergency medical situations. *Electronics*, vol. 11, no. 14, p. 2160, July 2022.
63. Á. Restás, Drone applications fighting COVID-19 pandemic—Towards good practices. *Drones*, vol. 6, no. 1, p. 15, January 2022.
64. P. Royo, À. Asenjo, J. Trujillo, E. Çetin, C. Barrado, Enhancing drones for law enforcement and capacity monitoring at open large events. *Drones*, vol. 6, no. 11, p. 359, November 2022.
65. J. Lee, Z. Zhong, K. Kim, B. Dimitrijevic, B. Du, S. Gutesa,, Examining the applicability of small quadcopter drone for traffic surveillance and roadway incident monitoring. In *Transportation Research Board 94th Annual Meeting*, Washington, DC, no. 15–4184, p. 15, January 2015.
66. I. Bisio, Z. Zhong, K. Kim, B. Dimitrijevic, B. Du, S. Gutesa, A systematic review of drone based road traffic monitoring system. *IEEE Access*, vol. 10, pp. 101537–101555, September 2022.
67. H. Gupta, O.P. Verma, Monitoring and surveillance of urban road traffic using low altitude drone images: A deep learning approach. *Multimedia Tools and Applications*, vol. 81, no. 14, pp. 19683–19703, January 2022.
68. P. Darby, V. Gopu, Bridge Inspecting with Unmanned Aerial Vehicles R&D, 2018.
69. H. Habeenzu, P.J. McGetrick, D. Hester, S.E. Taylor, Bridge Management Systems-A Review of the State of the Art and Recommendations for Future Practice. In *Bridge Maintenance, Safety, Management, Life-Cycle Sustainability and Innovations*, edited by H. Yokota, D.M. Frangopol. Boca Raton: CRC Press, Taylor & Francis Group, pp. 926–933, April 2021.
70. E.K. Mahama, *Developing Test and Evaluation Techniques for UAV-Enabled Bridge Inspection*. Doctoral dissertation, North Carolina Agricultural and Technical State University, 2021.
71. C. Barrado, J. Ramírez, M. Perez-Batlle, E. Santamaria, X. Prats, E. Pastor, Remote flight inspection using unmanned aircraft. *Journal of Aircraft*, vol. 50, no. 1, pp. 38–46, January 2013.
72. T. Whitley, A. Tomiczek, C. Tripp, A. Ortega, M. Mennu, J. Bridge, P. Ifju, Design of a small unmanned aircraft system for bridge inspections. *Sensors*, vol. 20, no. 18, p. 5358, September 2020.

73. P. Nooralishahi, C. Ibarra-Castanedo, S. Deane, F. López, S. Pant, M. Genest, N.P. Avdelidis, X.P.V. Maldague, Drone-based non-destructive inspection of industrial sites: A review and case studies. *Drones*, vol. 5, no. 4, p. 106, September 2021.
74. D.A. Rodríguez, C.L. Tafur, P.F.M. Daza, J.A.V. Vidales, J.C.D. Rincón, Inspection of aircrafts and airports using UAS: A review. *Results in Engineering*, vol. 22, p. 102330, May 2024.
75. W.C. Castelar, J. Pflughaupt, L. Moshagen, M. Kurenkov, T. Lewejohann, G. Schildbach, LiDAR-based automated UAV inspection of wind turbine rotor blades. *Journal of Field Robotics*, vol. 41, no. 4, pp. 1116–1132, June 2024.
76. Y. Zhu, H. Tang, Automatic damage detection and diagnosis for hydraulic structures using drones and artificial intelligence techniques. *Remote Sensing*, vol. 15, no. 3, p. 615, January 2023.
77. A. Varghese, J. Gubbi, H. Sharma, P. Balamuralidhar, Power infrastructure monitoring and damage detection using drone captured images. In *2017 International Joint Conference on Neural Networks (IJCNN)*, Anchorage, AK, pp. 1681–1687, IEEE, May 2017.
78. E. Papadopoulos, F. Gonzalez, UAV and AI application for runway foreign object debris (FOD) detection. In *2021 IEEE Aerospace Conference* (50100), Big Sky, MT, pp. 1–8, March 2021.
79. A. Parker, F. Gonzalez, P. Trotter, Live detection of foreign object debris on runways detection using drones and AI. In *2022 IEEE Aerospace Conference (AERO)*, Big Sky, MT, pp. 1–13, IEEE, March 2022.
80. D. Appelbaum, R.A. Nehmer, Using drones in internal and external audits: An exploratory framework. *Journal of Emerging Technologies in Accounting*, vol. 14, no. 1, pp. 99–113, Mar 2017.
81. F. Ullah, S.M. Sepasgozar, C. Wang, A systematic review of smart real estate technology: Drivers of, and barriers to, the use of digital disruptive technologies and online platforms. *Sustainability*, vol. 10, no. 9, p. 3142, September 2018.
82. A. Quattrini, A. Mascheroni, A. Vandone, M. Coluzzi, A. Barazzetti, F. Cecconi, T. Leidi, Real estate advisory drone (READ): System for autonomous indoor space appraisals, based on deep learning and visual inertial odometry. *IOP Conference Series: Materials Science and Engineering*, vol. 1226, no. 1, p. 012112, February 2022.
83. Q. Galvane, C. Lino, M. Christie, J. Fleureau, F. Servant, F.-L. Tariolle, P. Guillotel, Directing cinematographic drones. *ACM Transactions on Graphics (TOG)*, vol. 37, no. 3, pp. 1–18, Jul 2018.
84. R. Campbell, Drone film theory: The immanentisation of kinocentrism. *Media Theory*, vol. 2, no. 2, pp. 52–78, December 2018.
85. J. Kim, *Documentary's Expanded Fields: New Media and the Twenty-First-Century Documentary*. Oxford University Press, 2022.
86. E. Tuba, I. Tuba, D. Dolicanin-Djekic, A. Alihodzic, M. Tuba, Efficient drone placement for wireless sensor networks coverage by bare bones fireworks algorithm. In *2018 6th International Symposium on Digital Forensic and Security (ISDFS)*, Antalya, Turkey, pp. 1–5, IEEE, March 2018.
87. O. Zerlenga, V. Cirillo, R. Iaderosa, Once upon a time there were fireworks. The new nocturnal drones light shows. *Img Journal*, vol. 4, pp. 402–425, September 2021.
88. M. Royo-Vela, M. Black, Drone images versus terrain images in advertisements: Images' verticality effects and the mediating role of mental simulation on attitude towards the advertisement. *Journal of Marketing Communications*, vol. 26, no. 1, pp. 21–39, January 2020.
89. A. Pamnani, V. Parvathi, Innovations in contemporary marketing through artificial intelligence and robotic drones. *Innovations*, vol. 1, no. 1, pp. 34–43, July 2021.

90. AirVuz, *Drone Videos of Sports Cars*, June 2024 (available online: https://www.airvuz.com/collection/video/drone-videos-of-sports-cars?id=BkahBhC-Q).
91. G. Rohi, G. Ofualagba, Autonomous monitoring, analysis, and countering of air pollution using environmental drones. *Heliyon*, vol. 6, no. 1, p. e03252, January 2020.
92. A. Hossain, A. Hossain, M.J. Anee, R. Faruqui, S. Bushra, P. Rahman, R. Khan, A GPS based unmanned drone technology for detecting and analyzing air pollutants. *IEEE Instrumentation & Measurement Magazine*, vol. 25, no. 9, pp. 53–60, November 2022.
93. M. Durgun, Y. Durgun, Smart environmental drone utilization for monitoring urban air quality. *Environmental Research and Technology*, vol. 7, no. 2, pp. 194–200, June 2024.
94. R.R. Hemamalini, R. Vinodhini, B. Shanthini, P. Partheeban, M. Charumathy, K. Cornelius, Air quality monitoring and forecasting using smart drones and recurrent neural network for sustainable development in Chennai city. *Sustainable Cities and Society*, vol. 85, p. 104077, October 2022.
95. R. Camarillo-Escobedo, R. Camarillo-Escobedo, J.L. Flores, P. Marin-Montoya, G. García-Torales, J.M. Camarillo-Escobedo, Smart multi-sensor system for remote air quality monitoring using unmanned aerial vehicle and LoRaWAN. *Sensors*, vol. 22, no. 5, p. 1706, February 2022.
96. S. Hossain, A. Helal, M.S. Ahsan, K.M. Hasan, M. Maniruzzaman, M.E. Kabir, Design and development of an autonomous air quality monitoring drone. *Engineering Headway*, vol. 5, pp. 39–53, May 2024.
97. J. Burgués, S. Marco. Environmental chemical sensing using small drones: A review. *Science of the Total Environment*, vol. 748, p. 141172, December 2020.
98. UN (United Nation), *Secretary-General's Strategy on New Technologies*, September 2018 (available online: https://www.un.org/en/newtechnologies/#:~:text=The%20goal%20of%20this%20internal,their%20alignment%20with%20the%20values).
99. P. Royo, M. Perez-Batlle, R. Cuadrado, E. Pastor, Enabling dynamic parametric scans for unmanned aircraft system remote sensing missions, *Journal of Aircraft*, vol. 51, no. 3, pp. 870–882, May 2014.
100. P. Royo et al., The mapping of alpha-emitting radionuclides in the environment using an unmanned aircraft system. *Remote Sensing*, vol. 16, no. 5, p. 848, February 2024.
101. E. Salamí, C. Barrado, E. Pastor, UAV flight experiments applied to the remote sensing of vegetated areas. *Remote Sensing*, vol. 6, no. 11, pp. 11051–11081, November 2014.
102. S.M. Schoenung, R.T. Albertson, Remote sensing in the arctic using autonomous sensors developed under NASA's airborne science program for the international polar year. In *34th International Symposium on Remote Sensing of Environment*, Sydney, Australia, p. 4, April 2011.
103. J.C. Hodgson, *Using Drones to Improve Wildlife Monitoring in a Changing Climate*. Doctoral thesis, University of Adelaide, Australia, 2020.
104. G. Mahajan, Applications of drone technology in construction industry: A study 2012–2021. *International Journal of Engineering and Advanced Technology,* vol. 11, no. 1, pp. 224–239, October 2021.
105. L. Stannard, *6 Ways Drones in Construction Are Changing the Industry*. BigRenz Blog, February 2022 (available online: https://www.bigrentz.com/blog/drones-construction?srsltid=AfmBOorgMjRjXqS8h7WFqAfEtD8jVgfeTvJNlFcvmxh9ZA-Z4L6y0izP).
106. M.C. Tatum, J. Liu, Unmanned aircraft system applications in construction. *Procedia Engineering*, vol. 196, pp. 167–175, January 2017.
107. K. Asadi et al., An integrated UGV-UAV system for construction site data collection. *Automation in Construction*, vol. 112, p. 103068, April 2020.
108. S.S.C. Congress, A.J. Puppala, A road map for geotechnical monitoring of transportation infrastructure assets using three-dimensional models developed from unmanned aerial data. *Indian Geotechnical Journal*, vol. 51, no. 1, pp. 84–96, February 2021.

109. P. Kim, J. Park, Y. Cho, As-is geometric data collection and 3D visualization through the collaboration between UAV and UGV. In *Proceedings of the 36th International Symposium on Automation and Robotics in Construction (ISARC)*, Banff, AB, Canada, pp. 544–551, 2019.
110. D. Giordan et al., The use of unmanned aerial vehicles (UAVs) for engineering geology applications. *Bulletin of Engineering Geology and the Environment*, vol. 79, pp. 3437–3481, September 2020.
111. S.P. Bemis, S. Micklethwaite, D. Turner, M.R. James, S. Akciz, S.T. Thiele, H.A. Bangash, Ground-based and UAV-based photogrammetry: A multi-scale, high-resolution mapping tool for structural geology and paleoseismology. *Journal of Structural Geology*, vol. 69, pp. 163–178, December 2014.
112. M.J. Westoby, J. Brasington, N.F. Glasser, M.J. Hambrey, J.M. Reynolds, Structure-from-Motion'photogrammetry: A low-cost, effective tool for geoscience applications. *Geomorphology*, vol. 179, pp. 300–314, December 2012.
113. S. Bang, H. Kim, H. Kim, UAV-based automatic generation of high-resolution panorama at a construction site with a focus on preprocessing for image stitching. *Automation in Construction*, vol. 84, pp. 70–80, December 2017.
114. O.G. Ajayi, M. Palmer, A.A. Salubi, Modelling farmland topography for suitable site selection of dam construction using unmanned aerial vehicle (UAV) photogrammetry. *Remote Sensing Applications: Society and Environment*, vol. 11, pp. 220–230, August 2018.
115. S. Siebert, J. Teizer, Mobile 3D mapping for surveying earthwork projects using an unmanned aerial vehicle (UAV) system. *Automation in Construction*, vol. 41, pp. 1–14, May 2014.
116. M.H. Chae, S.O. Park, S.H. Choi, C.T. Choi, Commercial fixed-wing drone redirection system using GNSS deception. *IEEE Transactions on Aerospace and Electronic Systems*, vol. 59, no. 5, pp. 5699–5713, April 2023.
117. S. Razzaq, C. Xydeas, A. Mahmood, S. Ahmed, N.I. Ratyal, J. Iqbal, Efficient optimization techniques for resource allocation in UAVs mission framework. *PloS One*, vol. 18, no. 4, p. e0283923, April 2023.
118. W. Budiharto, Mapping and 3D modelling using quadrotor drone and GIS software. *Journal of Big Data*, vol. 8, pp. 1–12, December 2021.
119. H.W. Choi, H.J. Kim, S.K. Kim, W.S. Na, An overview of drone applications in the construction industry. *Drones*, vol. 7, no. 8, p. 515, August 2023.
120. J. Kim, S. Lee, J. Seo, D.-E. Lee, H.S. Choi, The integration of earthwork design review and planning using UAV-based point cloud and BIM. *Applied Sciences*, vol. 11, no. 8, p. 3435, April 2021.
121. P. Kavaliauskas, D. Židanavičius, A. Jurelionis, Geometric accuracy of 3D reality mesh utilization for BIM-based earthwork quantity estimation workflows. *ISPRS International Journal of Geo-Information,* vol. 10, no. 6, p. 399, Jun 2021.
122. W.J. Baker, C.L. Meehan, Spatial interpolation of UAV survey data for lift thickness determination during earthwork construction. In *Geo-Congress 2023*, pp. 336–346, 2023.
123. A. Ostrovskyi, I. Kolb, A. Vivat, V. Lozynskyi, V. Zhyvchuk, Simplified method of obtaining data for calculating the volume of earthworks based on aerial survey materials from UAVs. *International Conference of Young Professionals «GeoTerrace-2021»*, vol. 2021, no. 1, pp. 1–5, European Association of Geoscientists & Engineers.
124. T.S.N. Rachmawati, H.C. Park, S. Kim, A scenario-based simulation model for earthwork cost management using unmanned aerial vehicle technology. *Sustainability*, vol. 15, no. 1, p. 503, December 2022.
125. H.C. Park, T.S.N. Rachmawati, S. Kim, UAV-based high-rise buildings earthwork monitoring—A case study. *Sustainability*, vol. 14, no. 16, p. 10179, August 2022.

126. H. Hasegawa, A.A. Sujaswara, T. Kanemoto, K. Tsubota, Possibilities of using UAV for estimating earthwork volumes during process of repairing a small-scale forest road, case study from Kyoto Prefecture, Japan. *Forests*, vol. 14, no. 4, p. 677, March 2023.
127. D. Kim, K. Yin, M. Liu, S. Lee, V.R. Kamat, Feasibility of a drone-based on-site proximity detection in an outdoor construction site. In *Computing in Civil Engineering 2017,* Seattle, Washington, pp. 392–400, 2017.
128. J. Wu, L. Peng, J. Li, X. Zhou, J. Zhong, C. Wang, J. Sun, Rapid safety monitoring and analysis of foundation pit construction using unmanned aerial vehicle images. *Automation in Construction*, vol. 128, p. 103706, Aug 2021.
129. J.G. Martinez, G. Albeaino, M. Gheisari, R.R.A. Issa, L.F. Alarcón, iSafeUAS: An unmanned aerial system for construction safety inspection. *Automation in Construction*, vol. 125, p. 103595, May 2021.
130. M. Attaran, B.G. Celik, Digital twin: Benefits, use cases, challenges, and opportunities. *Decision Analytics Journal*, vol. 6, p. 100165, March 2023.
131. R. Samsami, A. Mukherjee, C.N. Brooks, Mapping unmanned aerial system data onto building information modeling parameters for highway construction progress monitoring. *Transportation Research Record*, vol. 2676, no. 4, pp. 669–682, April 2022.
132. S. Zhou, M. Gheisari, Unmanned aerial system applications in construction: A systematic review. *Construction Innovation*, vol. 18, no. 4, pp. 453–468, October 2018.
133. A. Sawhney, M. Riley, J. Irizarry, *Construction 4.0.: Introduction and Overview.* Routledge, 2020.
134. Z. Li, Q. Xu, L.M. Tam, A survey on techniques and applications of window-cleaning robots. *IEEE Access*, vol. 9, pp. 111518–111532, August 2021.
135. W. Reckling, J. Levine, S.A. Nelson, H. Mitasova, Predicting residential septic system malfunctions for targeted drone inspections. *Remote Sensing Applications: Society and Environment,* vol. 30, p. 100936, April 2023.
136. H. Liang, S.-C. Lee, W. Bae, J. Kim, S. Seo, Towards UAVs in construction: Advancements, challenges, and future directions for monitoring and inspection. *Drones*, vol. 7, no. 3, p. 202, March 2023.
137. T. Kirschstein, Energy demand of parcel delivery services with a mixed fleet of electric vehicles. *Cleaner Engineering and Technology*, vol. 5, p. 100322, December 2021.
138. G. Aiello, R. Inguanta, G. D'Angelo, M. Venticinque, Energy consumption model of aerial urban logistic infrastructures. *Energies*, vol. 14, no. 18, p. 5998, September 2021.
139. M. Torabbeigi, G.J. Lim, S.J. Kim, Drone delivery scheduling optimization considering payload-induced battery consumption rates. *Journal of Intelligent & Robotic Systems*, vol. 97, pp. 471–487, March 2020.
140. R. Pinto, A. Lagorio, Point-to-point drone-based delivery network design with intermediate charging stations. *Transportation Research Part C: Emerging Technologies*, vol. 135, p. 103506, February 2022.
141. R. Daknama, E. Kraus, Vehicle routing with drones. *arXiv preprint arXiv:1705.06431*, May 2017.
142. Z. Bi, X. Guo, J. Wang, S. Qin, G. Liu, Truck-drone delivery optimization based on multi-agent reinforcement learning. *Drones*, vol. 8, no. 1, p. 27, January 2024.
143. R. Kellermann, T. Biehle, L. Fischer, Drones for parcel and passenger transportation: A literature review. *Transportation Research Interdisciplinary Perspectives*, vol. 4, p. 100088, March 2020.
144. C. Schlag, The new privacy battle: How the expanding use of drones continues to erode our concept of privacy and privacy rights. *Journal of Technology, Law and Policy*, vol. 13, pp. 2–22, Spring 2013.
145. B. Rao, A.G. Gopi, R. Maione, The societal impact of commercial drones. *Technology in Society*, vol. 45, pp. 83–90, May 2016.

146. E. Vattapparamban, İ. Güvenç, A.İ. Yurekli, K. Akkaya, S. Uluağaç, Drones for smart cities: Issues in cybersecurity, privacy, and public safety. In *International Wireless Communications and Mobile Computing Conference*, Paphos, Cyprus, pp. 216–221, September 2016.
147. E. Serafinelli, Imagining the social future of drones. *Convergence*, vol. 28, no. 5, pp. 1376–1391, March 2022.
148. S.A. Applin, Deliveries by drone: Obstacles and sociability. In *The Future of Drone Use: Opportunities and Threats from Ethical and Legal Perspectives*, edited by B. Custers. The Hague, Berlin: Asser Press; Springer, pp. 71–91, September 2016.
149. E. Çetin, A. Cano, R. Deransy, S. Tres, C. Barrado, Implementing mitigations for improving societal acceptance of urban air mobility. *Drones*, vol. 6, no. 2, p. 28, January 2022.
150. A. Raghunatha, P. Thollander, S. Barthel, Addressing the emergence of drones–A policy development framework for regional drone transportation systems. *Transportation Research Interdisciplinary Perspectives*, vol. 18, p. 100795, March 2023.
151. B. Schäffer, R. Pieren, K. Heutschi, J.M. Wunderli, S. Becker, Drone noise emission characteristics and noise effects on humans—A systematic review. *International Journal of Environmental Research and Public Health*, vol. 18, no. 11, p. 5940, June 2021.
152. E. Lewis, J. Ponnock, Q. Cherqaoui, S. Holmdahl, Y. Johnson, A. Wong, H. Oliver Gao, Architecting urban air mobility airport shuttling systems with case studies: Atlanta, Los Angeles, and Dallas. *Transportation Research Part A: Policy and Practice*, vol. 150, pp. 423–444, August 2021.
153. A. Filippone, G.N. Barakos, Rotorcraft systems for urban air mobility: A reality check. *The Aeronautical Journal*, vol. 125, no. 1283, pp. 3–21. January 2021.
154. T., Jones, *International Commercial Drone Regulation and Drone Delivery Services*, RAND Corporation, 2017 (available online: https://www.rand.org/content/dam/rand/pubs/research_reports/RR1700/RR1718z3/RAND_RR1718z3.pdf).
155. ICAO, *ICAO Cir 328: Unmanned Aircraft Systems (UAS)*. Quebec, Canada, 2011.
156. EASA, *Easy Access Rules for U-space (Regulation (EU) 2021/664)*, 2024 (available online: https://www.easa.europa.eu/en/document-library/easy-access-rules/easy-access-rules-u-space-regulation-eu-2021664).
157. C. Lidynia, R. Philipsen, M. Ziefle, Droning on About Drones—Acceptance of and Perceived Barriers to Drones in Civil Usage Contexts. In *Advances in Human Factors in Robots and Unmanned Systems: Proceedings of the AHFE 2016 International Conference on Human Factors in Robots and Unmanned Systems (eds. P. Savage-Knepshield, J. Chen), July 27-31, 2016, Florida, USA*. Springer International Publishing pp. 317–329, 2017.
158. M. Jaller, C. Otero-Palencia, A. Pahwa, Automation, electrification, and shared mobility in urban freight: Opportunities and challenges. *Transportation Research Procedia*, vol. 46, pp. 13–20, January 2020.
159. B. Anbaroğlu, Parcel delivery in an urban environment using unmanned aerial systems: A vision paper. *ISPRS Annals of the Photogrammetry, Remote Sensing and Spatial Information Sciences*, vol. 4, pp. 73–79, November 2017.

11 Smart Digitalization Technologies for Future Resilient and Sustainable Energy Systems

Vladimir Terzija

11.1 INTRODUCTION

Modern electrical power systems are experiencing significant changes. They are integrated into larger systems, so-called integrated energy systems, in which, e.g., heat and gas networks are connected to the electricity one. Within the power system, components based on power electronics are introduced, which affects the dynamic performance of the system, and which directly affects the margins of system stability or its resistance. Finally, through the intensive system digitalization and the introduction of smart grid technology, e.g., phasor measurement units (PMUs) and cyber-secure communication channels, new opportunities for improving system monitoring, protection, and control are opened. The chapter addresses the above-mentioned issues and more specifically discusses the system resilience and system integrity protection schemes (SIPSs). In this context, it gives an example of the solution for the prevention of power system blackouts – intentional system splitting/islanding.

Massive integration of converter interfaced generation (CIG) and renewable energy sources (RESs) has significantly changed the nature and complexity of modern electrical power systems. This is obvious from Figure 11.1, in which different

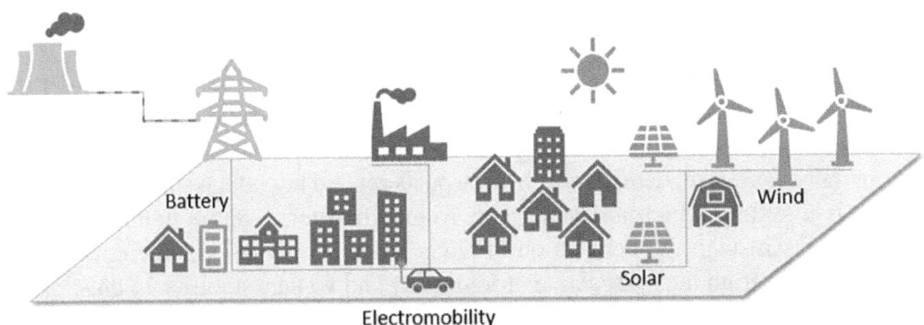

FIGURE 11.1 Modern electrical power system with converter interfaced generations and renewable energy sources.

DOI: 10.1201/9781003511298-11

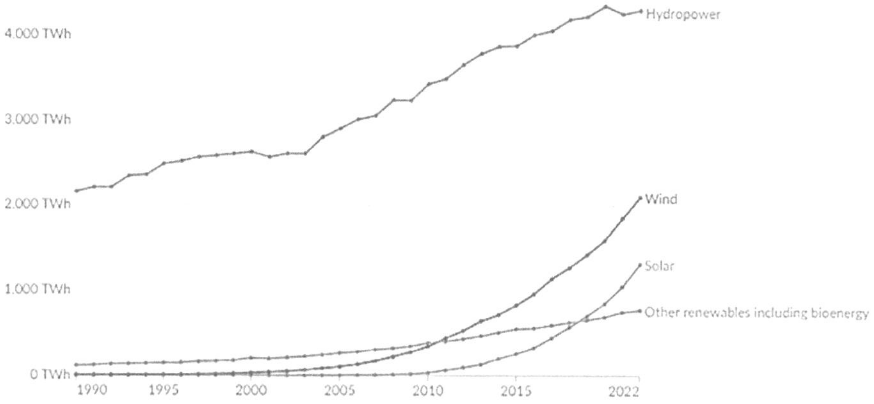

FIGURE 11.2 Modern renewable energy generation by source.

new system components are shown, e.g., wind farms, solar energy, battery, or electrical vehicles. Practically all components presented are connected to the grid over inverters/power electronics.

Massive integration of converter interfaced generation (CIG) and renewable energy sources (RESs) in a modern electrical power system.

Historically, the penetration level of CIG and RES has been permanently increased, which is depicted in Figure 11.2, in which the share of different renewable energy generation is depicted [1, 2].

Through the integration of CIGs and RESs, the system is changing its fundamental properties. These changes can be summarized in the conclusion that the entire system became weak or at least weaker than the system from the past. Here a definition of a strong system is a system whose stability cannot be significantly affected by different types of perturbations, e.g., sudden topological changes, faults, or loss of generation. The reasons for the power system to become weaker are power electronics-based system components, particularly generation. For example, Type-3 and Type-4 generators, connected to the grid, do not contribute to the system inertia and fault level [3–5]. From Figure 11.3 it can be concluded that only the stator of the generator is synchronously connected to the grid, whereas the rotor is decoupled from it through inverters.

System inertia is directly related to the system frequency stability, and under the term "inertia" we traditionally understand the rotational inertia, resulting from the rotation of rotors in synchronous generators and different types of AC motors. On the contrary, fault level determines those aspects related to voltage stability. A strong system is a system which is stable, both from a frequency and a voltage stability perspective. Through the connection of CIGs/RESs, the system is becoming weak and vulnerable from the perspective of frequency and voltage stability. In conclusion, system perturbations, e.g., faults, or sudden imbalances of active powers (P) caused by, e.g., sudden disconnection/connection of generators, can provoke instabilities and start cascading events, which can lead to power system blackouts.

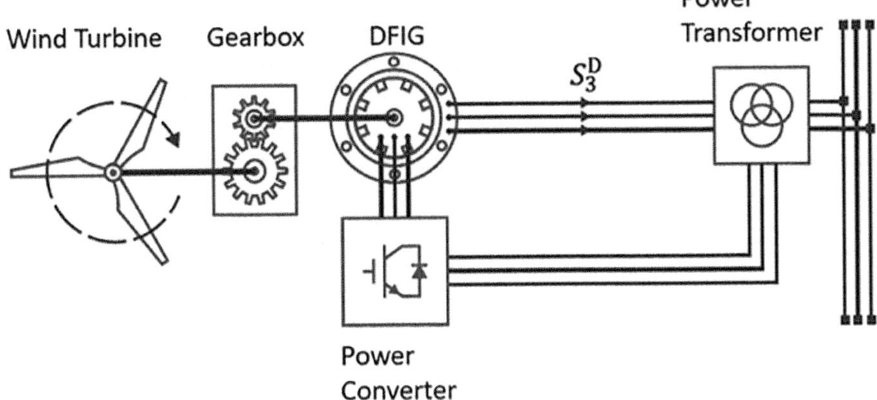

FIGURE 11.3 Doubly fed induction generator (DFIG) wind turbine, type-3 wind generator.

11.2 ENERGY SYSTEM RESILIENCE

Power system resilience is one of the key attributes which modern power systems have to possess and satisfy. Traditionally, the focus was on system security and reliability. Through the integration of different energy systems, e.g., heat, gas, electricity, or hydrogen, the importance of system flexibility became another important system attribute [6]. Figure 11.4 presents an example of such an integration.

Resilience is related to low probability but high-impact events, which as such can lead to catastrophic power system blackouts and a number of consequences to the system and the society. The system can be characterized as resilient if it can successfully go through large-scale perturbation and manage to retain its key security and stability, or operational, characteristics. In this context through the integration of different energy systems into an integrated energy system (Figure 11.4),

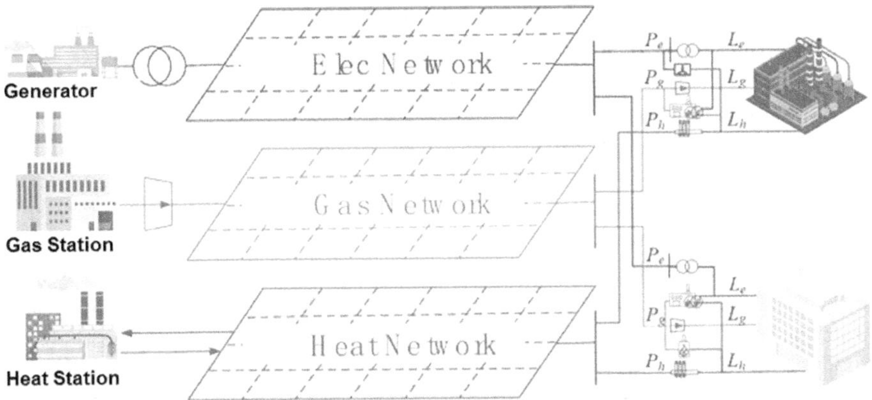

FIGURE 11.4 Integrated energy systems connected over different types of coupling technologies.

the system resilience can be further improved. Stankovic [7] defines power system resilience as "the system ability to withstand and reduce the magnitude and/or duration of disruptive events, which includes the capability to anticipate, absorb, adapt to, and/or rapidly recover from such an event." The term "disruptive events" refers to high impact and low-probability unexpected perturbation, including extreme natural disasters or man-made attacks. Typical examples are snow storms or cyber-attacks [8,9].

The system resilience can be improved through short-, mid-, and long-term planning. Different types of investing strategies can boost system resilience. This has to do with, e.g., planning actions 20 years ahead. However, system resilience can also be monitored in real time, subject to the accepted resilience definition and metrics. Finally, the system resilience can be significantly improved through adequate monitoring, protection, and control actions.

Large perturbations, e.g., simultaneous outage of a large number of system components, resulting from natural disasters and extreme weather conditions, traditionally cause so-called cascading events, i.e., sequential/parallel outage/loss of the system components, e.g., transmission lines, power transformers, generators, or demand. By assessing the resilience level in the grid, it can be concluded how critical the system "resilience state" is. This is addressed in Figure 11.5, in which the resilience level in a power system, connected to other energy systems, e.g., gas, heat, or hydrogen, is presented. The resilience trapezoids [10] changed over time are used to quantify the resilience level.

11.3 DIGITALIZATION AND DIGITAL SUBSTATION

The quality of the system monitoring, protection, and control significantly relies on the data obtained, on their quality and quantity, as well as on the speed how quickly data are collected from the physical process observed. Smart grid technology is accelerating the speed of the digitalization of modern electrical power systems. Novel sensors, like time-synchronized PMUs, or high-speed and cyber-secure communication channels strongly support advanced monitoring, protection, and control schemes. Here not only the quality of data but also cyber security is critically important for the success of new protection and control schemes. In Figure 11.6, a digital substation (right), presented next to the classical substation (left), is given [11]. The digital substation has non-conventional instrument transformers, precise conversion from analog to digital voltages and currents, fiber-optic communication infrastructure having no problems with electromagnetic compatibility, fast data transfer to higher hierarchical levels, enabling new protection solutions, both at the unit and system levels (SIPS), e.g., prevention of power system blackouts, as presented in the rest of the section. It can be observed that the information from the switchgear is transferred from the relay and substation control rooms to the central grid control room, in which the energy management system (EMS) is located. At this level, e.g., the system state is estimated.

Reliable information about the system state, which can be also linked to the entire situational awareness request, is a prerequisite for the optimal operation of the entire

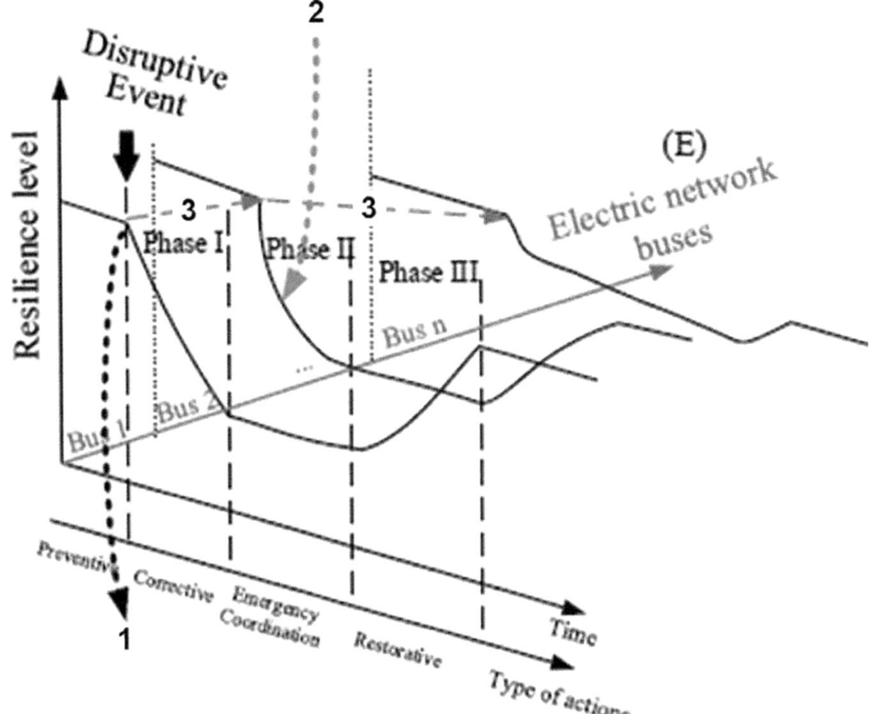

FIGURE 11.5 Resilience level changes over time in a power system connected to other energy systems/sectors.

power system, which is traditionally related to a series of actions undertaken under the umbrella of the EMS. In Figure 11.7, the major building blocks, creating a closed loop monitoring, protection, and control based on different sources of input data, are presented [12,13]. Here, time-synchronized PMU data are integrated with other data obtained over traditional SCADA (Supervisory Control And Data Acquisition), and those data are forecasted or obtained in any other way over EMS. A properly managed database is now a source from which key situational awareness attributes can be extracted, for example, security margins, the level of the system flexibility, or resilience. From the perspective of advanced protection and control applications, the system state, information about load dynamics, system inertia, fault level, etc. can contribute to the quality of the SIPS actions and mitigating measures necessary for detecting cascading events and preventing power system blackouts.

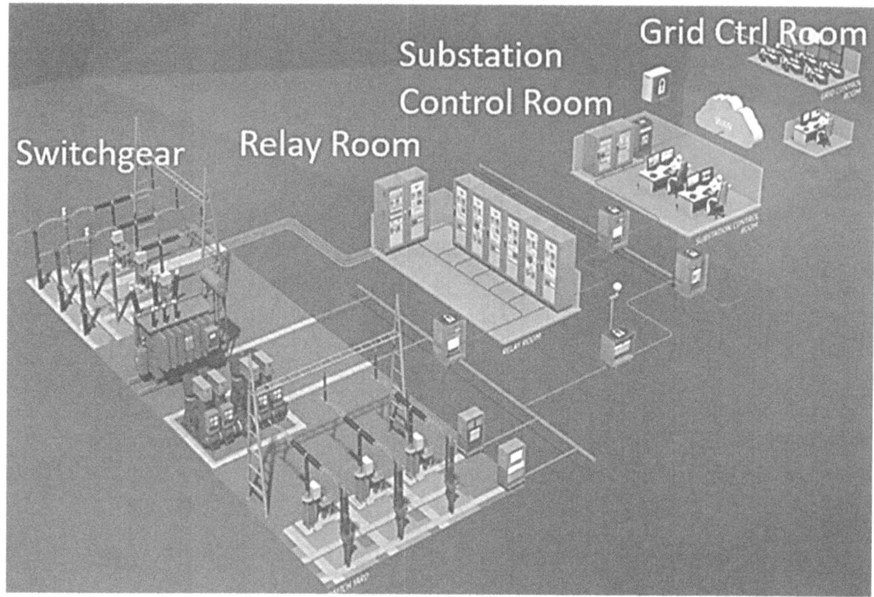

FIGURE 11.6 Digital substation (switchgear).

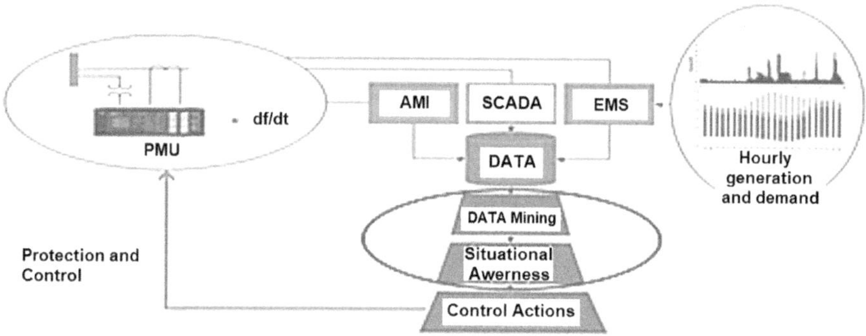

FIGURE 11.7 Building blocks, creating a closed loop power system monitoring, protection, and control.

An example of how PMUs can support advanced situational awareness, i.e., power system state estimation, is new installations worldwide, e.g., in the UK. In Figure 11.8, PMUs are collecting real-time data from different locations of the grid. Data are collected at different locations (data concentrators – DC) and centrally stored at the super data concentrator (super DC). Such a hierarchical structure can enable fast state estimation or protection and control functions which would have

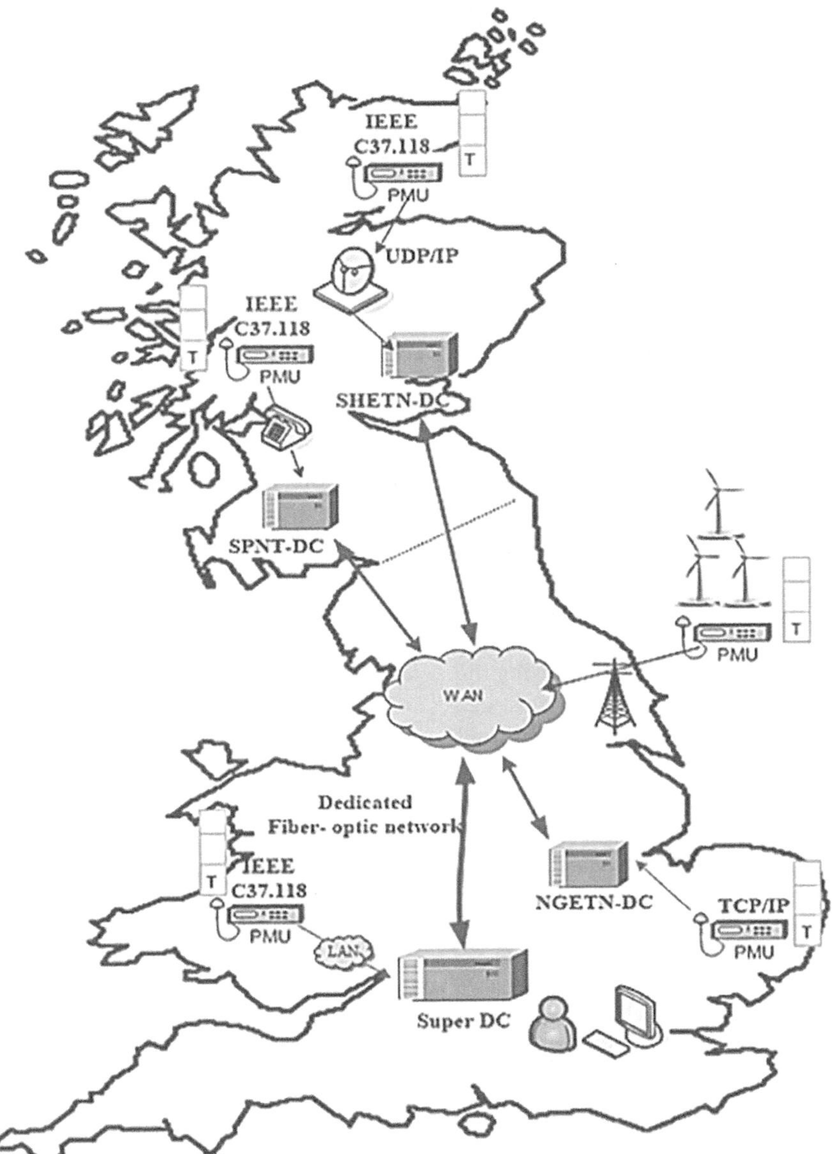

FIGURE 11.8 Integration of phasor measurement units for advanced monitoring, protection, and control of power systems with converter interfaced generations and renewable energy sources.

been impossible to be realized without such a smart grid technology supporting the overall system digitalization. Using PMUs and fast communication channels, the system state can be obtained within 0.1–1 seconds, whereas the traditional, purely SCADA-based estimator required more than 5–10 minutes to accomplish the state estimation procedure.

11.4 SYSTEM INTEGRITY PROTECTION SCHEMES

The role of power system protection is (i) to protect individual system components – unit protection, or (ii) to protect the entire power system – SIPSs. SIPS is defined as an automatic protection system designed to identify abnormal system conditions and to perform predictive or remedial actions necessary for maintaining system integrity. In the past, conventional protection devices were designed to protect individual elements of equipment from being damaged during faults by quickly detecting overcurrent conditions or other dangerous operating conditions and then selectively and quickly isolating the faulty equipment from the system. In contrast, SIPSs are designed to maintain the integrity of the entire power system by simultaneously monitoring and controlling multiple elements of the system. They integrate and analyze both local and system-level information against wide-area contingencies. This approach allows SIPS to improve the efficiency and security of system operation under specific conditions. SIPSs encompass a few different levels or schemes: special protection schemes, remedial action schemes, as well as additional schemes such as underfrequency, undervoltage, and out-of-step. It is commonly believed that modern SIPSs are derived from the coordination of different levels of local protection schemes.

Both unit protection and SIPSs can prevent the negative development of cascading events and support the system integrity. In Figure 11.9 a block diagram of a typical SIPS is presented. The following three key SIPS stages can be observed: (i) disturbance identification, which must be secure and robust; (ii) decision making, which must be reliable and fast enough; and (iii) action, which has to be reliably forwarded back to the system.

The quality of SIPS input data is critically important for its efficacy. Modern SIPSs rely on wide-area measurements and the availability of novel sensors and information and communications technology (ICT) technology. PMUs and fast and

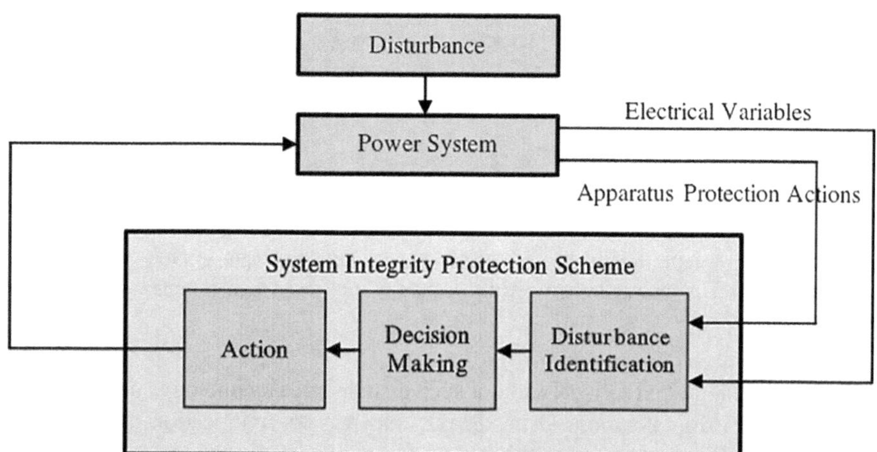

FIGURE 11.9 Block diagram of a typical system integrity protection schemes.

cyber-secure communication infrastructure are the backbone of modern wide-area monitoring, protection, and control systems [14]. Nowadays they are enablers of modern SIPSs. Intentional controlled islanding of a power system is an efficient SIPS for preventing power system blackouts caused by large system disturbances. This kind of SIPS limits the occurrence and consequences of blackouts by splitting the power system into a group of smaller, islanded power systems, called islands. The essence of an islanding solution is in determining a suitable set of transmission lines that must be disconnected to create a set of electrically isolated/independent islands. This section describes an algorithm for determining suitable islanding solutions.

The prerequisite for a successful system islanding is the selection of lines/cables, which must be opened to create independent islands. It is necessary to create stable islands, and in this context the instant at which the islanding is undertaken plays an important role in preventing blackouts.

However, the schemes for the prevention of power system blackouts rely on the quality of input data, as well as the system monitoring, which is traditionally linked to the issue of power system state estimation. Modern system estimation algorithms rely on PMUs integrated into the existing SCADA system (hybrid state estimator) or on PMUs only (PMU-based linear state estimator).

11.5 FUNDAMENTALS OF CONTROL ISLANDING

The complexity of electrical power systems with a high share of RESs is exposed in the fact that such technical systems are less stable compared to the old systems in which synchronous generators were the dominant source of generating power. Through the integration of wind farms the goal is to maximize utilization of the primary energy, and this can be done by connecting them over inverters (power electronics-based components). This leads to frequency and voltage problems and in general to an increased likelihood of power system blackouts. To cope with this risk, systems are equipped with system integrity protection mechanisms. One of them is intentional system splitting/islanding. Here instead of allowing the system to suffer from cascading events leading to the total system blackout, one intentionally split the system into self-sustained stable islands. Below, an advanced approach is presented. It relies on smart sensors and fast and cyber-secure communication infrastructure.

11.5.1 Intentional Control Islanding

To create stable islands, the islanding solution must satisfy a large number of constraints, such as load-generation balance, generator coherency, availability of transmission lines and their thermal limits, voltage stability, and transient stability. It would be far too complicated to search for a solution satisfying all these constraints. In extreme cases it is likely that such a solution does not even exist. Considering only a sub-set of these constraints, e.g. (i) load-generation active power balance and (ii) generator coherency, a set of feasible candidate islanding solutions could be created. This set of candidates can be coordinated with other corrective measures to find the final islanding solution that satisfies all possible constraints [15].

The existing islanding methods can be classified according to the objective function used for decision making, i.e., the creation of the islanding solution. Two main types of objective functions are used here:

a. Minimal power imbalance
b. Minimal power-flow disruption

Methodologies for minimal power imbalance minimize the power imbalance within the islands formed to reduce the amount of load that must be shed after system splitting.

Methodologies for finding islanding solutions based on the minimal power-flow disruption minimize the change of the power-flow pattern within the system following the system splitting.

The difference between power imbalance and power-flow disruption is that the power imbalance can be expressed by the algebraic sum of active power (considering the direction of power flow) on each disconnected transmission line, while the power-flow disruption can be expressed by the arithmetical sum of active power on each disconnected transmission line. Finding a solution with minimal power imbalance belongs to a class of so-called np-hard problems, which are computationally exceptionally demanding. This is the reason for designing new approaches which are practical and particularly capable of coping with large power networks and a requirement for real-time decision making. Most existing algorithms overcome this challenge by using heuristic search methods or by solving the problem by creating small-dimension network equivalents of the original large-size network. Heuristic search methods are usually quite flexible and have satisfactory computational efficiency. However, the solution quality cannot be guaranteed since these methods tend to converge to local, rather than global, solutions. Consequently, the islanding solution might be not effective enough.

In Section 11.5.2, the two-step spectral clustering controlled islanding (SCCI) algorithm will be presented and used to demonstrate its efficacy [16]. In the first step of the SCCI algorithm, the generator nodes are grouped using normalized spectral clustering. The results of this grouping serve as pair-wise constraints in the next step of the SCCI algorithm, in which every node is grouped based on constrained spectral clustering. This constrained spectral clustering uses power-flow data to produce an islanding solution with minimal power-flow disruption. Therefore, the two-step SCCI algorithm proposed here can identify, in real time, an islanding solution that has a minimal power-flow disruption and satisfies the constraint of generator coherency.

11.5.2 Controlled Islanding Problem

The problem of finding coherent generator groups is equivalent to an optimization problem of finding the weakest dynamic coupling between different generator groups, as shown in (11.1) [16].

$$\min S = \underset{V_{G1},V_{G2} \subset V_G}{\min} \left(\sum_{j \in V_{G2}} \sum_{j \in V_{G1}} \left(\frac{\partial P_{ij}}{\partial \delta_{ij}} \cdot \left(\frac{1}{H_i} + \frac{1}{H_j} \right) \right) \right) \quad (11.1)$$

Minimal power imbalance and minimal power-flow disruption, defined according to (11.2) and (11.3), respectively, can both be used as objective functions of controlled islanding. Each objective will produce a different solution with different advantages and disadvantages.

$$\min_{V_1, V_2 \subset V} \left(\left| \sum_{i \in V_1, j \in V_2} P_{ij} \right| \right) \tag{11.2}$$

$$\min_{V_1, V_2 \subset V} \left(\sum_{i \in V_1, j \in V_2} \left| P_{ij} \right| \right) \tag{11.3}$$

where P_{ij} denotes the P_{ij} value of the active power on the transmission line between node i and j.

The controlled islanding problem that is solved in this section consists of the minimal power-flow disruption objective function (11.3) and the generator coherency constraint (11.1). These two optimization problems are combined together to form the final SCCI algorithm expressed through (11.4). This is done by first solving (11.1), to find the set of coherent generator groups, and second by solving (11.2) subject to these generator groups [16].

$$\left[V_{G1}^*, V_{G3}^* \right] = \underset{V_{G1}, V_{G2} \subset V_G}{\arg\min} \left(\sum_{j \in V_{G2}} \sum_{i \in V_{G1}} \left(\frac{\partial P_{ij}}{\partial \delta_{ij}} \cdot \left(\frac{1}{H_i} + \frac{1}{H_j} \right) \right) \right) \tag{11.4}$$

$$\min_{V_1, V_2 \subset V} \left(\sum_{i \in V_1, j \in V_2} \left| P_{ij} \right| \right) \text{ subject to } V_{G1}^* \subset V_1, V_{G2}^* \subset V_2$$

Here, argmin stands for the argument of the minimum, i.e., is the node grouping that minimizes the objective function of (11.1). Spectral clustering is the tool used to solve the islanding problem. In the next section, the two-step SCCI algorithm will be presented.

The two-step SCCI algorithm can solve the optimization problem expressed in (11.4). Its solution is equivalent to determining a suitable islanding solution.

This solution can be found by constructing two graphs, based on the objective function and constraint from (11.4), and applying the SCCI algorithm to find the minimum cut of these two graphs.

In the first step of the SCCI algorithm, the dynamic graph is constructed. It only contains the generator nodes, and the edge weights of this graph are the synchronizing coefficients that describe the dynamic coupling between the nodes i and j. To satisfy the generator coherency constraint (11.1), the generator nodes are grouped using the normalized spectral clustering algorithm. These groups of generator nodes then serve as constraints for the second step of the SCCI algorithm.

In the second step of the algorithm, the static graph is constructed using power-flow data. It contains every node, and the edge weights are defined as the absolute value of the active power exchange between nodes i and j. The nodes are then grouped using constrained spectral clustering, which will be described in this section, to solve the optimization problem described in (11.4).

In Figure 11.10 a flowchart depicting the execution of the SCCI algorithm is presented [16].

Obviously, a prerequisite for a successful algorithm execution is the information about the system state. This can be achieved using PMU-based linear state estimators, which can be executed practically in real time. For this purpose, approximately 30% of nodes of the transmission grid should be covered by PMUs.

To demonstrate the advantages and efficacy of the proposed SCCI algorithm, the IEEE 118-bus test system was used.

After executing the first algorithm step, the following three coherent groups of generators are obtained (see Table 11.1):

After executing the second algorithm step, the splitting solution presented in Figure 11.11 was obtained. The two cutsets produced in the second step of the SCCI algorithm separated Group 1 from Groups 2 and 3 and then separated Group 2 from Group 3, respectively. Combined, these two cutsets form the final islanding solution marked in the figure.

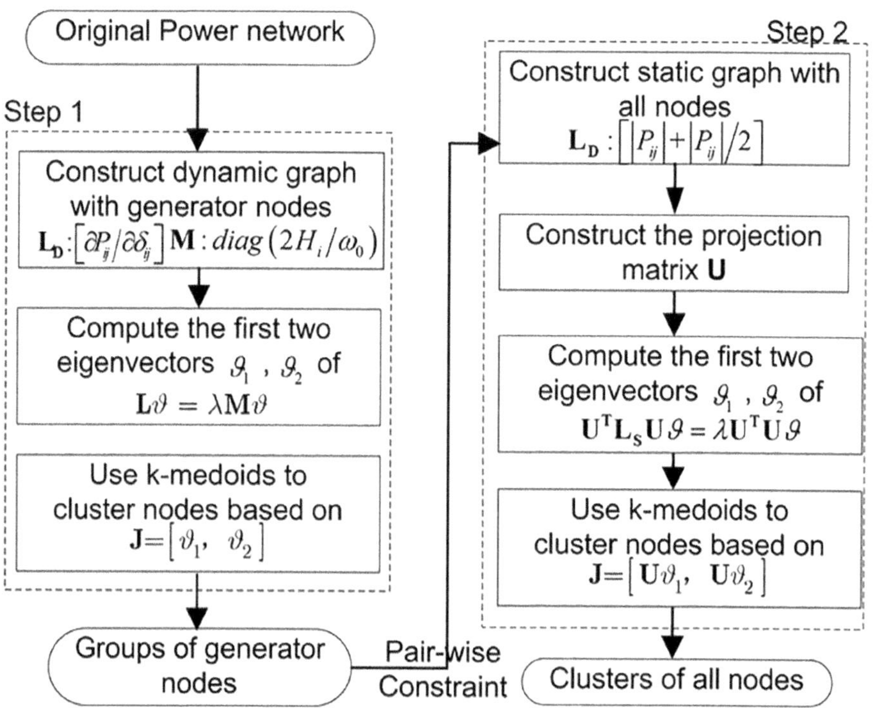

FIGURE 11.10 Flowchart of the spectral clustering controlled islanding algorithm.

TABLE 11.1
Generator Groups for the Assessed IEEE 118-Bus Test System

Group 1	Group 2	Group 3
10,12,25,26,31	46,49,54,59,61,65,66,69,80	87,89,100,103,111

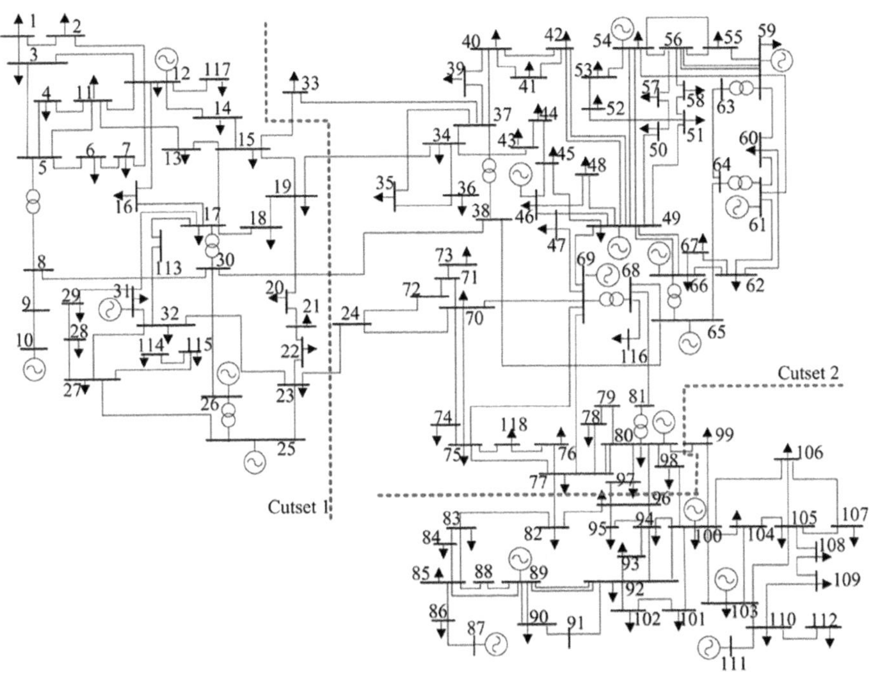

FIGURE 11.11 Single-line diagram of the IEEE 118-bus test network and the islanding solution.

As it was shown by Ding et al. [16], this method outperforms other methods aimed to be applied for controlled islanding. The method relies on understanding the system state and fast communication infrastructure needed to send centralized commands from the electricity smart meters (ESM) to circuit breakers through which the islanding is happening.

The above islanding scheme is obviously quite advanced, and it can be also extended to approaches focused on voltage stability. For this purpose, real-time information about reactive powers in the grid would be critical [17]. On the contrary, information about the power flow in the grid with, e.g., electric vehicles can be equally important to be known [18]. From the perspective of the understanding of the grid parameters measurement-based transmission line parameter estimation methods (e.g. Li et al. [19]) can improve the confidence in the system data and consequently protection and control schemes supporting the system resilience.

The proposed control and protection schemes must be properly tested using, e.g., hardware in the loop-testing facilities [20]. Furthermore, the presented islanding methodology should also involve the assessment of the instant/moment when the islanding should occur. Research on this topic is addressed by Ding et al. [21]. Next, the islanding has to be effective in low inertia power systems, in which the quality of the system dynamics is different and the application of islanding schemes even more challenging [22]. Last but not least, since the entire scheme relies on real-time data from the grid and commands to circuit breakers responsible for network splitting are sent over communication infrastructure (ICT), the security of this infrastructure must be extremely high and immune to potential cyber-attacks.

11.6 CONCLUSION

The chapter discusses the challenges with future power networks in which the penetration level of CIG and RESs is permanently increased so that networks become weak and prone to different types of instabilities. Digitalization and advanced technologies, based on novel sensor and communication infrastructure, can support solutions for the prevention of cascading events and power system blackouts. Discussing the structure of typical SIPSs, it was concluded that modern situational awareness tools, integrated into modern EMSs, can enable the next generation of SIPSs capable of preventing power system blackouts. The chapter also describes a two-step SCCI algorithm for determining islanding solutions for power systems. At the core of this algorithm is a single optimization problem that uses the minimal power-flow disruption as an objective function and considers ensuring generator coherency as a constraint. A more secure communication infrastructure (ICT) will enable practical solutions leading to advanced wide-area protection solutions. These will support even higher deployment of RESs and clean energy targets.

REFERENCES

1. M. Albrecht."Konzeptionierung und Erprobung von dezentralen Frequenzhaltungsmaßnahmen und leistungsflussorientierten Lastabwurfverfahren im Verteilnetz." (2021).
2. https://ourworldindata.org/renewable-energy (available online)
3. T. Bendong, J. Zhao, M. Netto, V. Krishnan, V. Terzija, and Y. Zhang. "Power system inertia estimation: Review of methods and the impacts of converter-interfaced generations." *International Journal of Electrical Power & Energy Systems* 134 (2022): 107362.
4. J. Zhao, Y. Tang, and V. Terzija. "Robust online estimation of power system center of inertia frequency." *IEEE Transactions on Power Systems* 34, no. 1 (2018): 821–825.
5. M. Sun, Y. Feng, P. Wall, S. Azizi, J. Yu, and V. Terzija. "On-line power system inertia calculation using wide area measurements." *International Journal of Electrical Power & Energy Systems* 109 (2019): 325–331.
6. P. C. Taylor, M. Abeysekera, Y. Bian, D. Ćetenović, M. Deakin, A. Ehsan, V. Levi, et al. "An interdisciplinary research perspective on the future of multi-vector energy networks." *International Journal of Electrical Power & Energy Systems* 135 (2022): 107492.

7. A. Stankovic. *"The definition and quantification of resilience."* IEEE PES Industry Technical Support Task Force: Piscataway, NJ, USA (2018): 1–4.
8. M. Panteli, and P. Mancarella. "The grid: Stronger, bigger, smarter? Presenting a conceptual framework of power system resilience." *IEEE Power and Energy Magazine* 13, no. 3 (2015): 58–66.
9. Z. Bie, Y. Lin, G. Li, and F. Li. "Battling the extreme: A study on the power system resilience." *Proceedings of the IEEE* 105, no. 7 (2017): 1253–1266.
10. J. Wang, P. Pinson, S. Chatzivasileiadis, M. Panteli, G. Strbac, and V. Terzija. "On machine learning-based techniques for future sustainable and resilient energy systems." *IEEE Transactions on Sustainable Energy* 14, no. 2 (2022): 1230–1243.
11. FITNESS project final report, *Network Innovation and Competition Scheme*, Ofgem, https://www.ofgem.gov.uk/sites/default/files/docs/fitness_submission.pdf
12. S. Clark, D. Wilson, N. Al-Ashwal, F. Macleod, P. Mohapatra, J. Yu, P. Wall, et al. "Addressing emerging network management needs with enhanced WAMS in the GB VISOR project." In *2016 Power Systems Computation Conference (PSCC)*, Genoa, Italy, pp. 1–7. IEEE, 2016.
13. P. Wall, N. Shams, V. Terzija, V. Hamidi, C. Grant, D. Wilson, S. Norris, et al. "Smart frequency control for the future GB power system." In *2016 IEEE PES Innovative Smart Grid Technologies Conference Europe (ISGT-Europe)*, Ljubljana, Slovenia, pp. 1–6. IEEE, 2016.
14. V. Terzija, G. Valverde, D. Cai, P. Regulski, V. Madani, J. Fitch, S. Skok, M. M. Begovic, and A. Phadke. "Wide-area monitoring, protection, and control of future electric power networks." *Proceedings of the IEEE* 99, no. 1 (2010): 80–93.
15. V. Terzija, "Adaptive underfrequency load shedding based on the magnitude of the disturbance estimation", *IEEE Transactions on Power Systems* 21, no. 3 (2006): 1260–1266.
16. L. Ding, F. M. Gonzalez-Longatt, P. Wall, and V. Terzija. "Two-step spectral clustering controlled islanding algorithm." *IEEE Transactions on Power Systems* 28, no. 1 (2012): 75–84.
17. J. Dragosavac, Z. Janda, J. V. Milanovic, D. Arnautovic, and B. Radojicic. "On-line estimation of available generator reactive power for network voltage support." In *8th Mediterranean Conference on Power Generation, Transmission, Distribution and Energy Conversion (MEDPOWER 2012)*, Cagliari, IET, 2012.
18. R. Azizipanah-Abarghooee, V. Terzija, F. Golestaneh, and A. Roosta. "Multiobjective dynamic optimal power flow considering fuzzy-based smart utilization of mobile electric vehicles." *IEEE Transactions on Industrial Informatics* 12, no. 2 (2016): 503–514.
19. C. Li, Y. Zhang, H. Zhang, Q. Wu, and V. Terzija. "Measurement-based transmission line parameter estimation with adaptive data selection scheme." *IEEE Transactions on Smart Grid* 9, no. 6 (2017): 5764–5773.
20. J. Shair, X. Xie, Y. Li, and V. Terzija. "Hardware-in-the-loop and field validation of a rotor-side subsynchronous damping controller for a series compensated DFIG system." *IEEE Transactions on Power Delivery* 36, no. 2 (2020): 698–709.
21. L. Ding, Y. Guo, P. Wall, K. Sun, and V. Terzija. "Identifying the timing of controlled islanding using a controlling UEP based method." *IEEE Transactions on Power Systems* 33, no. 6 (2018): 5913–5922.
22. L. Ding, Z. Ma, P. Wall, and V. Terzija. "Graph spectra based controlled islanding for low inertia power systems." *IEEE Transactions on Power Delivery* 32, no. 1 (2016): 302–309.

12 Blockchain Technology Integration in Industry 4.0

Challenges and Potential Solutions

Zoran S. Bojkovic and Dragorad A. Milovanovic

12.1 INTRODUCTION

This chapter aims to inform readers of the latest developments in the field of blockchain technology (BCT) and the potential for future blockchain (BC) integration into Industry 4.0. Under these circumstances, BC can be a useful basis for transaction-based applications and interaction, and it can assist in resolving the majority of security flaws and tracing capabilities. To facilitate smooth data sharing, BC can serve as a single platform for device interoperability and communication [1]. BC has found successful use because of its decentralized trust-building capabilities in various industries. It should be mentioned that this could change how we handle and protect data. Additionally, it has established itself in developing various new complex decentralized software systems. Key characteristics and applications of BC are as follows:

- **Decentralization:** Operates without a central authority, while transactions are verified by a network of nodes. The absence of a central point of failure, combined with hashed blocks and transaction ledgers, keeps data safe, secure and reliable over the network.
- **Immutability:** BC's core data is based on a chain of immutable blocks that contain important data. All transactions that take place in the BC are verified with the consensus mechanisms. Data added to the BC cannot be altered or deleted, while each transaction is cryptographically linked to the previous one forming a chain. An attacker would have to attempt to compromise all nodes without the network.
- **Security:** BC places a strong emphasis on robust security measures and enhanced privacy features. It explores advanced cryptographic techniques, zero-knowledge proofs and privacy-preserving protocols to ensure the confidentiality of sensitive data, while maintaining transparency and integrity of the BC. BC is designed to tolerate malicious behavior as long as certain thresholds of participants are correct.

DOI: 10.1201/9781003511298-12

- **Supply Chain Management:** Enables transparent and traceable supply chains.
- **Crypto Currencies:** Underlying technology for Bitcoin and Ethereum.

The BCT helps in moderating the solution for secure and trusted networks to keep information in the form of blocks. A BC consists of a chain of blocks. They contain details of stored transactions within the network. A header and a body belong to each of the previous blocks. The header includes the identifier of the previous block, while the body contains stored transactions. BC is a type of distributed ledger technology (DLT) which was introduced in 2008. The technology has undergone several significant transformations, with each generation introducing new capabilities, while addressing the limitations of its predecessors [2].

- Blockchain 1.0 is the seminal BC attached to cryptocurrency applications. The initial phase was primarily associated with the creation of the first cryptovalute Bitcoin in 2009 and other digital currencies together with decentralized peer-to-peer (P2P) electronic cash systems. It means that BC 1.0 focused on enabling secure and transparent transactions without the need for intermediaries. However, transactions were allowed to be recorded on only a BC, a type of public ledger, and they were restricted to financial transactions.
- Blockchain 2.0 generation started with Ethereum back in 2013. The first generation's inefficient mining and limited scalability led to the idea of extending the use of BCT beyond money. As a result, second-generation BC 2.0 emerged, built on novel ideas for smart contracts and proof-of-work consensus techniques. The 2015 debut of the digital currency Ethereum played a crucial part in popularizing this idea. This makes it possible to automate and carry out the formation of agreements without the need for middlemen. Additionally, it made it easier for developers to create a variety of decentralized apps, including supply chain management (SCM), identity verification and finance, on top of BC platforms.
- Blockchain 3.0 is an evolution of the technology into more aspects of social life with improved performance, interoperability and intraoperability. Scalability has been a persistent challenge for BCT, while interoperability solutions enabled different BCs to communicate and share data securely. BC is mostly based on proof of work and takes hours to confirm transactions. Blockchain 3.0 utilizes Proof of Stake and Proof of Authority consensus mechanisms to enable enhanced speed and computing power for smart contracts with no separate transaction fees. It hence eliminates the dependency on miners to verify and authenticate transactions and instead uses inbuilt mechanisms. They are thus extremely fast to allow thousands of transactions per second. However, the third generation of BC also has several disadvantages like bug fixing or updating due to their decentralized nature. The consensus mechanisms applied are comparatively complicated.

- Blockchain 4.0 makes it possible for many platforms to be seamlessly integrated and function cohesively under one roof to meet the needs of business and industry. It might also spread to other thriving technologies, including artificial intelligence (AI). As a technological advancement, it surpasses the capacities of earlier BC generations by integrating essential features including scalability, interoperability, security and related consensus processes.

The last two generations of BC are still in their infancy and undergoing several modifications. The usage of a particular generation thus depends on the application domain it has to sustain with its corresponding consensus models and respective parameter specifications [3–5].

From industrial application perspectives, smart factories, smart products, supply chains and smart solutions are some of the quality drivers and enablers that have been applied to develop BCT for their specific services. Industry 4.0 is the first great transformation that has progressed from a technology focused to an advanced one and which relies on different cornerstones including the openness of knowledge, assistance and interconnections. Industry 4.0 technologies provide a critical perspective for future innovation and business growth.

Potential solutions are based on advanced digital services and technological advancement with the fusion of 5G (mobile network), Internet of Things (IoT), AI, cloud computing (CC) and BC. Each of these technologies is coming forward with new engineering challenges in understanding the requirements.

12.2 INTEGRATION OF ADVANCED TECHNOLOGIES

Technologies such as IoT, AI, big data (BD) and other advanced upcoming technologies are being used to implement Industry 4.0. The area of study that deals with the extraction of information and draws useful insights from IoT data using various algorithms and scientific methods is called data science. To ensure security requirements and enhance industrial IoT (IIoT) development, BCT is used as a decentralized and distributed approach. Also, it is used as an immutable storage system. The major challenges that restricted the use of BCT for IoT are scalability and transaction fees. In smart industries, connected manufacturing processes depend on different machines that interact using AI automation systems by capturing and interpreting all data types. AI provides appropriate information to take decision making and alert people of possible malfunctions. Industries will use AI to process data transmitted from IoT devices and connected machines based on their desire to track entire end-to-end activities [6].

Industry 4.0 is used to point out the business and chain manufacturing development. To improve security, privacy and data transparency for small and large enterprises, BC can be incorporated. Not only is it a decentralized technology used to share, replicate and synchronize data across different geographical locations, but trusted transactions in any untrustworthy environment are guaranteed. Numerous advantages of BC such as security, trust open source, traceability, transparency and many more make it very popular to apply in different Industry 4.0 sectors. Various industries (healthcare, agriculture, supply chain and logistics, business, tourism, hospitality) together with their various operations have completed digital transactions in Industry 4.0. As core functionalities of Industry 4.0, SCM and demand-driven supply chain are considered to be a core functionality. On the contrary, the existing security mechanisms are either based on centralized architecture or consume high computation power. To resolve this issue BCT has been proven to be the best solution that depicts the new business aspects.

12.2.1 Blockchain Technology

BC is a P2P distributed ledger including an ever-growing list of blocks to enable non-confident parties to maintain a set of global identical states. Each block is generally composed of two parts: block header and block body. Block header constrains and identifier of the previous block. On the other hand, the block body stores a series of encrypted data. All data are spread and verified across the networks in a P2P manner. The valid data will be added into blocks and attached to the BC with distributive consensus, which exploits the whole storage/computing power of networks to guarantee the immutability of this data. BC leverages the integration of advanced technologies such as IoT, AI and BD. The integration enables the development of autonomous systems, smart cities (SCs) and machine-to-machine transactions, expanding to potential applications and impact of BCT. Each of these technologies is coming forward with new engineering challenges, e.g. understanding the requirements related to trust and ethics, system resilience toward unpredictable market behavior or development of ultra-long-term software systems. BC enables technological transformation by enhancing digitalization aspects from a software enabler standpoint. BCT helps in moderating the solution for secure and trusted networks, to keep information along with distributed networks in the form of blocks.

BC has varying benefits such as security, privacy, trust and transparency. The data and information are communicated among the users in a secure network using hash functions. Hash is a type of signature which uses cryptographic primitives and is immutable. BCT that maximizes full potential is transparency and validation information, consensus, robustness and incorruptibility.

A feature that makes BCT robust is the lack of a centralized approach. It means that no single point of failure can disrupt the normal functionality. It implies that regular functionality cannot be interfered with by a single point of failure. Since information is dispersed throughout the network, it is nearly impossible to tamper with or change it. Massive processing power is required to override the entire network in the event that hackers try to change the data. The process of coming to an agreement is referred to as consensus, and it is necessary for any legitimate transaction between parties to begin. With the participation of every node, transparency is attained and maintained, leading to an increase in data integrity. In a BC, digital data is dispersed throughout the network to assist participants in verifying the data without requiring any centralized authority. On the other side, any change in the information can be viewed by all users. Hence, the transparency of the information available in the BC is achieved. Concerns over possible tampering are eliminated.

In a technology environment, the BC is used as a permanent distributed directory to record all value transactions. This enables decentralization and confidence in the transaction ecosystem. Participants initially reviewed all transactions connected to the BC. This technology forms the database for transaction information to be stored in a decentralized and transparent manner. The database is run by a network of computers (nodes), so there is no single point of failure, and information can be accessed in real time [7].

There are a few issues in BCT. Namely, there is a growing need to address these issues and challenges to optimally utilize the capacity of the technology across different sectors. Some of the issues to be mentioned are the size of operations, security and regulation, cost of environment, complexity and slow speed.

- For any addition to the chain of blocks, all the previous blocks need to be validated and require the consensus of the network. All such data require huge storage capacities and pose a challenge to the scalability of BC, which has been a persistent challenge for BCT. As the number of transactions increases, the scalability of BC networks becomes crucial.
- Although BC is a very secure technology, it is not completely free from attacks. The attacker may manipulate the entire network [8]. Public and private keys dictate the security of BC. Public BCs also called permission-less BC are decentralized. Anyone is allowed to maintain a replica of this BC, is allowed to maintain a replica of this BC and is also engaged in validating new blocks. For example, Bitcoin and Ethereum are public or open BC.
- BCT requires huge computational powers which results in huge energy consumption. A huge requirement for energy poses a lot of challenges to the environment. A very complex technology is not easily understandable by anyone because BC is essentially based on encryption and cryptographic algorithms. In that way, it is difficult to implement in small organizations which do not have enough skills and resources to understand and implement the technology.
- As a distributed technology, BC requires the consensus of the user in the network for any changes. It takes time to obtain the confirmation of the users leading to the show process [9].

12.2.2 BC and Industrial IoT

The increasing amount of data in diversified applications has presented significant security and privacy challenges for IoT as a ubiquitous networking paradigm [10]. By sending data to private BC networks, IoT devices can produce tamper-proof records of shared transactions. Technology and business will have a powerful future when BC and IoT are combined since they improve data security, transparency and trust. With its cutting-edge characteristics like decentralization, BC has emerged as a crucial component of the IoT, where data created by the devices is distributedly stored as blocks [11,12]. BC represents one of the most promising candidate technologies to establish a secure and distributed ecosystem for IoT. Integration and management/

deployment operation with the corresponding challenges of BC-based IoT are summarized as follows:

- layer of integration, energy constraints, supportability and computability and heterogeneity
- failure of recovery, monitoring and audibility, extensibility and flexibility

Since IoT and BC are developed by different communities with distinct design goals, technical issues arise during the attempt to integrate the two systems into one coherent design seamlessly.

It should be noted that many different standards and designs exist for IoT systems. Large-scale practical applications are likely to use multiple IoT devices, possibly from different vendors, at the same time. Also, as most IoT devices are fragile and resource constrained, deploying and operating IoT systems smoothly is an important element to be addressed. The integration of BC and IoT is of mutual interest because BC achieves the desired resilient, distributed P2P network as well as the ability to interact with others in a controlled manner.

IoT is an important carrier of BC which not only makes BC more ubiquitous but also improves the security of IoT. Because IoT has high dynamics, configurations of the consensus in BC are required. For example, an IoT node is usually realized by various applications in different BC. This means that the corresponding IoT node should be switched frequently to cross-consensus in different BCs [13].

BC is treated as a convenient form of distributed P2P encryption storage application. This provides an innovation in networking and computing models. It can widely be used in security and computing models. It can be widely used in security and trust-critical environments, such as finance and industry. In the BC, the transaction party is the entity that records, deposits and stores transaction information. The block packet by a node can be successfully verified by each node and added to the BC. Each block in the BC contains a large amount of transaction information, which is organized in a specific structure.

In fact, it is necessary to implement cross-chain collaboration and interaction among different BC, especially in the era of the Internet of Everything (IoE). At the same time, IoT has been widely used in environment monitoring, intelligent transportation, e-Health and so on. BC enables trustless networks that provide secure P2P transactions in IoT without a trusted intermediary. The secure and unchangeable storage in BC generates the reliability and traceability of the data in IoT. Moreover, BC-based IoT eliminates a single point of failure in the centralized networking structure of IoT. This leads to the position where BCs can be deployed at the edge of the network and enhance the security of IoT. The applications of IoT have high dynamics. On the other side, IoT node is usually neutralized by many various applications, such as transportation control, weather forecast and environmental monitoring. The dynamic management capabilities of software-defined networking (SDN) and network function virtualization (NFV) provide the possibility to reconstruct the implementation architecture in BC, CC and IoT [14–16]. The critical roles that BC plays in the domain of IoT applications are as follows:

- **Providing High Scalability:** The consensus around BC initiates high scalability, allowing latencies to be mitigated [17,18].
- **Preserving Full Data Privacy:** Prevention of connecting transactions to specific individuals. Cryptographic methods and strategies for identity protection in BC [19].
- **Orchestration of Connected IoT Devices:** BC coordinates the administration of small contracts and IoT technology in a particular setting [20].
- **Ensuring Interoperability:** Integration with current systems is guaranteed, together with starting activities with other links, connected to the application on the same chain [21].

12.2.3 BC AND AI

It is well known that BCT has scope for handling huge data records and transactions. Also, it can manage data for huge size and scale of operations. The issues and limitations depend on the application and availability across different sectors. To achieve this objective, the ideal integrating BC with AI has the potential to other several benefits, and, as these technologies continue to evolve, new and innovative applications will continue to emerge.

The two technologies, AI and BC, when integrated provide a robust platform for secure communication among the users in the network system. The two technologies complement each other. Namely, BC helps AI to boost the credibility, integrity and trust of the users, while AI provides the necessary intelligence to BC, especially eliminating the security risks. Achieving the integration of BC and AI in all sectors requires huge efforts in successful implementation [22]. Basically, AI can work on the data stored in BC and identify the patterns which might have security concerns. On the other side, BC will verify the authenticity of such identified records and take necessary actions. The integration of BC and AI helps in addressing various security concerns. As there is no central server, the manipulation of systems is not possible in integration. The risk of hiding is managed in the integration model as the communication system is resistant to manipulation, while the anomalies could be kept hidden in the presence of the advanced cryptographic algorithms built into the integration environment. The integration of BC and AI may offer the following capabilities:

- effective models for capturing and storing huge data
- effective systems for ensuring the authenticity of records and verification processes at the global level
- robust systems to ensure compliance and regulations
- governance of network system in a more transparent manner
- various types of advanced analytics which help in predictions

As previously stated both technologies BC and AI require data, algorithms and modeling to achieve the objectives. Some of the benefits of integration are security and protection, efficiency, trust, management, privacy, storage, computation, authenticity, monetarization of data, automation, transparency, fairness, autonomy and power management. Challenges in the integration of BC and AI are data input concerns, data collaboration, learning transfer, huge complexity, scale of issues and lack of standards.

Data collaboration means that all of the traditional systems need to interact with BC and AI tools, which is of great importance for the required data and information. Additionally, it will take some time to build the necessary infrastructure, which might serve as a foundation for much more complexity in the future. The term *scale issues* refers to the consensus attribute in BC DLT, which creates a significant scaling barrier because the network's consensus process is necessary for the passage of any record, data or transfer. The consensus property guarantees the accuracy and reliability of the data. Naturally, this will cause operations and procedures to move slowly. Due to the fact that records must be kept for every extra block, there are enormous storage needs, which affect the scalability. AI also needs huge data for better analysis and results. Unfortunately, there are no fixed global standards available for BC and AI integration. AI integration is dependent on the data available in BC to produce meaningful outputs. If there is any issue either in the quality or in the quantity, the integration suffers from the change of providing enough quality data. Taking into consideration that both BC and AI technologies are quite complex working with different algorithms and languages, it can be difficult to transfer the output of integration to another machine. Thus, new protocols and algorithms must be built to ensure smooth learning transfer from the integrated framework to other machines.

The synergy that BC and AI create in the digital age is a significant answer to the needs that are arising. Only when users reach an agreement within the network can a new version of a block be added to the current chain, which affects performance and simultaneously erodes user confidence and integrity. To get the optimal answer, there should be an ideal trade-off between these attributes. In the ever-changing technology landscape, laws are also desperately needed. Regarding technologies, there are a few regulations, including AI and BC. Numerous business associations that conduct research on the creation, and uses of BC and AI ought to improve the regulatory framework to bring and ensure more security as well as trust among users. The way ahead consists of making permanent awareness of the benefits of the integration of BC and AI. That is the main reason that these two technologies can be conducted to get optimum integration.

12.2.4 BC and 5G/6G

Driven by recent advances in BCT and the radically expanded capacity of 5G networks, 5G services are anticipated to significantly boost Industry 4.0 and accelerate development. Great strides have been made in defining methodologies, architectures, standards and solutions in the past few years. More intriguingly, BC facilitates worldwide accessibility, maybe lowers transaction costs and delays communication between users while assisting in the establishment of secure communication. However, there are new issues that need to be considered as a result of the extensive use and implementation of these services, including network reliability, decentralization, control, transparency, hazards associated with data interoperability, data immutability and privacy. Moreover, the new requirements might not be adequately addressed by current procedures.

Using BCT to meet 5G network needs for decentralization, transparency, anonymity, immutability, traceability and resilience is a creative option. Existing 5G services and apps that provide spectrum sharing, data sharing, network virtualization, resource management, network slicing, privacy and security may benefit from the use of BCT. But the BC is also more complex because of a number of other issues, including latency and expensive computers. To continue with additional development and implementation, it is critical to investigate the challenges we encounter in this integration.

In addition to facilitating more effective resource sharing, enhancing trusted data interaction, securing access control, protecting privacy and offering tracing, certification and supervision functionalities for 5G and beyond networks, BCT can address a number of issues pertaining to trust and security in communications networks. The idea of trust in relation to communications, which includes human–computer interaction, has been discussed [23–27]. A key component of a 5G heterogeneous infrastructure that enhances efficiency and security across a range of resources is distributed trust. However, a number of trust-related problems that are frequently disregarded in network design stand in the way of this objective. Novel BCT offers a possible remedy. BC, with its decentralized structure, can create collaborative trust. Trust among separate network entities enables efficient resource sharing, trusted data interaction, secure access control and privacy protection for wireless networks.

BCT promises low-cost 5G networks, strong security, seamless interoperability and privacy. The challenges of the current deployment of 5G mobile networks are addressed by a number of field experiments and proof of concept (PoC) projects. The registration and administration of data from several networks and Industry 4.0 is the fundamental testing problem. This technique aims to improve the reliability and transparency of data gathered from various networks. Based on preliminary trials, major network operators are evaluating the use of BC in select applications. On the basis of the substantial study done thus far, it is anticipated that BC will reach its full potential in the near future.

A significant vertical market for 5G applications is manufacturing. The need to oversee automated, real-time production activities that necessitate exact orchestration is great. Imagine a factory of the future where robotics, drones and cars all move wirelessly, sending and receiving data as they enter and exit the building. For certain enterprise-sized operations, this kind of automation necessitates not just extremely dependable low-latency capabilities but also the capacity to oversee up to one million devices per square kilometer.

The next 6G generation of wireless communication systems is expected to address the issues of slow information speed that have become more pressing as 5G networks have grown in the number of data applications [28]. When combined with current candidate schemes, a few critical 6G technologies will ensure the desired quality of experience to achieve ubiquitous wireless connectivity for the IoE, which includes digital smart sectors and the telecom sector [29]. The current industry is being transformed by the IoT into smart infrastructure with 6G data-driven architecture [30].

Due to its decentralization, transparency, limited spectrum resources, intrinsic privacy and security, inadequate interoperability, secrecy and emergence as a smart application domain, BCT has drawn a lot of interest, including industrial IoT and Industry 4.0. The mismatch between the requirements of many data-intensive disruptive IoT applications and 5G network capabilities steered the demand for decentralized BCT-based 6G architecture.

To create an open ecosystem of industrial solutions that can be implemented by various sectors, the ETSI (European Telecommunications Standards Institute) industry specification group (ISG) on permissioned distributed ledger (PDL) analyzes and provides the foundations for the operation of PDLs. This encourages the application of these technologies and, as a result, helps to consolidate the trust and dependability of information technologies supported by global, open telecommunications networks [31]. Permission-less or PDLs relate to the conditions under which a node must be authorized to verify transactions and enter them into the ledger. With Bitcoin serving as a prime example, permission-less ledgers have drawn the greatest public attention; nonetheless, PDLs are more suited to address most of the use cases of interest to industrial and governmental institutions.

ETSI PDL specifications PDL-003 and PDL-012 contain the reference architecture definition. Key concerns like reputation, redaction, immutability and interoperability are addressed in great detail, as are particular applications like SCM and wireless networks. PDL-020 and PDL-025 discuss and specify wireless consensus in important Wireless IoT automation, while PDL-021 looks into 3GPP use cases as wireless networks become an essential component of PDL implementations. A methodology for technological assessment and demonstration using proofs of concept has been built in addition to textual deliverables. The ISG has completed three successful PoCs to date and has a solid relationship with research activities.

12.3 CHALLENGES AND POTENTIAL SOLUTIONS

Today, in the era of 5G wireless intelligent connectivity has given the basis for Industry 4.0 with technologies such as BC, IoT, AI and CC. BC provides an important role in understanding smart technology and real-time recording of transactions and data. There are a whole series of possibilities as an intelligent network, a higher degree of automation and interparty frictions, thereby reducing costs and accelerating operations due to network flexibility [32,33]. Industry 4.0 represents a kind of transformation that has advanced from different technologies including knowledge, assistance and interconnections. BCT has begun to hit high on the consumer and has quality influenced the production sector. Thus, there has been much development in the industry, which will involve an improved degree of privacy [34,35].

The BC is entirely owned by the organization, which frequently makes it superior to other data storage methods. The ledger would be accessible to any node taking part in the transaction via a variety of devices. Decentralization and network-wide autonomy are therefore necessary to build the transition ecosystem's credibility. Participants review every transaction related to the BC. A type of database that stores transaction data in an open, decentralized fashion already exists. Companies are more interested in decentralization, disintermediation and the production of commodities on demand than in software, algorithms and automation robotics. Industry 4.0 will be stimulated by the integration of AI and BC, which will have a substantial impact on many elements that solve a variety of issues in a modern industrial environment [36,37]. The transition to Industry 4.0 with BC has brought a range of technologies. To minimize energy consumption, streamlined manufacturing processes are detected.

A wide range of health and education services are available, as well as government agencies, logistics and transportation. The sub-spheres that facilitate and enable BC capabilities for usage in Industry 4.0, specifically, are also very important since they offer data privacy, efficient healthcare, simple data maintenance and time and money savings. The primary concepts are practicing Industry 4.0 deployment in many sectors more precisely. When examining BC's function in Industry 4.0's IoT integration, it's important to remember that IoT stands for the Internet of Things, which is a network of physically connected objects (such as actuators, sensors and cars) that simultaneously record transactions and gather data via the Internet. A comparison of industrial equipment using BC and IoT is shown in Table 12.1. The BC can provide a service layer that will be included in the IoT framework. The BC agent is given access to a service under the BC that interacts with and collects data from sensors in the IoT. The primary purpose of a BC is to enable and verify transactions in a distributed, P2P network. Enhancements to privacy and security, as well as the IoT system interoperability, are complementary to BC technologies and their integration in Industry 4.0.

12.3.1 BC in Industry 4.0

Numerous academic and industrial BC-based initiatives exist to support various IoT applications, including BC virtualization, network security, scalability and computation offloading. IoT security can be improved using a few robust sensor nodes and

TABLE 12.1
Comparison of BC and IoT for Industrial Equipment

Items	IoT	BC
Privacy	Lack of privacy	Ensure the privacy of the participating node
Bandwidth	Many IoT devices have low resource and bandwidth capacities	Utilization of a significant amount of bandwidth
System structure	Centralized	decentralized
Scalability	The number of connected objects is very high	Poor scalability on a large device
Resources	Restriction of resources	Consuming resources
Latency	Low latency	time is a major factor in block mining
Security	A major concern is security	Enhanced safety features

networking interface modules/nodes with BC deployment at the network's edge. Technology applications are quite dynamic [38,39]. Furthermore, a variety of applications, including weather forecasting, environment monitoring and transportation control, typically centralize an IoT node. For instance, the smart factory's operation and training are smoothly integrated with the trading process. Low-complexity consensus techniques are used in smart factories to deliver industrial services with low latency. To maintain transaction trust in Industry 4.0 trade systems, BC apps usually empty the proof-of-work consensus process. An Industry 4.0 node cannot be realized in a smart factory or effectively integrated into trading if it is unable to switch between the aforementioned multiple consensuses. Similar to this, security surveillance and intelligent transportation systems can both make use of a roadside camera [40]. SDN and NFV have the potential to reconstruct the BC implementation architecture in the IoT due to their dynamic management capabilities [41].

The decentralized and transparent nature of BCT renders it a powerful instrument for tackling issues related to Industry 4.0. Two powerful technologies are IoT and BC. Because of this, when coupled, they may achieve security and trust in a variety of applications, making them useful for a wide range of use cases and industries. Modern digital services are made possible by the concept of intelligent connectivity, which combines 5G, IoT, AI, CC and BC. This accelerates technological advancement. Thus, intelligent connection is defined as the final intersection of BC, AI and IoT, where the intersection denotes a cooperative effort between the participating technologies working together [42]. This is supported by the gathering of copious amounts of excellent input, which enables customized goods and services to be developed to increase the number of clients and target real-world issues [43].

The integration of BC and IoT technologies in Industry 4.0 has led to robust distributed applications including SCs, healthcare, education and transportation; BC should be consolidated with 5G and IA to tackle the challenges associated with digital transformation in Industry 4.0.

The use of emerging technologies has been impacting BC integration in Industry 4.0. DLTs such as BC have emerged as a digital innovation which provides new business chances. On the contrary, standards for DLT offer coordination between different DLT platforms and external digital platforms. Unfortunately, although considerable progress on the interoperability of DLT has been made, legacy systems and public and private BCs cannot seamlessly communicate with each other. Current solutions are not standardized and do not support the possibility of seamlessly transmitting data [44].

At the moment, BC is the most widely used DLT used in sustainable SCs. Like a long list of data records, data transactions of records in BC are kept within the ledger in the form of a BC of blocks. In BC, the network's active node authenticates a new transaction as soon as it occurs. Following authentication, each transaction is given a unique hash ID, which is subsequently recorded in the distributed ledger along with the new transaction. Note that every transaction block in the distributed ledger is distinguished and synchronized by a unique ID included in each block of a BC. This ID is known as the hash. Following the addition of a new data transaction to the ledgers, it cannot be altered or removed [45].

IEEE SA (Standard Association) has been engaging in BC standardization by handling different activities in multiple industrial sectors including Industry 4.0. Also, the future directions committee has endorsed the creation of the IEEE Blockchain Institute (BCI) to be the focal point for all BC activities and projects [46]. In the energy sector, IEEE P2418.5 BC provides a consistent, open and interoperable reference model for BC. It also contained three domains, including Industry 4.0 and related services, that serve as references for BC application cases in the oil and gas, electrical power and renewable energy sectors. The overall goal was to design a technology and open protocol layered framework, supply technology and give system interfaces for BC applications within the energy sector, all based on the reference architecture for interoperability. In conclusion, IEEE P2418.5 assists in assessing and offering standards for interoperability, smart contract evaluation, the consensus algorithm and several BC deployment models for the energy sector.

ISO/TC307 international technical committee has found in 2021 various advisory groups, study groups and working groups (WGs) to promote BC standardization in interoperability, security, use cases, privacy, smart contract and offer directions. Besides working on development standards, for example, the TC68, which comprises a technical committee for financial services, and TC46, which involves a technical committee for information and documentation, there exists JTC1, which is a technical committee for information technology. WGs such as the SG6 in the future aim to provide a governance guide research report for BC to exploit the relationship between the strategic deployment of BC (including market, business goals and benefits) and BC users and stakeholders, proposing a reference model for system lifecycle consensus and management. Group SG7 under TC307 focuses on conducting more research on inter-system and inter-chain interoperability solutions, thereby providing a standardization framework for enabling interaction between different digital technologies [47].

A BC community group led by W3C has been established to create message format standards for BC that are based on ISO 20022. The purpose of W3C pertains to BC and includes creating standards for message formats, offering rules for the use of stocks, including side chains and private BC and assessing cutting-edge technology and creative applications, such as interbank communications. W3C offers guidelines for using storage, side chains and content delivery networks, as well as support message format specifications for BC based on ISO20022 [48].

12.3.2 BC in SCs

The world is currently shifting toward highly advanced, technologically reliant and digitalized city planning and administration. SC are urban areas that are technologically advanced, inventive and reliant on technology. By resolving unique problems and enhancing daily operations, a significant portion of managing and building SCs plays a crucial role in both the present and the future. The 5G initiative has grown and approached businesses and industries, facilitating extensive digital commerce. Given the harsh living conditions for city dwellers, this reality has given rise to the idea of innovative SC architecture [49]. The ubiquitous presence of IoT devices, diversified network systems, scalable data storage capacities and the deployment of more specialized services form the basis of SCs. As a result, local governments, businesses and private entities will have a continuous view of their properties and manage access.

It is advantageous to construct distributed edge computing systems to process data on a big scale [50]. From a security perspective, BC makes sense for creating decentralized security mechanisms that connect edge devices, IoT devices and city dwellers. This would allow for the handling of data processing, exchange and business transactions on the BC platform [51]. In terms of lower latency, power consumption, better customer support, faster user reaction, privacy and security guarantees and more, the use of decentralized BCT offers benefits over centralized designs using a single cloud platform [52].

There are many ways in which BC makes SCs smarter.

Improved Cybersecurity: These days, one of the biggest issues facing SCs is cybersecurity. This challenge has drawn a lot of attention from the cybersecurity research community in recent years. BC seems to give the security and data architecture based on BD and AI to improve SCs concerns cybersecurity, among the several technologies suggested to handle the topic. Effective solutions for cyber-secure SCs can be achieved by designing and executing the corresponding analytics. SCs are becoming more connected as a result of providing more digitalized services, but they also increase the risk of cyberattacks and hazards [53].

The following are use cases of using BC for SCs that help in improving cybersecurity:

- End-to-end encryption, secure communication and authentication made possible due to BC can help cybersecurity with AI and IoT devices.
- The integrity of the updates can be verified using BC to prevent malicious software from being installed.
- BC validation techniques, like cryptography, can be used for identifying theft.

Enhanced Healthcare: SCs are growing more interconnected and vulnerable to cyber-risks and cyberattacks as a result of providing more and more digitalized services [54]. These days, one of the biggest issues facing improved healthcare is cybersecurity. The cybersecurity research community has focused a lot of attention on this problem in recent years. BC seems to provide data architecture and security based on AI, BD and BC to enhance cybersecurity in SCs. To overcome these obstacles and offer an effective solution from cyber-secure SC healthcare, BCT integration is crucial.

The BC has revolutionized a number of sectors, including manufacturing, banking, e-commerce, education and even the healthcare sector. Additionally, cryptographic algorithms are used to preserve data integrity. It is hence the suitable choice for healthcare applications. The immutability of data saved in BC is one of the many BC features that the healthcare sector is interested in. Essential BC features like decentralized management data authenticity, resilience, improved security and—above all—the restoration of character rights serve as the foundation for the evolution. To raise the bar on current practices for medical data concealing, sharing, processing, analysis and outcome-based classification, BC is being used to develop creative and cutting-edge solutions. BC is able to offer healthcare administration that is safe, open and unchangeable. To guarantee accessibility and interoperability, it can establish a single source of truth. It should be mentioned that BC affects stakeholders in the health ecosystems widely. BC provides safe, interoperable medical record management solutions that guarantee data integrity and optimize workflows for the healthcare sector. Authorized healthcare practitioners can securely access patient records housed on the BC, enhancing patient care and facilitating improved coordination among various medical groups.

BCT improves data protection, makes consent management easier and streamlines the processing of insurance claims and bills. Additionally, it can be used to establish smart contracts, which will facilitate and streamline the process of transferring payments from donors to receivers. Wireless healthcare monitoring technologies have been developed in the past ten years. One of the main areas where IoT infrastructure and solutions are extensively employed to support the greatest patient experience, precise diagnosis and prompt treatment of patients with pre-existing disorders is smart healthcare. Nonetheless, a centralized monitoring unit that may be susceptible to single-point failures is frequently in charge of handling diverse difficulties. Hospitals can collaborate by using Global BC. In this instance, patients are not restricted to a particular group of physicians and hospitals in the area. For small issues that require fewer doctor consultations and iteration, the BCs are tailored to the patient's location. Every model that has been put forth in the public domain promises a decentralized structure that makes information sharing and exchange as well as information integration among all users easier.

Due to the ease of management and operation of healthcare facilities, intelligent connections throughout the healthcare sector can help give more effective preventive care at a reduced cost. Additionally, intelligent connection may enable remote diagnosis and treatment. For instance, this could redefine healthcare access, which is currently limited by the geographical locations of medical specialists.

Along with basic health metrics, a key component that 5G technology integrated equipment might measure is the mobility of patients, staff and healthcare specialists throughout the hospital [55,56]. The Internet of Body refers to the partnership between 5G and the healthcare sector. But the most exciting aspect of 5G is its potential for remote operations. Healthcare will be transformed by the tactile Internet since it will make it possible to operate on patients who are physically present somewhere else.

Transportation Management: The idea of a smooth and effective transportation management system that meets citizen needs while reducing environmental impact is at the heart of SC development. Intelligent and sophisticated technologies, like IoT, AI, CC and data-driven solutions, are revolutionizing urban transportation and affecting urban ecosystems in ways that have resulted in delays or waste at any point. Regarding the waste management sector, the desirable characteristics of BCT include trust, decentralization, better security and data privacy. Within the BC network, the accountability feature indicates who is in charge of a specific action. BC improves autonomous vehicle communications, vehicle tracing and logistics through intelligent transportation management. It responds to data interchange in transportation networks that is transparent and safe. Issues including air pollution, traffic jams, poor infrastructure and a poorly run public transportation system have come together to pose serious obstacles. Furthermore, transportation infrastructures are under immense pressure due to the world's rising urbanization. To solve these problems, smart transportation management must support intelligent and sustainable mobility solutions that increase accessibility, shorten commute times and enhance overall quality of life.

Real-time data is used by intelligent transportation management systems to optimize traffic flow, cutting down on traffic and trip times. Increased productivity and economic gains result from this efficiency. Greenhouse gas emissions and carbon footprint are considerably decreased by smart transportation management, which also encourages the use of electric vehicles, shared mobility services and redundancy in traffic congestion. A smooth and multimodal travel experience is delivered by smart transportation, which combines multiple forms of transportation. Computers can seamlessly transition between shared mobility services, personal vehicles and public trust, hence decreasing reliance on private automobiles. Last but not least, intelligent transportation management systems provide enormous amounts of data that can be examined to learn important lessons about transportation trends, optimize outcomes and schedule infrastructure development, all of which help decision-makers make well-informed choices. Road transportation management has greatly improved and more efficient if every vehicle and individual in the environment can seamlessly communicate among themselves. Furthermore, traffic management system integration in the 5G area has the potential to significantly worsen driving conditions by providing vehicles with real-time updates on when to slow down [56]. Thus, 5G's reader and computer vision features will enhance and streamline traffic systems even more. Reliability in autonomous vehicles will be facilitated by 5G. The on-board computer chip must be capable of receiving data from other road-related incidents, as these together comprise the 5G network.

Better Waste Management: In SCs, BC can support the upkeep of a hygienic atmosphere. It offers real-time tracking of several elements of improved waste management, including transportation and unchangeable data about the volume of waste collected and the recycling process. Also included is the promotion of cleanliness. Waste can be defined as any type of stuff that the owner no longer wants. Wastes are acknowledged as one of the main issues facing our society, and they ought to be managed by qualified professionals. The significance of waste management services becomes evident only in the presence of technical or societal issues. By selecting the proper approaches, an integrated waste management system provides a complete approach to waste management by choosing the appropriate techniques and systems according to the situation of the waste region.

For effective implementation, there are several criteria involved in transforming an urban area into a smart zone. The three main parameters are people, data and technology. The possibility exists for SCs to address the environmental issue brought on by inappropriate trash disposal. Enhancing human health, preserving the healthy ecology and cutting down on air pollution are the ways to do this. Nonetheless, the technologies, methods and systems used in waste management today are centralized. However, a sizable percentage of the current waste management systems in SCs will include characteristics like audit, security, traceability and operational transparency. Consequently, BCT plays a crucial role in waste management in SCs since it can achieve the required objectives in a decentralized, reliable and secure manner. By safely storing and carrying out transactions, BC, a decentralized technology, can help to secure data and transactions. It uses a P2P architecture to handle and store data in a transparent, safe, dependable and trustworthy manner [57,58].

Energy generation stations are frequently the recipients of industrial waste produced by SCs. Features for tracking and trailing help confirm the legitimacy of information and procedures related to the gathering, handling and transportation of garbage from SCs. An SC that is safe and free of pollution can be achieved by effectively channeling waste material [59,60]. The policies, practices and guidelines that must be adhered to during the collection, transportation, segregation and recycling of trash are outlined in the waste management-related certificates and documentation. The waste disposal recycling form, waste declaration form, hazardous goods shipping document, waste document shipping plan, waste management inspection plan document and so forth are among the important records and certifications that are kept up to date by the waste handling organizations during waste management [61]. The current centralized-based systems' lack of transparency regarding waste management resources and operations can lead to less reliable, ineffective, fault-tolerable and unsafe systems. Therefore, choices about the allocation of resources based on data that may be unstructured or malleable may prove to be ineffective and expensive. The authorities are driven to employ BCT for assurance of compliance actions because of its artificial decentralization, traceability through time-stamped transactions and use of consensus protocols. Penalties, including license cancellation, are enforced on individuals, organizations and businesses that fail to comply with approved waste management standards. Everyday waste collection in SCs involves planning and optimizing truck routes while taking socioeconomic and environmental aspects into account. The planning of the road to gather SCs waste is affected by several parameters such as traffic jams, fuel costs, amount of waste segregation time and improved waste handling efficiency, and presenting an auditable way to issue incentives to the owners of the robots involved in the waste management operations is of great importance. The existing systems responsible for managing waste in SCs are inefficient in holding individuals or organizations.

Simplified Education: Simplifying the educational procedures is one of the best applications of BC for SCs. Educational institutions have a massive amount of student data to manage. In a similar vein, data transfers between several institutions become laborious. The issue of building a centralized, immutable data base can be resolved with the use of BCT. Information from many academic institutions can be readily accessed and shared via the BC network. This will facilitate the involvement of administrative activities. The BCT is being implemented by the education sector in SCs in a variety of potential areas. The key benefits of the many applications of BC in education will be increased security and transparency. Additionally, it enhances processing speed, lowers costs, boosts efficiency, enhances traceability, increases efficiency, reduces costs and improves processing speed.

One of the most important inventions, the BCT, is well known for its ability to provide SCM security. Industry 4.0's education sector in SCs must also take advantage of BCT's advantages. Educational institutions are currently thinking about using this application to enhance student–teacher cooperation. Moreover, cloud storage identity management, digital degrees and certifications and e-transcripts will all be used. The educational sectors can make the most of BC's technological applications. It streamlines verification procedures and improves the integrity of educational data [62]. This will enable academic institutions to evaluate and determine which BC applications for Industry 4.0 SC education will be advantageous for their particular setups. Additional study on advancement is concentrated on the application of BC with the confidential documents.

Mixed reality (XR) in learning environments has the advantage of promoting a deeper intuitive understanding of the subject matter. Using visors, sensors and headsets, teachers can use a variety of hypothetical scenarios to convey information to students in a virtual setting, adding realism to the learning process [63].

Advancements in education and on-the-job training that would have sounded exciting a few decades ago are now possible thanks to 5G. For instance, recent studies on the relationship between digital technology and education are producing exciting developments in both classroom instruction and on-the-job training. The integration of digital technologies into education has resulted in revealing discoveries. Additionally, 5G will assist employees in various Industry 4.0 industries in becoming experts in their field and expanding their skill set. Applications for mixed reality in the workplace have the potential to boost worker engagement, reduce training costs, facilitate hearing more quickly and easily and offer additional career advancement chances. Industry 4.0 facilitates a more seamless transition from theoretical ideas to real-world applications, such as manufacturing and logistics, with augmented reality and virtual reality technology. It should be mentioned that XR headsets can provide step-by-step instructions in real time and allow lecturers to evaluate practical sessions. Additionally, 5G will make real-time haptic feedback possible for applications where actions are dictated by mobility, with achievable latencies of only 20 ms.

Increased Energy Saving: BCT has the potential to revolutionize a number of industries by providing dependable, economical and efficient substitutes for conventional techniques. Authenticity and dispersed data security are combined with the advancement of renewable energy in the future. P2P energy trade and the encouragement of the production of renewable energy are made possible by the transparent and decentralized energy markets that BC is instrumental in fostering. The ability of BC-based platforms to trace the source and effects of energy services should also be mentioned to guarantee transparency and confirm clean energy. It is imperative to overcome the operational risks associated with smart contrasts to fully realize the benefits of BC in security registration and settlement. Conversely, smart contracts facilitate grid management, optimize energy distribution and automate energy transactions. BC can help integrate renewable energy sources, such as solar and wind power, into the grid by enabling the development of local energy markets. This may facilitate the switch to low-carbon emissions and aid in the reduction of carbon emissions [64]. BC can be applied in a number of ways to reduce energy savings in SCs. A BC-based network, for instance, can be used to track how much energy each citizen uses. To exchange incentives, residents can also swap additional electricity with other participants.

Efficient Mobility: One essential component of sustainable transportation is efficient mobility. It involves transporting passengers and cargo with the least amount of damaging greenhouse gas emissions. Recall that efficiency is all about minimizing waste and loss while attaining travel goals and maximizing freight and cargo handling to generate profits for the transportation industry as a whole. The world's fastest-growing economies are investing in SC transportation infrastructure. Smart mobility is an innovative approach to communications that offers residents several, reasonably priced modes of transportation.

In SCs, BCT has the potential to improve transportation services. It could assist in building a vehicular point-to-point network. A network like this can provide effective vehicle tracking, offer a safe registration platform for both drivers and vehicles and confirm with drivers regarding important changes. For instance, utilizing BCT can improve the sale or recovery process and minimize car thefts by enabling safe vehicle owner data tracking. BC is being used in SC implementation by several nations worldwide. BCT is being adopted by others to improve security, improve healthcare and streamline educational procedures. These days, BC is utilized in every aspect of SC operations.

12.4 CONCLUDING REMARKS

BC and transactions, together with many other non-BC computing and data resources, are components of broader heterogeneous distributed systems. The idea of a BC is compatible with other off-chain transactions related to BD, the cloud continuum, machine learning, IoT and other organizational contexts. Following the appropriate technologies' integration with BC, the advantages become noteworthy and of shared interest. The new technologies offer users a multitude of global production and supply chain opportunities. The best method to integrate Industry 4.0 with emerging technologies is to perform more research, which will require increased efforts in technology integration. Not only does BC integrate with Industry 4.0 in the finance system, but it also offers information.

The goal of Industry 4.0, an industrial revolution, is to increase the automation of business and manufacturing processes by integrating all previously improved technologies. Industry 4.0 encompasses a wide range of applications for BD and analytics, including SCs, energy, SCM, logistics and transportation, security in the context of autonomous vehicles, government services and more. BCT for safe social media computing, when combined with Industry 4.0, addresses a broad spectrum in the smart areas of the next generation of society and trends toward the healthcare potential. As Industry 4.0 progresses toward Industry 5.0, more innovative BC applications should be anticipated.

REFERENCES

1. D.A. Milovanovic, T.P. Fowdur, Z.S. Bojkovic, 5G Multimedia communication and Blockchain technology: emerging potential and challenges, in *Blockchain Technology for Secure Social Media Computing* (R.S. Bhadoria, N. Saxena, B. Nagpal, Eds), IET, 2023. pp. 139–155
2. R. Colomo-Palacios, M. Sánchez-Gordón, D. Arias-Aranda, "A critical review on blockchain assessment initiatives: a technology evolution viewpoint", *Journal of Software: Evolution and Process*, vol. 32, no. 11, pp. 1–11, Wiley, 2020.
3. M.N.M. Bhutta, A.A. Khwaja, A. Nadeem, H.F. Ahmad, M.K. Khan, M.A. Hanif, H. Song, M. Alshamari, Y. Cao, "A survey on blockchain technology: evolution, architecture and security", *IEEE Access,* vol. 9, pp. 61048–61073, 2021.
4. C. Fan, S. Ghaemi, H. Khazaei, P. Musilek, "Performance evaluation of blockchain systems: a systematic survey", *IEEE Access,* vol. 8, pp. 126927–126950, 2020.
5. A.F. Monrat, O. Schelén, K. Andersson, "A survey of blockchain from the perspectives of applications, challenges, and opportunities", *IEEE Access,* vol. 7, pp. 117134–117151, 2019.
6. T.P. Fowdure, L. Babooram, M. Indoonundon, A.P. Murdan, Z. Bojkovic, D. Milovanovic, Enabling technologies and applications of 5G/6G-powered intelligent connectivity, in *Driving 5G Mobile Communications with Artificial Intelligence Towards 6G* (D.A. Milovanovic, Z.S. Bojkovic, T.P. Fowdur, Eds), CRC Press, Taylor & Francis Group, 2023, pp. 355–402.
7. M.S. Ali, M. Vecchio, M. Pincheira, K. Dolui, F. Antonelli, M.H. Rehmani, "Applications of blockchains in the Internet of Things: a comprehensive survey", *IEEE Communications Surveys & Tutorials*, vol. 21, no. 2, pp. 1676–1717, 2019.
8. F. Casino, T.K. Dasaklis, C.A. Patsakisa, "A systematic literature review of blockchain-based applications: current status, classification and open issues", *Telematics and Informatics*, vol. 36, pp. 55–81, Elsevier, 2019.
9. H. Watanabe, S. Fujimura, A. Nakadaira, Y. Miyazaki, A. Akutsu, J.J. Kishigami, "Blockchain contract: a complete consensus using blockchain", *In Proceedings of the IEEE Global Conference on Consumer Electronics (GCCE 2015),* Osaka, pp. 577–578
10. Z. Zheng, S. Xie, H. Dai, X. Chen, H. Wang, "An overview of blockchain technology: architecture, consensus, and future trends", In *Proceedings of the IEEE International Congress on Big Data (BigData Congress 2017),* Honolulu, HI, USA, pp. 557–564
11. L. Tseng, L. Wong, S. Otoum, M. Aloqaily, J.B. Othman, "Blockchain for managing heterogeneous Internet of Things: a perspective architecture", *IEEE Network*, vol. 34, no. 1, pp. 16–23, 2020.
12. W. Baiod, J. Light, A. Mahanti, "Blockchain technology and its applications across multiple domains: a survey", *Journal of International Technology and Information Management*, vol. 29, no. 4, pp. 78–119, 2021.
13. S. Hakak, W.Z. Khan, G.A. Gilkar, M. Imran, N. Guizani, "Securing smart cities through blockchain technology: architecture, requirements and challenges", *IEEE Network*, vol. 34, no. 1, pp. 8–14, 2020.
14. L. Lee, M. Azamfar, J. Singh, "A blockchain enabled cyber-physical system architecture for Industry 4.0 manufacturing systems", *Manufacturing Letters*, vol. 20, pp. 34–39, 2019.
15. W. Lin, H. Huang, H. Yin, G. Min, Y. Yuan, D. Wu, "Scalable blockchain-based data storage in Internet of Things", *IEEE Communication Magazine*, vol. 62, no. 1, pp. 40–45, 2024.
16. R. Tapwal, P.K. Deb, S. Misra, S.K. Pal, "Shadows virtualization for interoperate computations in IIoT environments", *IEEE Transactions Computers*, vol. 72, no. 3, pp. 868–879, 2023.

17. J. Lu, J. Shen, P. Vijayakumar, B.B. Gupta, "A blockchain-based secured data storage protocol for sensors in the Industrial Internet of Things", *IEEE Transactions Industrial Informatics*, vol. 18, no. 8, pp. 1422–1431, 2022.
18. J. Wu, M. Dong, K. Ota, J. Li, W. Yang, "Application aware consensus management for software-defined intelligent blockchain in IoT", *IEEE Network*, vol. 34, no. 1, pp. 69–75, 2020.
19. C. Xu, K. Wang, M. Guo, "Intelligent resource management in blockchain-based cloud data", *IEEE Cloud Computing*, vol. 4, no. 6, pp. 50–59, 2017.
20. S. Luo, M. Dong, K. Ota, J. Wu, J. Li, "A security assessment mechanism for software-defined networking-based mobile networks", *MDPI Sensors*, vol. 15, no. 12, pp. 31843–31858, 2015.
21. J. Wu, M. Dong, K. Ota, J. Li, Z. Guan, "Big Data analysis-based secure cluster management for optimal control plane in software-defined networks", *IEEE Transaction Network and Service Management*, vol. 15, no. 1, pp. 17–38, 2018.
22. F. Wen, L. Yang, W. Cai, P. Zhou, "DP-hybrid: a two layer consensus protocol for high scalability permissioned blockchain", In *Proceedings of the International Conference on Blockchain and Trustworthy Systems,* Dali, China, 2020, pp. 57–71.
23. C. Benzaïd, T. Taleb, M.Z. Farooqi, "Trust in 5G and beyond networks", *IEEE Network*, vol. 35, no. 3, pp. 219–222, 2021.
24. P. Gorla, V. Chamola, V. Hassija, D. Niyato, "Network slicing for 5G with UE state based allocation and blockchain approach", *IEEE Network*, vol. 35, no. 3, pp. 184–190, 2021.
25. Y. Zhang, K. Wang, H. Moustafa, S. Wang, K. Zhang, "Blockchain and AI for beyond 5G networks", *IEEE Network*, vol. 34, no. 6, pp. 22–23, Nov/Dec. 2020.
26. Y. Liu, F.R. Yu, X. Li, H. Ji, V.C.M. Leung, "Blockchain and machine learning for communications and networking systems", *IEEE Communications Surveys & Tutorials*, vol. 22, no. 2, pp. 1392–1431, 2020.
27. A.K.M.B. Haque, M.O.M. Zihad, M.R. Hasan, 5G and Internet of Things—integration trends, opportunities, and future research avenues, in *5G and Beyond* (B. Bhushan, S.K. Sharma, R. Kumar, I. Priyadarshini, Eds), Springer, 2023. pp. 217–245.
28. X. Ling, J. Wang, Y. Le, Z. Ding, X. Gao, "Blockchain radio access network beyond 5G", *IEEE Wireless Communications*, vol. 27, no. 6, pp. 160–1682, 2020.
29. T. Maksymyuk, J. Gazda, M. Volosin, G. Bugar, D. Horvath, M. Klymash, M. Dohler, "Blockchain-empowered framework for decentralized network management in 6G", *IEEE Communications Magazine*, vol. 58, no. 9, pp. 86–92, 2020.
30. A. Jahid, M.H. Alsharif, T.J. Hall, "The convergence of blockchain, IoT and 6G: potential, opportunities, challenges and research roadmap", *Journal of Network and Computer Applications*, vol. 217, p. 103677, 2023.
31. M. Dohler, D.R. Lopez, C. Wang, *Blockchains in 6G: A Standardized Approach to Permissioned Distributed Ledgers*, Routledge, 2024.
32. D. Perez, J. Xu, B. Livshits, "Revisiting transactional statistics of high-scalably blockchains", In *Proceedings of the ACM Internet Measurement Conference,* 2020, New York, NY, pp. 535–550.
33. J.B. Bernate, J.L. Canovas, J.L. Hernandez-Ramos, R.T. Moreno, A. Skarmeta, "Privacy-preserving solutions for blockchain: review and challenges", *IEEE Access*, vol. 7, pp. 164908–164940, 2019.
34. C. Pahl, N.E. Ioini, S. Helmer, B. Lee, "An architecture pattern for trusted orchestraction in IoT edge clouds", In *Proceedings of the International Conference on Fog and Mobile Edge Computing FMEC,* 2018, Barcelona, Spain, pp.63–70
35. B. Pillai, K. Biswas, V. Muthukkumarasamy, "Cross-chain interoperability among blockchain-based systems using transaction", *Knowledge Engineering Review*, vol. 35, p. e23, 2020.

36. Z. Zhang, X. Song, L. Liu, J. Yin, Y. Wang, D. Lan, "Recent advances in blockchain and artificial intelligence integration: feasibility analysis, research issues, applications, challenges and future work", *Security and Communication Networks*, vol. 3, p. 9991535, 2021.
37. S. Khan, R. Singh, and Kirti, "Critical factors for blockchain technology implementation: a supply chain perspective", *Journal of Industrial Integration and Management*, vol. 7, pp. 479–492, 2021.
38. U. Bodkhe, S. Tanwar, K. Parekh, P. Khanpara, S. Tyagi, N. Kumar, M. Alazab, "Blockchain for Industry 4.0: a comprehensive review", *IEEE Access*, vol. 8, pp. 79764–79780, 2020.
39. W. Viriyasitared, L.D. Xu, Z. Bi, A. Sapsomboon, "Blockchain-based business process management (BPM) framework for service competition in Industry 4.0", *Journal Intelligent Manufacturing*, vol. 31, pp. 1737–1748, 2018.
40. T. Kumar, H. Erkki, E. Muneeb, M. Ahsan, P. Pawani, A. Ijaz, L. Madhusanka, B. An, Y. Mika, "Block edge: blockchain-edge framework for industrial IoT networks", *IEEE Access*, vol. 8, pp. 154166–154185, 2020.
41. D. Mihta, S. Tanwar, U. Bodkhe, A. Shukla, N. Kumar, "Blockchain-based royalty contract transactions scheme for Industry 4.0 supply-chain management", *Information Processing Management*, vol. 58, no. 4, p. 102586, 2021.
42. D.V. Lypnytsky, "Oportunities and challenges of blockchain in Industry 4.0", *Economy of Industry*, vol. 1, no. 85, pp. 85–100, 2019.
43. C. Zhang, Y. Chen, "A review of research relevant to the emerging industry trends: Industry 4.0, IoT, blockchain, and business analytics", *Journal of Industrial Integration and Management*, vol. 5, no. 1, pp. 165–180, 2020.
44. R. Belchior, A. Vasconcelos, M. Correia, T. Hardjono, "Hermes: fault-tolerant middleware for blockchain interoperability", *Future Generation Computer Systems*, vol. 129, pp. 236–251, 2022.
45. F. Ezzi, A. Jarbouri, K. Monakhar, "Exploring the relationship between blockchain technology and corporate social responsibility performance: empirical evidence from European firms", *Journal of the Knowledge Economy*, vol. 14, pp. 1–22, 2022.
46. D. Blowmik, F. Temmermans, *JPEG White paper: towards a standardizaed framework for media blockchain and distributed framework for media blockchain and distributed ledger technologies*, Technical Report WGIN84038, 2019.
47. V. Cali, C. Lima, X. Li, Y. Ogushi, "DLT/Blockchain in transactive energy use cases segmentation and standardization framework", In *Proceedings of the IEEE PES Transactive Energy System Conference TESC,* 2019, Minneapolis, MN, pp. 1–5.
48. Z. Li, "Standardization of blockchain and distributed ledger technologies – a legal voice the data protection low perspective", In *Proceedings of the EURAS Annual Standardization Conference – Standards for Digital Transformation: Blockchain and Innovation (The European Academy for Standards),* Aachen, Germany, 2020.
49. K.E. Skouby, P. Lynaggard, "Smart home and smart city solutions enabled by 5G, IoT, AAI and CoT services", In *Proceedings of the International Conference on Contemporary Computing and Informatics IC3I,* 2014, Mysore, India, pp. 874–878.
50. S.Y. Nikonei, R. Xu, D. Nagothu, Y. Chen, A.J. Aved, E. Blasch, "Real-time index authentication for event-oriented surveliance video quality using blockchain", In *Proceedings of the International Conference Smart Cities ISC2,* 2018, Kansas City, MO, pp. 1–8.
51. R. Wang, W.-T. Tsai, J. He, C. Liu, Q. Li, E. Deng, "A video survelieance system based on permissioned blockchains and edge computing", In *Proceedings of the International Conference on Big Data Smart Computing BigComp,* 2019, Kyoto, Japan, pp. 1–6.
52. A. Daminianou, C.M. Angelopoulas, V. Katus, "An architecture for blockchain over edge-enabled IoT for smart circular cities", In *Proceedings of the International Conference on Distributed Computing in Sensor Systems DCOSS,* 2019, Santorini, Greece, pp. 465–472.

53. H. Sharma, E. Podoplelova, G. Shapovalov, A. Tselykh, A. Tselykh, "Sustainable smart cities: convergence of artificial intelligence and blockchain", *MDPI Sustainability*, vol. 13, no. 23, p. 13076. 2021.
54. G. Hameed, Y. Singh, S. Haq, B. Rana, Blockchain-based model for secure IoT communication in smart healthcare, in *Emerging Technologies for Computing, Communication and Smart Cities* (P.K. Singh, M.H. Kolekar, S. Tanwar, S.T. Wierzchon, R.K. Bhatnagar, Eds), Lecture Notes in Electrical Engineering, vol. 875, Springer, April 2022.
55. PWC, *5G in healthcare*, 2020.
56. GSMA, *Intelligent connectivity – haw the combination of 5G, AI, BigData and IoT is set to change everything*, 2022.
57. Y. He, H. Li, X. Cheng, Y. Liu, C. Yang, L. Sun, "A blockchain based truthful incentive mechanism for distributed P2P applications", *IEEE Access*, vol. 6, pp. 27324–27335, 2018.
58. R. Wasim Ahmad, K. Salah, R. Jayaraman, I. Yaqoob, M. Omar, "Blockchain for waste management in smart cities: a survey", *IEEE Access*, vol. 6, pp. 131520–275540, 2021.
59. CPCB, *Guidelines on information of eWaste management rules*, 2021.
60. T.K. Dasaklis, F. Casino, C. Patsakis, "A traceability and auditing framework for electronic equipment reverse logistics based on blockchain: the case of mobile phones", arXiv 11556, 2020.
61. G. Ongena et al., "Blockchain-based smart contract in waste management: silver bullet?", In *Proceedings of the Bled eConference,* 2018, Bled, Slovenia.
62. N. Joshi, *Revolutionizing waste management with blockchain technology*, 2020.
63. R.W. Ahmad, H. Hasan, I. Yaqoob, K. Salah, R. Jayaraman, M. Omar, "Blockchain for aerospace and defense: oppurtunities and open research challenges", *Computers & Industrial Engineering*, vol. 151, p. 106981, 2021.
64. A. El Koshiry, E. Eliwa, T. Abd El-Hafeez, M.Y. Shams, "Unlocking the power of blockchain in education: an overview of innovations and outcomes", *Blockchain: Research and Applications,* vol. 4, no. 4, pp. 1–19, Elsevier, 2023.

13 Exploring 5G/6G Energy-Efficiency in Mobile Communications for Sustainable Future

Dragorad A. Milovanovic and Zoran S. Bojkovic

13.1 INTRODUCTION

Worldwide deployment of fifth-generation (5G) mobile networks achieves significant breakthroughs in terms of latency, data rates, mobility, and a massive number of connected devices as well as marks the beginning of a true digital society [1]. Beyond 2030, the sixth-generation (6G) wireless communication network will integrate terrestrial, aerial, and maritime communications into a robust network which would be more reliable and fast and can support a massive number of devices with ultra-low-latency requirements. Wireless networks have been mainly designed to optimize performance metrics such as the data rate, throughput, and latency. In parallel, after intense research stimulated by operational, economic, and environmental considerations, energy efficiency (EE) has emerged as a key criterion in the design of mobile communications [2]. 5G is the first wireless technology designed to be energy efficient and sustainable with an industry target of 90% improvement in 4G spectral efficiency [3,4]. However, 5G systems serve an unprecedented number of devices, providing ubiquitous connectivity as well as innovative and rate-demanding services [5–9].

Numerous studies and technological advancements in EE and renewable wireless communications and networking have been made to reduce the high power consumption of wireless systems. Academic, industry, and government institutions must come up with plans that maximize 5G networks' EE. In the 5G decade, wireless communication networks are predicted to significantly increase in terms of both power consumption and carbon emissions. The objective is to preserve coverage, capacity, and service quality while reducing energy usage through energy saving (ES). It is imperative to maintain constant or even lower mobile network energy usage while significantly increasing network capacity. The key is to make effective use of the available spectrum at moderate power levels. To achieve optimum ESs, power management is designed to adjust to the current and predicted loads of network traffic. A key requirement for 5G network development and standardization technical specifications is EE. The 3GPP, ETSI (European Telecommunications Standards Institute),

DOI: 10.1201/9781003511298-13

and ITU have developed key performance indicators (KPIs), which come in many varieties. Consolidation of resources, selective connections, and virtualization are the fundamental concepts of achieving an energy-efficient 5G network architecture. All measurement profiles, metrics, techniques, and test profiles must be defined to evaluate the performances [10,11].

It is a significant challenge developing 5G energy-efficient technologies than it was for previous communication network generations. Since EE directly affects the economic and environmental challenges faced by wireless networks, it is the primary component of sustainable mobile communication:

- Present networks are designed to increase transmit power to maximize capacity. However, this kind of strategy is unsustainable, considering how quickly the number of connected devices is increasing. Operating costs will rise to an unacceptably high level if the communication capacity is increased by continuously using more energy. Therefore, increasing only the transmit power will not result in the anticipated capacity increase in current wireless communication systems.
- Conventional carbon-based energy sources are the primary source of electricity for present wireless communication systems. Comparable to the carbon emissions of the avionics industry, actual information and communication technology (ICT) systems account for 5% of global CO_2 emissions. In the wireless community, there is universal agreement on one important point: the 1,000× capacity increase should be accomplished at a power consumption that is comparable to or less than that of current networks.

Power-saving features, new architecture, and protocols can all significantly increase the EE of 5G mobile networks. In developed markets, energy accounts for more than 20% of network OPEX. Base stations (BS) and radio access networks (RANs) can use up to 75% of the energy. In particular, there are a number of significant opportunities to enhance the EE of macro base stations: reduce energy consumption (EC) while the BS is not transmitting data, reduce EC as a result of auxiliary equipment, and increase hardware efficiency. The highest possible data rate or channel capacity is determined by the power of the transmitted signals. Channel capacity is increased by raising the power and bandwidth at the same time. However, raising the maximum power beyond a certain threshold, causes excessive power consumption, more heat generation and dissipation from BSs and mobile or stationary terminals, and a number of health risks. Increasing power indefinitely is therefore not practical for a number of reasons [12,13].

EE is currently a key consideration in mobile communications architecture. The majority of strategies that are useful for improving wireless networks' EE fall into the following general categories:

- **Network Planning and Deployment:** It is necessary to deploy infrastructure nodes to maximize the covered area per consumed energy, rather than just the covered service area. In addition, the use of BS ON–OFF switching algorithms and antenna MUTING techniques in a short time scale to adapt to the network traffic conditions can reduce EC. If the idea of dense networks is to densify the number of infrastructure nodes, the idea of massive multiple input multiple output (MIMO) is to densify the number of deployed antennas. However, massive MIMO systems also come with several challenges and impairments. Offloading techniques are another key 5G strategy to boost the capacity and EE. In particular, device-to-device communications and mmWave cellular methods/strategies can be envisioned.

- **Resource Allocation:** Allocating the system radio resources to maximize EE rather than throughput is one way to make a wireless communication system more efficient. It has been demonstrated that this strategy offers significant benefits at the expense of a minor throughput loss. Reducing EC is a different, though less effective, strategy for ES. Despite their apparent similarity, the two strategies typically result in differing resource allocations. The energy cost is represented by the amount of consumed energy related to the amount of data reliably transmitted in the time interval. Several performance functions that are dependent on the signal-to-noise ratio (SNR, SINR) of the single communication link have been used, including system capacity/achievable rate, throughput, and outage capacity, all measured in (bits/s). Information transmission efficiency is expressed in (bits/joule), which is the measure of how efficiently each Joule of energy is used. Maximizing EE is also possible, given all realistic limitations. In addition to the maximum power constraint, quality of service (QoS) constraints has been implemented recently. It is necessary to combine the used energy reduction problem with certain minimum QoS limitations to ensure throughput, capacity, or outage capacity. This approach results in consuming the minimum amount of energy required to maintain a given minimum system performance. Power minimization, however, is a specific instance of EE maximization that is constrained by QoS requirements.
- **Energy Harvesting and Transfer:** Harvesting energy from the environment and converting it to electrical power applies to both renewable and clean energy sources like sun or wind energy and to the radio signals present over the air. The random quantity of energy available at any one time is the primary design difficulty in wireless communication.
- **Hardware Solutions:** The concept is about designing the hardware devices for wireless systems explicitly accounting for its EC and use of simplified transmitter/receiver structures. For mmWave communications, given the required large number of antenna elements, the implementation of digital beamforming poses serious complexity, EC, and cost issues. The concept of cloud-RAN (C-RAN) is based on software implementation of the RAN and network function virtualization (NFV) in remote data center. This enables a great deal of flexibility in the network, thus leading to substantial savings as far as both deployment costs and EC are concerned. The idea is to use transmitter/receiver structures that are simplified and to explicitly account for energy usage when designing hardware for wireless systems. Digital beamforming solution for mmWave communications presents significant challenges in terms of complexity, EC, and cost due to the enormous number of

antenna elements required. The software implementation of the RAN and NFV in remote data centers forms the foundation of the C-RAN concept. This gives the network a great deal of flexibility and saves significant EC and implementation costs.

Furthermore, emerging techniques and new energy models should be used for energy-efficient optimization. Numerous academic articles on artificial intelligence/machine learning (AI/ML) intelligent wireless engineering systems have been published in recent years. AI and communication will converge, with AI either enhancing the communication network's capacity for data transmission or the communication network supporting AI applications. The efficiency of algorithms, data, and the amount of computing power accessible for training are the main drivers of AI advancement. An AI/ML model is trained by a training dataset and evaluated by a test dataset.

A set of training data is utilized to iteratively adjust the hyper-parameters to construct the model. In general, the size of the training dataset, the number of hyper-parameter experiments, and the cost of running the model multiply each other to determine the computing cost of an AI model. Elapsed real time, number of parameters, and floating point operations should be measured to compare the EC of different models [14].

Physical layer solutions based on AI/ML can improve 6G networks' EE by reducing transmit power by up to 50% compared to 5G for the same bandwidth and data rate. To realize the vision of an intelligent and linked world that 6G aims to bring about, networks need to be designed from the ground up with inherent AI capabilities that facilitate AI workflows. Globally, next-generation 6G networks are anticipated to provide more environmentally friendly and socially seamless wireless connectivity. 6G could be a potential solution for integrating digital technologies in the context of Industry 4.0 and beyond, supporting the sustainable evolution of industries [15–20].

In the first part of the chapter, we have discussed the growing need for EE in mobile networks. Network consumption models, KPIs, and related metrics as well as efficient power management are presented to evaluate new aspects of 5G EE. The recent results of the development study on scenarios and 5G requirements, ES opportunities and potential solutions, standardized technical specifications of ES features, and methods for performance evaluation are reviewed in the second part. In the third part of the chapter, we have analyzed recent trends in the field of mobile communications which indicated a paradigm shift from network efficiency to sustainable communication. New aspects of 5G-Advanced toward 6G sustainable networks are presented.

13.2 POWER CONSUMPTION IN MOBILE NETWORKS

The evolution of mobile networks from the introduction of the first-generation (1G) systems until today 5G and 5G advances toward 6G clearly indicate a growth of installed network equipment and carried traffic. Despite more efficiency of the new generations, EC unavoidably increases so that economic and environmental impacts cannot be neglected. More consumption means more costs for the mobile networks operators and a greater carbon footprint. The communications industry has to therefore develop strategies to optimize the EE of 5G networks, without compromising performance. However, developing energy-efficient technologies is a significant challenge.

The power consumption of wireless networks has been increasing steadily with every generation. In principle, the analog 1G network had the lowest power density and the lowest power consumption on the side of the BS and on the side of the user device. Moving to 2G with improving coverage area (CA), 3G with improving system capacity, and 4G networks, the largest increase occurred in the power density. The power at the BS also increased. The increasing trend of EC is shown in Figure 13.1 [21–23].

A BS is the most energy-consuming part of a mobile network; still, the EE of a BS is very low. A comparison of the power consumption between 4G and 5G is shown in Table 13.1. This analysis shows a massive increase in power consumption, though based on a recent assessment, 5G networks are 90% more efficient than their predecessors in data rate bits/kW.

The development of 5G has already reduced the static power consumption where networks had to transmit radio signals more frequently in 4G or even continuously in 3G even if there were no active data transmission. With regard to 5G, the most important design decision was to depart from the standard reference signals transmitted. Using power-efficient waveforms and eliminating all transmissions when there is no actual user data to be transferred are the next steps in network power

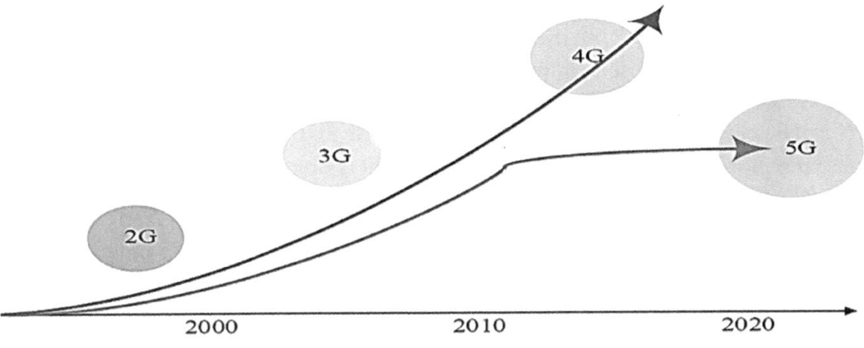

FIGURE 13.1 Increasing trend of power consumption of mobile networks.

TABLE 13.1
Comparison of Power Consumption of 4G LTE and 5G NR Base Stations (BS)

Generation	Bandwidth (MHz)	Transmission Power (W)	Baseband Power (W)	Radio Unit Power (W)	Max Data Rate (Mbps)
4G	20	40	150	950	120
5G	100	240	220	4,080	2,000

efficiency. In terms of energy consumed per traffic unit, 5G is 10× more energy efficient than 4G thanks to these and other energy-saving technologies.

The network EC model is established as part of the 5G-Advanced studies, which should allow vendor-independent analysis of the impact of network power consumption when adding support for a new channel or signal to be conveyed or new procedures between network parts. The EC model allows one to evaluate, for instance, the network-side cost of providing a unique signal to low power wake-up radio at the user equipment (UE) side. The network scheduler should ideally send data only very sparingly when there is a low traffic load.

5G BSs enable optimization of both the number of antenna ports active at the network side and the number of frequencies to be active for the coverage and traffic support required at a given moment in the area of interest. In the case of an antenna array, the network might afford to employ only a subset of antenna ports, thereby lowering the number of ports, especially under low traffic loads. This is anticipated to be made possible primarily by AI/ML-based control, which will enable dynamically coordinated energy-saving actions across several network levels to reduce network EC.

5G-Advanced systems aim to achieve high EE. The research community is increasingly exploring and developing technology enablers for 6G including radio components and network architectures. It has become clear that one solution fits all is no longer feasible as innovation for new business models, introduction of novel use cases and vertical-specific KPIs diversify the future systems. Different verticals require a very different set of capabilities. Selected verticals (Industry 4.0, future mobility, energy, eHealth, finance and banking, public safety, and agri-sector) are presented in Table 13.2 with the technical ITU-R WP5D KPI.

Starting with 5G, research on EE has been published in a large number of articles over the last several years. In general, most researchers agree that the techniques for ES can be classified into several categories. However, the distinction between different types of ES techniques is not always completely obvious and there are techniques that can easily belong to multiple categories or stand outside of the categories:

TABLE 13.2
6G Vertical-Specific KPI Toward 2030 and Beyond

Vertical	Capacity	Density	Reliability	Mobility	Energy Efficiency
Industry mMTC	<10 Gbps	$100/m^3$	10^{-6}	240 km/h	High
Industry eURLLC	<100 Mbps	$10/m^3$	10^{-9}	240 km/h	Nominal
Mobility	1 Tbps	$100/m^3$	10^{-7}	1,200 km/h	Nominal
Energy	<100 Mbps	$10/m^3$	10^{-6}	N/A	Nominal
eHealth	<10 Gbps	$1/m^3$	10^{-9}	240 km/h	High
Finance	<100 Mbps	$1/m^3$	10^{-9}	Low	High
Public safety	<10 Gbps	$1/m^3$	10^{-7}	240 km/h	Nominal
Agri-sector	1 Gbps	$100/km^2$	10^{-7}	240 km/h	Nominal

- Deploying energy-efficient architecture based on optimization of the cell size, relaying or cooperative communications
- Utilizing energy-efficient resource management may include techniques such as joint power and resource allocation, utilization of mMIMO or packet scheduling
- Introducing energy-efficient radio technologies could include heterogeneous network deployment or the utilization of higher frequency (mmWave and TeraHertz) bands

Relaying is used by the energy-efficient design to divide interior and outdoor network cells, minimize the communication distance between user devices, and introduce sleep modes. The BS has been shown to be lightly loaded approximately 80% of the time, but it nearly always uses peak power. This is true, at least in the case of components like air-conditioning and power-amplifying circuits. This necessitates the implementation of a number of mechanisms to switch OFF the BS during periods of low traffic. Exploiting the dynamics of the traffic load over time and space is necessary. However, interactive monitoring of power consumption, traffic load, and other parameters is needed, which then allows for the activation and deactivation of nodes based on a number of pre-set rules. It is even feasible to incorporate several sleep depths (depths) with varying durations and activation and deactivation profiles into BS designs through flexible reference signal design. These depths can range from 100 μs to several seconds. This results in energy-saving levels that are flexible to varying demands on data traffic.

Future networks may implement smart energy resource management as a way to dynamically balance energy availability and demand. Smart grids are electrical networks that minimize costs and preserve the stability and reliability of the grid while more effectively matching the supply and demand of electricity in real time. Random traffic arrival, dynamic connection quality, on-site power storage, wireless energy transfer, on-grid energy price, and grid load are just a few of the variables that a smart grid-enabled network must concurrently handle.

Radio technology NOMA (Non-Orthogonal Multiple Access) is a promising candidate for improving the overall throughput in the network under the condition of a large number of users sharing the same resources. It has been speculated that EE can also be dynamically improved in NOMA. This would be achieved by dynamically regulating the bandwidth to comply with both constraints of spectral efficiency and EE simultaneously, which is possible because the extra user interference, while degrading the performance, decreases the energy usage. EE improvement can also be observed if the network resources and infrastructure are shared by different network operators or under the condition of a large number of users sharing the same resources. Radio technology NOMA dynamically regulates the bandwidth to comply with both constraints of spectral efficiency and EE simultaneously.

Another radio technology massive MIMO supports the bulk of ES on the side of the user device, rather than on the side of the BS. The same signals are delivered over various pathways between the transmitter and reception antennas in MIMO. The same time-frequency resource can serve a large number of users by densifying the number of antennas on both sides. Collaboration also gives mobile network operators the power to plan and carry out future network deployments according to local network demands. Additionally, if each operator deployed their equipment independently, the issues of either over-provisioning or under-utilization of the deployed network infrastructure would be avoided.

13.2.1 Energy Models, KPI and Related Metrics

To estimate the EC of 5G networks, a model of the power consumption of BSs is developed. The EE of a network depends on low EC when the traffic load is low. When the traffic load is high, the EE is characterized by efficient data transmission (average spectral efficiency SE). However, there are fundamental and practical limits, so the performance-energy trade-off is unavoidable. EE optimization methods based on single-criterion maximize either spectral efficiency or EE to obtain better average performance. Multi-criteria optimization is approaching near fundamental limits of peak performance in joint optimization of both spectral efficiency and EE implying the use of trade-offs since the optimum is not unique. Additionally, it is essential that 5G wireless networks enhance EE with guaranteed QoS/Quality of Experience (QoE) for time-sensitive multimedia services. The key to increasing EE is to optimize the use of radio-spectrum per unit of electricity under the limit of average consumption in a certain time interval and maximum consumption [24–28].

EE metric is a mathematical value of the ratio of utility and consumed electricity. In this case, utility represents any significant communication parameter, probability of downtime, and number of BSs regardless of QoS. For example, in the case where power is considered at the equipment level, there is a relationship between output/input power and electricity consumption. The most commonly used metrics are:

- Energy per bit of information [J/b] or [W/bps] represents the energy required to transmit 1 bit of information.
- Power per unit area [W/m^2] refers to the average power required to establish a connection on the surface of the CA.

In Table 13.3, key power- and energy-related definitions are shown.

TABLE 13.3
Relevant Basic Metrics

Metric	Unit	Note
Power consumption	W or J/s	Amount of energy that is transferred or converted per time unit
Energy consumption	kWh or J	Amount of power used over a time period—1 kWh corresponds to 3.6×10^6 J
Energy efficiency	% or label or kWh/unit	Ratio of output of performance, service, goods, or energy, to the input of energy
Power efficiency	%	Output power/input power
Energy performance	Mbits/kWh	Ratio between the produced task or work and the consumed power for producing this task or work over a time period

EE KPIs are defined by various organizations (3GPP, ETSI, ITU), and there are various scopes in the application. This can include the entire network, a subset of the network for example the RAN, individual network entities, as well as specific communication sites that have networking and on-site devices. KPI can also be categorized according to the operator's network life cycle phase they may apply: during the design phase (comparison network elements from various vendors), during the build phase (several design options), and during the run phase (assess the EE of the live end-to-end network, or for sub-networks, or for single network elements or telecom sites) [27].

- KPI network elements are expressed in terms of data volume (DV) divided by the EC of the considered network elements. In the case of RANs, an EE KPI variant may also be used, expressed by the CA divided by the EC of the considered network elements.
- A quantitative analysis of the EE can be performed using a KPI network. This KPI network is particularly useful when the effects of different EE solutions or mechanisms are not clearly revealed with a qualitative analysis.
- KPI deployment scenario indicates the definition to target a given network environment for measuring the EE.
- KPI telecom sites are defined in relation to the management of energy for data centers and sites of operators but can also include other types of sites. These KPIs allow having a global vision about the technical environment impact on EE based on goals such as EC and reuse, renewable energy and operation efficiency.
- KPI virtualized network functions (VNF) are evaluated according to hardware EC, resource consumption, and utilization. The measurement methods are intended to be used to assess and compare the EE of the same functional components independently in lab testing and pre-deployment testing.

Existing measurement methods for the calculation of EE KPIs for mobile networks are based on the collection, on a per-network node basis, of DV measurements (collected via operations, administration, and management (OAM) as performance measurements), and EC measurements (collected using power meters or information from invoices provided by power suppliers, via built-in or external sensors). The methodology varies depending on whether the measurements are made in live networks or in testing laboratories. Potential measurement methods for 5G networks rely on the collection of related measurement data based on reporting methods, measurements (counters), and built-in or external sensors. For virtualized parts of BSs, the EC of the VNFs is the EC of the server(s) on which the VNF(s) run, minus the EC of the subject servers when they are in idle mode [29–32].

From the RAN side, the EE KPI is also represented with different metrics related to the focus of the subcomponents of the network. The overall EE consists of three factors: power efficiency of the site infrastructure, power efficiency of the BS equipment, and energy performance of the air interface (Figure 13.2). P_{AC} is the AC input from the grid, P_{BS} is the DC input power to the main equipment (BS), P_{output} is the cabinet-top power output of the BS antenna, and S_{pi} is the service provided

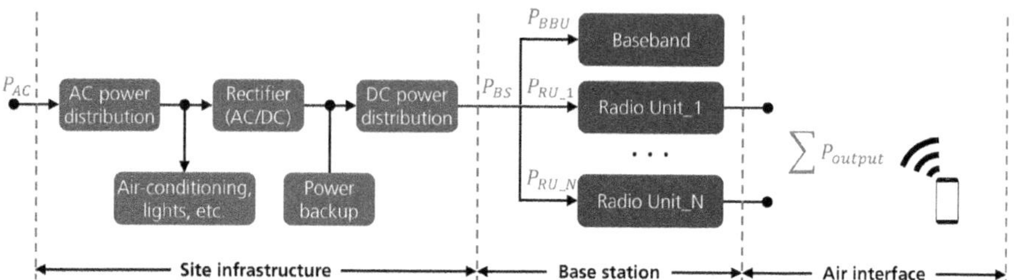

FIGURE 13.2 Overall radio access network (RAN) energy efficiency: power efficiency of site infrastructure, power efficiency of BS equipment, and energy performance of air interface.

by the BS (delivered bits, coverage, or number of subscribers served by the BS). The overall energy performance is represented by $EP = S_{pi}/P_{AC}$, $SiteEE = P_{BS}/P_{AC}$, $BaseStationEE = \Sigma P_{output}/P_{BS}$, $RadioEE = \Sigma S_{pi}/\Sigma P_{output}$. This ratio integrated over a time period is expressed then in [Mbits/kWh] or [Mbits/J].

Furthermore, EE KPI is also represented with different metrics focusing on different performance values. Most of them are defined in ITU-T L.1330. A list of relevant metrics is shown in Table 13.4.

Designing wireless networks on the basis of EE metrics is an extremely challenging task, as traditional power allocation schemes lead to continuous transmission using the maximum power available. However, this view has begun to change recently as bit/joule EE, defined as the amount of information that can be reliably transmitted per joule of consumed energy, has emerged as a KPI for 5G networks.

Detailed EC reporting is expected to become a regular element of telecom equipment as EC becomes a primary concern for mobile network service providers, notably in terms of operating expenses (OpEx). Telecom standards specify energy-related metering functions.

Standardization organizations ETSI Technical Committee Environmental Engineering and ITU-T Study Group 5 Environment and climate change are in the past 5–10 years specifying a series of standards related to the energy management of ICT systems and equipment in relation to the provision of EE KPIs and assessment principles.

Documents ETSI TR 103542 and ITU-T L.1310 Supplement 36 analyze the EE issues for the 5G systems with a focus on methods and metrics to measure EE. The document relies on the existing standards for legacy radio systems, ES 202706/ITU-T L.1310 for single BS measurements in a laboratory environment, and ES 203228/ITU-T L.1331 for access network aggregate measurements of EE.

- Recommendation ITU-T L.1310 contains the definition of EE metrics test procedures, methodologies, and measurement profiles required to assess the EE of telecommunication equipment. The document aims to define the topology and level of analysis to assess the EE of mobile networks, with a focus on the radio access part of the mobile networks and specifically the

TABLE 13.4
Mobile Network (MN) Energy Efficiency Metrics

Metric	Unit	Note
Energy consumption of mobile network (EC_{MN})	kW	Sum of the energy consumption of each piece of equipment included in the mobile network (ITU-T L.1331, ETSI ES203228)
Mobile network data energy efficiency ($EE_{MN,DV}$)	bit/J	The ratio between the data volume DV_{MN} and the energy Consumption EC_{MN} when assessed during the same time period
Mobile network coverage energy efficiency ($EE_{MN,CoA}$)	m^2/J	Ratio between the area covered by the mobile network and the energy consumption when assessed during one year (handling low data volumes, in particular in rural or deep rural areas)
Power usage effectiveness (PUE)	%	Energy efficiency of a Data Center (ISO/IEC 30134-2:2016)
Renewable electricity ratio	%	Percentage out of the total energy consumed by fixed and mobile networks coming from renewable sources
Energy consumption per customer	kWh/client	Quantitative value of the network energy efficiency per client
Energy intensity (energy consumption per data volume)	kWh/TB	Quantitative value about the network energy efficiency per data unit (TByte)
Network carbon intensity (NCIe)	$kgCO_2$/TB/kWh	ITU-T SG5 is currently working on two work items: ratio between total carbon emission and total data volume, and ratio between EC_{total} and total data volume

radio BSs, backhauling systems, radio controllers, and other infrastructure radio site equipment.

- In relation to the RAN aspects, ETSI Technical Report 3GPP TR 38.913 contains scenarios and requirements for next-generation access technologies, which can be used as not only guidance to the technical work to be performed in 3GPP RAN WGs but also input for ITU-R to take into account when developing IMT-2020 technical performance requirements. Qualitative inspection is suggested, for EE but also quantitative analysis. The first part of the report is dedicated to a summary of the possible 5G scenarios. In the second part, the report deals with the KPIs to be used to evaluate the performance of the new network in these scenarios. Two KPIs are relevant for the EE estimation area traffic capacity and user-experienced data rate. Network EE must be considered as a basic principle in the NR design of the new 5G systems.
- In relation to the service and systems aspects, ETSI Technical report 3GPP TR 21.866 deals with 3GPP system EE requirements and principles and

identifies four of them: high-level requirements, where the landscape for EE in mobile radio systems is set up; architectural requirements, where the architectural approaches to save energy are mentioned; functional requirements, where the possible impacts of EE on network performance is dealt with; EE control principles, where a list of possible actions to increase EE is given.

3GPP 5G system simultaneously represents an evolution of the current legacy systems and a revolution to satisfy the new needs of the innovative services offered by the inclusion of new vertical areas in the telecommunication environment. In addition, this two-facet aspect of 5G is reflected in a time-wise approach that starts with the 5G Phase 1 Release 15 NR system, based on an evolution of LTE, and 5G Phase 2 Release 16 that is about the new vertical services and applications [33–37].

- The metrics and methods described in ETSI ES 203228/ITU-T L.STP for the legacy networks are considered valid for 5G Phase 1 and an update. It is worth noting that 5G EE already includes some remarks on virtualization. The first phase of 5G starts earlier based on the eMBB services. In this sense, the first phase is quite an evolution of the legacy 4G networks, with an architecture similar to those already in place. Differences could be limited to the wider adoption of virtualization and orchestration in the core network and to wider usage of small cells to have the required network densification. Consequently, 5G network could be analyzed in terms of the EC and the EE aspects in the same way, or very similarly, to what is already specified for 4G networks in ITU-T L.1331. The single nodes are measured referring to ETSI ES 202 706-1 and ETSI ES 202 706-2 for both static and dynamic operations. Indeed, the capacity and coverage definitions, as given in ITU-T L.1310 and ITU-T L.1331, still hold for this phase of 5G, as well as, even more, for the EC measurement. A challenge arises from the increasing use of multi-radio equipment. Specific 5G BSs will be installed for dense urban high-capacity sites. In most other cases, 5G will be collocated at existing sites. Most of today's new BSs can be configured to operate with different technologies (LTE) simultaneously and even multi-band BSs are now available. Consequently, many new 5G BS will be multi-standard capable. There is currently no precise method to measure the fraction of energy consumed by a multi-standard BS for the different standards. The efficiency KPIs in ITU-T L.1331 can only be applied per technology, if separate equipment is used. In the case of multi-standard equipment, the measured KPI provides the average site efficiency measured over all equipment.
- The 5G Phase 2 impacts heavily the specifications to measure EE and requires an extensive update of them, in tight cooperation with the standard bodies that outline the new systems, especially 3GPP RAN and ITU-R. The objective is for example to leverage 3GPP SA5 work dealing with EE-related analytics. The second phase of 5G coincides with Release 16 and onwards in 3GPP and is expected in 2020. The focus is on new features for Industrial Internet of Things as well as enhancements for massive MIMO

and network slicing (NS). Each slice of the network has a different architecture in terms of access and core parts. Slice is a new network for the sake of EC and efficiency, being made of different real and virtual components. One possible approach could then be to introduce a different measurement method for each slice in the network. The approach in ITU-T L.1331 could still be applicable, but new network elements could be introduced in the formulas to evaluate the EC, according to the network layout of every single slice. Probably also the measurements of the capacity should be accorded to the single slice under test and its typical throughput values. The same for the coverage, that of course will be very different if sensors/actuators or vehicles or other network elements are analyzed. Also the ITU-T L.1310 and the standards ETSI ES 202 706-1 and ETSI ES 202 706-2 will have to be extended to the new network elements in 5G Phase 2.

13.2.2 Efficient Power Management

ITU-R performance requirements specify that the EC for the RAN of IMT-2020 should not be greater than IMT networks deployed today, while delivering the enhanced 5G capabilities. The network EE should therefore be improved by a factor at least as great as the envisaged traffic capacity increase of 5G relative to 4G for enhanced mobile broadband. 5G mobile networks have the capability to support a high sleep ratio and long sleep duration and other ES mechanisms for both network and device are encouraged. Network EE is the capability to minimize the RAN EC in relation to the traffic capacity provided. Device EE is the capability to minimize the power consumed by the device modem in relation to the traffic characteristics. EE of the network and the device should support two aspects [38,39]:

- Efficient data transmission (average spectral efficiency) in a loaded case
- Low EC when there is no data

Low EC when there is no data is estimated by the sleep ratio. The sleep ratio is the fraction of unoccupied time resources (for the network) or sleeping time (for the device) in a period of time corresponding to the cycle of the control signaling (for the network) or the cycle of discontinuous reception (DRX) (for the device) when no user data transfer takes place. Furthermore, the sleep duration, the continuous period of time with no transmission (for network and device) and reception (for the device), should be sufficiently long. This requirement is defined for evaluation in the eMBB mobile broadband usage scenario.

The management services required for the assessment of the EE of 5G networks are performance management services, management services for network function provisioning, and management services for fault supervision. 5G control framework for self-managed automated EE control processes should include:

- Policy management defines and manages the EE control policies related to the EC status and control operations at the network, equipment, and site levels. It translates the policy information into configurations at the EE optimization entities at network, equipment, and site levels, where applicable. The policy may be adjusted according to achievable EE KPI and the variations of QoS/QoE.

- Control and coordination are designed for power-saving operations across all the relevant elements at network, equipment, and site levels. The operating conditions such as traffic load and density at the network and equipment levels and operational conditions such as temperature and humidity at the site level are monitored and reported. According to the EE control policy and the current status, EE control operations are activated/deactivated in each of the EE optimization entities. The embedded energy metering function in each EE optimization entity collects the necessary statistics such as EE KPI and QoS/QoE and then reports the information to the EE profiles management. The EE optimization entity can be a logical or physical component to execute the EE policies and the corresponding EC optimization operations.
- Profiles management monitors, collects, processes, stores, and provides EE-related information and statistics including the profiles for traffic, operating conditions, the corresponding achievable EE KPI, and the variations of QoS/QoE in reference to the predefined QoS/QoE requirements from the embedded energy metering functions in each of the EE optimization entities. The information is assessed and then sent to the EE control and coordination for adjusting the power-saving control operations and possible policy adjustment by the EE policy control.

13.2.2.1 Enhancing EE in NR

The RAN domain is the primary focus for energy-saving efforts, as it accounts for 73% of total EC in mobile networks. Key techniques to enhance RAN's EE include adaptations of radio network parameters in time, frequency, spatial, and power domains in response to traffic variations. These adaptive capabilities play a pivotal role in optimizing power consumption while maintaining network performance.

- In the time-domain, micro-sleep techniques are very effective means to conserve energy by deactivating and reactivating power amplifiers during idle periods. The lean carrier design of 5G new radio (NR) enabled improved sleep opportunities compared to LTE as it removed the always-on cell-specific reference signaling of LTE.
- In the frequency domain, carrier aggregation can expand coverage using lower frequency bands while efficiently boosting capacity by moving user-plane data to higher bands. Furthermore, innovations such as spectrum sharing enable single service providers to deploy new frequency bands on existing infrastructure without sacrificing performance or environmental sustainability goals.

Adopting adaptive capabilities based on load and radio conditions across various radio domains is essential for modern BSs and RAN sites geared toward enhanced network performance while reducing power consumption and in turn environmental impact. Furthermore, the 3GPP enablers are continuously developed in 5G-Advanced and expected in the 6G design to enable radio adaptations to be more dynamic and at a more granular level than in today's networks so as to exploit further ES potentials.

EC in the RAN is a significant focus area for EE. Various techniques are being employed, including optimizing sleep modes for gNB, uplink wake-up signals (WUSs), dynamic transmission power adjustment, multi-carrier energy conservation, dynamic UE group switching, and dynamic bandwidth adaptation. These techniques reduce EC in RANs, contributing to a more sustainable and cost-effective operation. EC is increasingly being considered a core performance criterion alongside security, privacy, and complexity, particularly for best-effort communication. The exposure of EC and efficiency information is also a critical aspect, providing customers with insights into network services under varying energy conditions. Application EE monitoring and temporary coverage layer pooling are strategies aimed at optimizing energy usage. Energy is emerging as a key criterion for 5G environment adaptation, enabling operators to achieve significant ESs while maintaining service quality.

Flexible and scalable system design in 5G NR enables different standardized power-saving techniques to adapt to various traffic loads and traffic types in the networks. Specifically, the NR standard supports flexible reference signal design (avoiding always-on reference signal) and flexible muting of resources and ensures forward compatibility for energy-efficient network implementation, the DRX mechanism, and the inactive radio resource control (RRC) state.

It is expected that the data transmission duration can be significantly shortened by intelligent technologies in 6G. As a result, a device may be able to stay longer in an operating mode when it is not actively accessing or interacting with the network. This would make it feasible to operate a system with native power saving, which is especially important for energy-efficient devices and environmentally friendly networks. Effective transmission channels can be designed in such a way that control signaling can be optimized and the number of state transitions or power mode changes can be minimized to achieve maximal power saving for devices and network nodes.

There are several methods for enhancing EE in RANs. Implementing some of the following techniques can significantly enhance EE in RANs, reducing overall power consumption and contributing to a more sustainable and cost-effective network operation.

- **Optimizing Sleep Modes for gNB:** To bolster EE, it's crucial to allow gNBs to utilize deeper sleep modes by adjusting the transmission patterns of downlink common and broadcast signals, including Synchronization Signal Block (SSB), SI, paging, and cell common Physical Downlink Control Channel (PDCCH). Similarly, the transmission pattern and availability of uplink random access opportunities should be adapted. Eliminating unnecessary transmissions and receptions of UE-specific channels during low or no activity periods significantly reduces EC. This enables gNBs to spend more time in sleep mode, particularly valuable in low-traffic areas.
- **Uplink WUSs:** Minimizing gNB EC involves reducing the time spent in an active state when channels or signals are not required. UEs can send uplink

WUSs to prompt the gNB to transition from low-power mode to an active state for transmitting or receiving signals, conserving energy.

- **Dynamic Transmission Power Adjustment:** Improving network EE is achieved by dynamically adapting the transmission power of downlink signals and channels through UE feedback and configuration adjustments. This adaptive approach applies to Physical Data Shared Channel (PDSCH), channel state information (CSI)-RS, DMRS, and broadcast channels/signals. Power offset values between various signals and channels can be updated via lower-layer signaling, enabling the network to reduce power for unnecessary signals and channels, leading to ESs.
- **Multi-carrier Energy Conservation:** ESs can be enhanced by transmitting SSBs and System Information Block (SIBs) on only one carrier in a two-carrier deployment, leveraging the SSB-less operation feature. This not only reduces EC for the gNB but also benefits UEs, as they no longer need to transmit or receive signals on both carriers.
- **Dynamic UE Group 5G Primary Cell (Pcell) Switching:** The gNB can efficiently conserve energy by dynamically switching UEs between different Pcells using the UE group Pcell switching feature. The Pcell is the cell that the UE exchanges RRC signaling messages with. This prevents UEs from having to transmit or receive signals on all Pcells, contributing to ESs.
- **Dynamic Bandwidth Adaptation:** EE is also improved by dynamically adapting the bandwidth part (BWP) for UEs through the BWP adaptation feature. By allowing UEs to transmit and receive signals only on the required bandwidth, unnecessary EC is reduced.

13.2.2.2 Exposure of EC and Efficiency Information

Regardless of the RAN deployment options, the 5G network collects energy usage data and makes it available to clients upon request or subscription. Based on the energy states of the network, operators can provide different energy-related Service Level Agreements (SLAs), giving consumers better insight into network services under various energy conditions. The customer, approved third parties, or the operator can configure network performance statistics and link them to information about energy use. EC monitoring is critical to the deployment of 5G Non-Public Networks (NPN) private networks, particularly in RAN sharing settings. Customers that subscribe to the 5G Core Network receive EE data, which enables them to monitor and modify their EC, while operators are able to enforce policies relating to energy. New needs that come with this use case include EC monitoring and the establishment of energy credit limits for services.

- **Application EE Monitoring:** Next-generation mobile communication systems are gearing up to accommodate increasingly resource-intensive services, including immersive experiences and digital twins. Such use cases as well as the pervasive use of artificial intelligence and machine learning (AI/ML) in future network architectures will result in heightened EC for both devices and network infrastructure. To uphold EE, we can employ application layer adaptation strategies, such as dynamically adjusting service levels based on the expected EC within a specific geographical area.

- **Application Service EE (AEE)** comes to the forefront. Within this context. AEE can be effectively monitored and forecasted within the 5G system, empowering service providers to undertake well-informed measures to optimize energy usage. This process encompasses the consideration of pre-conditions, the management of service flows, and the evaluation of post-conditions, in addition to potentially necessitating the introduction of new prerequisites to ensure enhanced EE across mobile communication systems.
- **Temporary Coverage Layer Pooling for ESs:** In the pursuit of ESs within mobile networks, the implementation of coverage layer pooling can be a valuable strategy. This approach involves deactivating certain RAN nodes during periods of low usage, leaving only one active coverage layer. Agreements between operators can enable one network to provide coverage for subscribers from other networks during low-load periods, leading to network-wide ESs. This strategy can also be extended between NPN operators and public land mobile network operators. The 5G system should support the temporary pooling of coverage layers, enable operators to serve users from other networks, and allow UEs to display their home operator network name during coverage layer pooling.
- **Energy as a Service Criterion for 5G Environment Adaptation:** Operators and cloud/data service providers face the dual challenge of reducing carbon emissions while offering optimal service plans to end-users. By diligently monitoring and controlling network functions based on EC, operators can optimize energy usage without compromising service quality. The 5G system supports individual network function monitoring, registration, discovery, selection, load balancing, and overload control based on EC. This enables operators to adapt, migrate, and scale network functions in line with EE requirements and strategic plans. By measuring and managing EC, operators can achieve significant ESs while maintaining high service quality, thereby accommodating diverse end-user service demands and advancing environmental sustainability.

13.3 ES TECHNIQUES AND PERFORMANCES

This section introduces ES techniques elaborated in 3GPP standardization, from the UE and the network site, and methods for performance evaluation.

EE of mobile networks has long been an important aspect of 3GPP; already the second LTE release in Release 9 introduced signaling among eNodeBs to exchange messages related to network ES. This mechanism allows eNodeBs to inform neighbors when they deactivate capacity-booster cells, and, conversely, neighbor eNodeBs can request cell activation. Nonetheless, the overall network EC is still rising due to the massive growth in mobile data, which is not being compensated by energy-saving measures. In addition to sustainability, network EC accounts for a large portion of OpEx in mobile networks. 3GPP continues actively pursuing network ESs for these two reasons. The definition of EE energy efficiency, KPIs, and measurement techniques as a means of characterizing energy-saving use cases and solutions were recently specified. Part of the industry-wide 5G activity, this initiative spans an ecosystem that also includes output relevant to ES from other major actors like as GSMA, ETSI, ITU-T, and NGMN.

More recently, a new study item (SI) on network ESs was started on the RAN. The RAN consumes the main share of energy used to maintain a mobile network. Therefore, techniques for reducing power consumption on RAN should provide the highest benefit in terms of ES. The definition of the BS consumption model, evaluation methodology, and appropriate network ES strategies are all included in this SI. The results are recorded at TR 38.864. The EE of a RAN or UE is commonly determined by dividing its performance metric by its EC. Depending on the kind of network entity it applies to, the performance is defined. The 3GPP SA5 Group has expanded its scope to include the entire 5G system, defining EE KPIs for the 5G core network and NS in the TS 28 series. It also specifies requirements, solutions, and protocol-specific definitions for management, orchestration, and charging.

13.3.1 3GPP Specifications and Methods

3GPP continuously pursues faster, more reliable, and sustainable mobile networks. The evolution 5G-Advanced promises groundbreaking improvements in Release 18. The further advance continues into Release 19 and Release 20, focusing on expanding AI/ML use cases and exploring 6G possibilities. This evolution builds upon the foundation established in previous releases shown in Figure 13.3 [40–43].

The development of 5G-Advanced is made possible by cooperative efforts and standardization within 3GPP. To create a single, widely recognized standard, a wide range of stakeholders—including network operators, equipment producers, technology suppliers, and vertical industry partners—have collaborated. Ensuring 5G network interoperability, scalability, and future-proofing has been the goal.

3GPP has established a parallel workflow methodology, which allows for the simultaneous creation of numerous releases. The strategy for 5G-Advanced is to roll it out over several releases, with Release 18 serving as the initial target release. The fundamental specification for 5G-Advanced RAN scoping was finished by the end of 2023, with scoping starting in the second half of 2021. The UE conformance testing specification work that comes after in RAN5 and the performance portion of RAN4 are not included in this timetable.

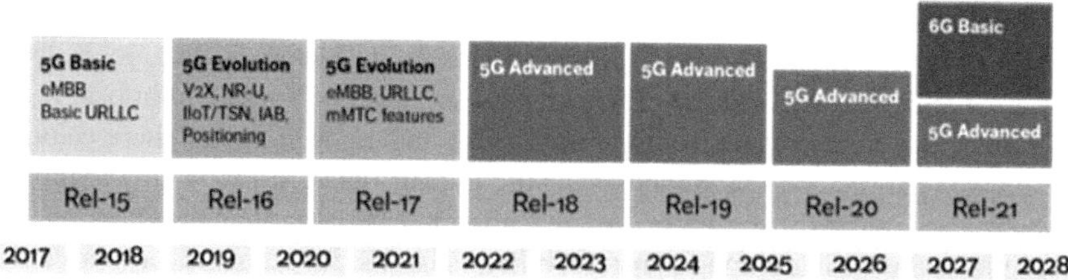

FIGURE 13.3 3GPP evolution toward 5G-advanced and 6G.

With regard to 3GPP advancements, the organization has been focusing on key technologies that will help to enable 5G-Advanced in the future and create the groundwork for the transition to 6G. These developments began with Release 18 and will continue into future releases. Important technologies involve integrating AI/ML into the network seamlessly, not just at the Physical Layer (PHY), but also at other layers of the network. Increasing overall efficiency and network capabilities is the goal.

13.3.1.1 AI/ML Performance and Usability

3GPP sustained advance AI technologies beyond Release 18 are based on the potential of AI for continuous network optimization and performance enhancement. As approaching Release 20, 6G studying AI/ML as an integral component of the system (Figure 13.4). Advanced technologies such as distributed learning, in conjunction with deeply embedded AI, will significantly boost performance and usability, marking 6G as the first generation of data-driven mobile networks [44–46].

In Release 18, AI/ML further spread to the new RAN, with 3GPP considering UE and network-side models, fostering enhanced capabilities such as beam management, CSI enhancements, and improved positioning. This exploration aims to establish a generalizable ML framework, facilitating ML applications across the air interface and preparing for 6G. As we look ahead to Release 19 and beyond, comprehensive incorporation of AI/ML within the air interface, RAN, and overall system architecture is anticipated. This integration of AI and ML represents a paradigm shift in mobile communication technology, particularly in the context of 6G's data-driven networks. Furthermore, the life cycle management of AI/ML functionality is crucial, focusing on features, functionality, and models. This process involves data collection, model training, functionality registration, inference operation, monitoring, and activation/deactivation, ensuring the viability of ML solutions across diverse scenarios and configurations. Trustworthy AI has become an essential concern as ML proliferates across mobile networks. It encompasses principles like explainable, fair, and robust AI, integrated into the design to ensure transparency, unbiased outcomes, and reliability. These principles apply throughout the ML process, including data processing, training, testing, and inference. Ensuring trustworthiness is pivotal for

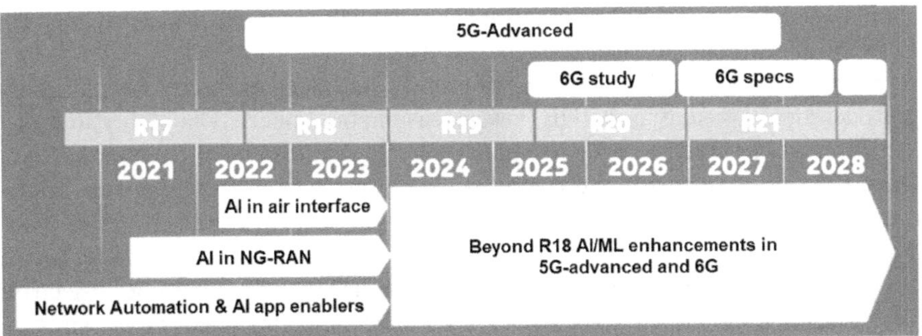

FIGURE 13.4 The evolution of artificial intelligence/machine learning (AI/ML) in 3GPP.

diverse use cases, and mechanisms must be tailored to individual requirements to maintain the desired levels of reliability and integrity.

Initially, network automation was the domain where AI/ML into 3GPP started. The arrival of AI/ML was indicated in Release 15 with the introduction of the network data analytics function, which offered network slice analysis capabilities. It then expanded its scope in Release 16 to encompass data gathering and exposure in the 5G core and enabled UE application data collecting in Release 17. The next-generation RAN, or NG-RAN, was also integrated with machine learning. The RAN intelligence enabled by ML is governed by the concepts described in the technical paper TR 37.817. It investigates the many use cases and solutions for ML-enabled RAN in addition to researching the functional framework. The first set of use cases includes ESs for networks.

By carefully offloading traffic inside a layered structure, this technique uses cell activation and deactivation to reduce energy usage over the whole RAN. Cells may be turned off and linked UEs moved to another target cell if the expected traffic volume drops below a predetermined threshold. Release 18 expands the specifications for AI/ML into the NR air interface, whereas Release 17 ML coverage was extended to include descriptions of the RAN intelligence concepts. The focus on this area of the air interface specifically addresses ML solutions that require interactions between network infrastructure and UE.

Future AI/ML use cases will be addressed in the air interface, RAN, and system architecture in Release 19. As we approach Release 20, 6G will be studied with AI/ML as an integral component of the system. Advanced technologies such as distributed learning, in conjunction with deeply embedded AI, will significantly boost performance and usability, marking 6G as the first generation of data-driven mobile networks.

13.3.1.2 Network ES

In Release 18, a SI on network ES was addressed to identify and evaluate potential techniques, which are summarized in a technical report TR 38.864. Following the SI, still in Release 18, a work-item phase is starting in 2023Q1 where some of the solutions for network ES shall be standardized. Furthermore, such a SI may lay the foundation for enhancements in further 5G NR releases and eventually also in 6G. The SI started by defining a BS EC modeling, an evaluation methodology, and the KPIs to be evaluated. Also, several techniques that can potentially improve the energy footprint of 5G NR networks were identified and evaluated.

Multiple transmission states, reception states, and sleep modes may be taken into consideration, just like in the UE power consumption model. The ability to switch off different hardware components translates into different sleep modes. Moreover, there is a transition period and transition energy involved in switching between the various sleep stages. For example, deep sleep, light sleep, micro-sleep, active downlink transmission, and active uplink reception are the various sleep modes and active states of a BS, according to the current agreement in TR 38.864. The micro-sleep has zero transition time to enter the active modes and simply indicates no transmission or reception. The transition timings from the deep sleep and light sleep modes to the micro-sleep or active modes are based on the shutdown times of multiple hardware components. Furthermore, the bandwidth, transmission power, power amplifier efficiency, and proportion of active antenna units are all taken into account in the appropriate scaling of the active transmission and reception models.

To maximize the benefits of switching to sleep modes and reducing the resources in active modes, network ES approaches concentrate on increasing savings during periods of idle, low load, or moderate load. The methods that the SI covers fall into the following categories: time, frequency, spatial, and transmit power resource optimization. Due to its improved spectrum efficiency, 5G NR generally has substantially higher EE than LTE when operating at high loads and in full-scale active modes. However, a very large capacity is not required in every cell at every instant of the day. As a result, there are numerous opportunities for savings, such as at night when traffic may be much lower than during the day. Reducing the always-on common signaling from BSs to maximize the possibility of successfully entering sleep modes is one of the most attractive approaches in the context of 5G standardization.

13.3.1.3 UE Power-Saving Techniques

The sustainability of mobile services depends both on infrastructure and UE EC. Furthermore, battery lifetime is a significant dimension of QoE. From the very beginning (Release 15), 5G NR includes features for UE power saving such as DRX and the new RRC Inactive state. In LTE, only two main UE operational states for UE were defined: RRC connected and RRC idle. The uprise of always-on apps and dynamic webpages has led 4G to an inefficient operation from a power and signaling perspective where the UE is constantly changing between these two states only to transmit small amounts of data. For this reason, NR introduced a new state called RRC inactive. The transition from RRC inactive to RRC connected involves less signaling and the connection can be resumed more quickly leading to less UE power consumption. Despite better efficiency per bit and enhanced design in NR the power consumption at UE is always a growing concern. Due to the increased bandwidth, more antennas, and higher data rates, a 5G NR UE may actually consume more power than a typical LTE UE. In early deployments also the UE needs to maintain both a LTE and a NR connection. For all these reasons, 3GPP has continued to improve UE power consumption also on Release 16 and 17.

- DRX is a method wherein a UE turns off its RX chains for network-configured time intervals in order to save energy. When deterministic traffic patterns are present, this feature maximizes power consumption gains. In contrast, it is ineffective in a scenario with sporadic traffic. More dynamic power savings are made possible by the addition of a WUS, which enables the UE to decide instantly whether to monitor the DRX during its on-period or to return to sleep. Depending on the number of antennas being utilized, MIMO can increase date rates. However, to feed more antennas, more active transceiver chains are required, which also increases power consumption. Release 16 addresses this issue by introducing an adaptation to the downlink antenna count, allowing the UE to minimize the number of in-use layers and so lower its power consumption. Analogously, it is possible to exactly choose how many antennas to use in the uplink, which lowers transmission power at the UE side and conserves energy at the UEs. Measurements are frequently required of UEs and serve as an input for Radio Resource Management (RRM) processes. To reduce power usage, Release 16 may decrease the frequency of these RRM measurements. RRM relaxation is the term for this technique since it relaxes the measurement requirements. Generally speaking, a number of factors, such as UE mobility, UE signal quality toward a serving cell, and beam status of a serving cell, can be taken into account to relax RRM measurements. Surprisingly, the RRM relaxation interval is mostly influenced by the UE's proximity to the serving cell and users with reduced mobility.

- Release 17 extended the RRM relaxation criteria and made them applicable to the RRC-connected mode. Additionally, Release 17 allows for a relaxation of metrics to link and beam failures. WUS for RRC-connected mode is applicable in Release 16; however for RRC idle, a similar idea called paging early indication (PEI) is introduced in Release 17. To be sufficiently responsive to incoming messages from the network, such as an incoming call, the UE must keep an eye on paging occasions (POs). However, UE EC is high because of waking up and checking PO messages. With the help of the PEI, the UE can detect paging messages faster. If there are no messages to decode, the UE can wake up and go back to sleep faster. Additional power-saving features of Release 17 include small data transmission in RRC inactive, extended paging/DRX cycles, reference signals for quicker synchronization in idle/inactive, and dynamic aperiodic skipping of control channel monitoring in RRC connected.

13.3.2 Performance Evaluation Metrics

In 3GPP performance metrics are critical metrics that evaluate network or user performance from a system- or link-level view. 3GPP strives to specify evaluation methodology and KPIs by which different approaches proposed by the companies are compared, and a decision is made based on the approach achieving the best performance.

The focus of current 5G networks is on EC and efficiency, although not as a primary performance criterion. Ongoing efforts have been dedicated to enhancing EE in system operations. However, as we evolve toward 5G-Advanced features, EE should stand as a core principle alongside security, privacy, and complexity [47–56].

3GPP Release 18 decided to investigate the paths assisting the RAN with EC reduction while considering much better traffic capacity. For this reason, 3GPP has started a study to improve the power consumption model that is currently required for the UE to be used in the network. Furthermore, the need for updates on both qualitative and quantitative KPIs was covered. Designing a network with the capacity to effectively transport traffic to the UE and the flexibility to cease transmission when no more traffic needs to be transmitted while taking network availability into account was the stated goal. In light of this, it was decided that the energy-saving gain for BS consumption evaluation should be accompanied by at least the following KPIs: user-perceived throughput, access delay, and latency.

Network traffic emerges as the primary driving force and a crucial parameter for optimizing network components, enabling adaptation of radio resources across time, frequency, space, and power domains in response to changes in traffic load.

UE were extensively studied and several performance metrics were specified to compute the overall gain from a system or link-level perspective. More specifically, the following performance metrics are defined:

- UE power-saving gain is computed in the power consumption model for a specific scenario.
- Latency, scheduling delay, and user throughput, o wherein system performance considering UE power-saving techniques is evaluated.
- False alarm rate and miss detection rate metrics indicate errors that occur when a procedure or signal is triggered.

13.3.2.1 Support for Multimedia Delivery

Using EC as a novel service criterion for unconstrained mobile telecommunication services ultimately leads to a more resource-efficient network. However, the 5G technology effectively facilitates services; nonetheless, EC is not included as a service-level metric. One method that can assist in reducing EC from the content provider to the mobile device is media delivery service optimization. One significant use case for mobile networks is media distribution. The EC of streaming video is impacted by both the user and the network. Users are discovering they may stream videos to their smartphones for entertainment and educational value, as well as participate in video-conferences from the most unexpected locations when new mid-band Time Division Duplex (TDD) spectrum is deployed for 5G [57–59].

Advanced network-side compression and user-side decompression are necessary for streaming video. Initiatives that are relatively recent and concentrate on the technical supply chain that supports streaming services and end-to-end EE. There is a set of best practices for the optimal use of compression technologies in streaming video processes, maximizing their energy economy while preserving a good QoE for viewers by default.

When operators shape video, they gain twofold benefits: they expand the capacity of their networks and enhance user QoE by extending battery life. Shaping also benefits third-party streaming providers by enabling them to effectively accommodate a larger number of users at a reduced energy cost.

13.4 6G GREEN NETWORKS

In this section, we give an overview of the current state of the art of sustainability in future mobile networks. In achieving the UN Sustainable Development Goals (SDGs), information and telecommunication industry ICT plays an important role. An increasing number of countries implement national laws to monitor and develop the EC toward more sustainability.

Metering and continuously improving the EE of mobile networks becomes highly important for operators.

Environmental sustainability refers to the ability of both the network and devices to minimize greenhouse gas emissions and other environmental impacts throughout their life cycle. Important factors include improving EE, minimizing EC, and the use of resources, for example optimizing for equipment long life, repair, reuse, and recycling. EE is a quantifiable metric of sustainability. It refers to the quantity of information bits transmitted or received, per unit of EC (bit/joule). EE is expected to be improved appropriately with the capacity increase to minimize overall power consumption. Therefore, the main strategy for de-carbonizing the operational emissions of the mobile industry requires complementary and urgent implementation of EE measures as well as switching to renewable or low-carbon sources of energy to power mobile networks [60–64].

Future mobile systems should challenge sustainability as a fundamental goal. It is anticipated that ITU IMT-2030 will contribute to addressing the demand for greater social, economic, and environmental sustainability. The implementations are expected to be designed to achieve the least possible environmental impact and to use resources efficiently by minimizing power consumption, using energy efficiently, and reducing greenhouse gas emissions. By taking crucial factors like reusing, repairing, repurposing, or recycling into account, circular economy concepts help preserve and recover value from resources and extend lifetime. Furthermore, IMT-2030 might make deployments and operations more efficient, which would enhance environmental sustainability and the affordability required to promote social sustainability. It is anticipated that IMT-2030 technologies take into account low power consumption and EE from the standpoints of the network and the user device.

13.4.1 Contribution of Standardization Bodies to Sustainability

Many standard developing organizations are working on EE and eco-design of the network equipment:

- ISO defines a methodology of a life cycle assessment (LCA) as also aspects of the environmental impact of a company, a product, or a service. It addresses all environmental aspects throughout the entire life cycle of products or services: from raw material acquisition through production, use, end-of-life treatment, recycling, and final disposal. The ISO Standards provide requirements and guidelines for LCA-inclusive tools, performance evaluation as audits, and reporting issues. Standards cover a large range of activities: quality management ISO 9000, environmental management ISO 14000, health and safety ISO 45001, energy management ISO 50001, food safety ISO 22000, and IT security ISO/IEC 27001.
- ETSI focuses on the technical committee environmental engineering to define the environmental aspects of telecommunication infrastructures and equipment. One engineering aspect related to sustainability relies on eco-environmental matters like EE, environmental impact analysis, and alternative energy sources. This includes the life cycle assessment of ICT goods, networks, and services; methods to assess the EE of wireless access

networks and equipment, core networks, and wireline access equipment including efficiency metrics/KPI definition for equipment and network; network standby mode for household and office equipment; circular economy standard for ICT solutions.

- ITU-T Study Group 5 works on standards and guidelines for the assessment of mobile network EE, smart sustainable cities, management of batteries, e-waste as also on energy-efficient blockchain systems, and circular economy. The telecommunication standardization sector ITU-T develops international standards known as Recommendations which act as defining elements in the global infrastructure of ICTs. The Study Group SG5 works on the environment and circular economy topics. SG5 is responsible for studies on methodologies for evaluating ICT effects on climate change and publishing guidelines for using ICTs in an eco-friendly way. Moreover, studies on design methodologies to reduce ICTs and e-waste's adverse environmental effects are also part of its mandate. ITU-T SG5 and ETSI-TC EE are working together to develop technically aligned standards on EE, power feeding solutions, circular economy and network efficiency KPI, and eco-design requirements for ICT, with the aim to build an international eco-environmental standardization [65–70].
- 3GPP has been constantly addressing EE aspects of UEs. In addition, further power-saving aspects for UEs were defined in the study and work items related to devices with reduced capabilities RedCap used for wearables or industrial wireless sensors. 3GPP started in the 2Q 2022 Release 18 with a SI on network ES to address RAN EC.

As 3GPP technological trends continue to evolve, spectrum and sustainability become more important considerations. The significance of the UN SDG is acknowledged by 3GPP. The 3GPP is actively developing standards for EE, resource efficiency, circularity, and social responsibility. This comprehensive strategy highlights 3GPP's dedication to setting the pace for worldwide technological trends while advancing systems, connectivity, and diverse communication modalities. Their strategy can reduce resource consumption, reduce the negative environmental effects of network operations, and increase global resource efficiency. These efforts extend beyond telecommunications, as 3GPP is making standards to enable wireless technologies to promote sustainability across industries.

An important area of focus is energy usage, which accounts for a sizable portion of network operators' operating costs. The RAN and its active antenna unit active antenna unit (AAU) are the main sources of EC within the network architecture. It is imperative to create and improve energy-saving methods with targeting for certain deployment scenarios and investigate the possibility of supporting UE.

3GPP has been implementing standards for communication services, increasing transparency by reporting energy usage to clients, studying requirements to close current gaps, and investigating security, billing, and privacy issues in the pursuit of EE in 5G networks. The RAN, which uses the most energy, has received a lot of attention from 3GPP. These initiatives align with broader sustainability goals and serve to reduce the carbon footprint of mobile communications.

13.4.2 Assignment of Sustainability Aspects to Different 6G Subsystems

This section deals with studies and projects addressing relevant results of research work on sustainability issues of mobile network components that could be relevant for future 6G activities [71].

- **Waveform Design for Network EE:** While the techniques considered presently by 3GPP for the SI on network ES do not yet consider fundamental alterations to the waveform design, future releases toward 6G should revisit it for the prospect of reducing EC active transmission and reception.
- **Radio Unit, MIMO, and Reconfigurable Intelligent Surface (RIS):** Regarding the main equipment, the radio unit or antenna unit is the main energy-demanding component. The hardware chain's energy-hungry part in the antenna unit is the power amplifier. Massive MIMO-based antenna units were launched with 5G, and ultra-massive MIMO antennas are planned for 6G in response to the need for larger transmission capacity and new frequency bands. The 6G vision presents additional subjects that further affect the radio portion's EC, such as the deployment of cell-free networks, ultra-massive MIMO antennas, and THz frequency ranges, which are likewise linked to higher bandwidths. The suggested approach to energy-saving benchmarking involves designing hypothetically 6G model networks and analyzing them under various circumstances to determine compromises based on various situations and extremes. The RIS is an additional area of study. The negative consequences of naturally occurring wireless propagation can be eliminated by mobile network operators (MNOs) by RIS controlling the radio waves' scattering, reflection, and refraction characteristics. The random channel (provided by the environment) could be changed into a programmable channel via RIS. Further investigations are required to evaluate their potential in use with respect to network ESs.
- **Orchestration Architecture of the Network:** The strategy for saving energy in part-load operation is to actively shift computing resources from selected system components to free or partially used resources. The purpose of this is to enable optimal utilization. Subsequently, idle systems are switched off in a controlled manner as long as they are not required. Various possible solutions are to be analyzed.

13.4.3 Green Design for Future Networks

In today's world, green is now linked to sustainability, reduction of pollution and carbon footprints, slowing down climate change, and other related aspects. To put it simply, going green means making efficient use of the energy that is available. However, starting with 5G, there will be a massive increase in both the volume of information transmitted and the data rate at which it occurs. The data rates of 6G will be several orders of magnitude higher than even those of 5G, while the data rates of 5G are at least 100 times higher than those of 4G. Every wireless generation sees an increase in the number of users at the same time, with a large proportion of users these days being machines rather than humans. As a result, with each wireless generation, the amount of energy needed to produce and transmit all the anticipated information increases significantly. Future wireless networks are therefore likely to be anything but environmentally friendly. In fact, future networks could become the highest energy consumers (and carbon generators) of the future.

It is obvious that this scenario is not practical or sustainable. The solution is, however, simple: the energy required to transmit a piece (bit) of information must be decreased so that the overall power consumption of future networks, despite the increased traffic flow, mostly stays on par with that of previous generations. Energy-per-bit needs to decrease at least at the rate at which the data throughput increases. In general, this has not been the case to date and the rate of efficiency improvement has been slower than the traffic growth rate. However, it is widely accepted that if future wireless networks maintain the status quo of the said EC, they will be considered *green*. Naturally, the concept of energy-per-bit is fairly abstract. Instead, the idea is that the energy required to transmit and receive chunks of information is decreased, leading to an average EC decrease in the whole network. In practice, this is achieved typically in two ways:

- Reducing the conventional EC by ES within the architecture and hardware of the network, both on the side of BS and on the side of the remote user
- Harvesting the freely available (renewable) energy to reach a net effect of low or even zero EC.

In a hybrid approach, the overall net effect of energy reduction can also be achieved if some of the energy that would normally be wasted during information generation or transmission is harvested elsewhere, typically on the side of the receiver.

As the world moves toward sustainable energy solutions, alternate energy sources are becoming increasingly important. Solar and wind power are two of the most prominent renewable energy sources that can significantly reduce carbon emissions.

- Solar power can be utilized in various ways, such as designing and installing enough solar capacity for excess electricity to be sent back to the grid during peak generation hours. Deploying solar panels at cell sites with excess real estate could also significantly reduce carbon emissions. Renewable energies can lead not only toward net-zero emissions but also operational expenditure reductions.
- Wind energy is another path toward reducing carbon emissions. Wind turbines generate renewable energy by harnessing wind power. There are two main types of turbines: horizontal-axis wind turbines (HAWTs) and vertical-axis wind turbines (VAWTs). HAWTs are more efficient than VAWTs but require more space for optimal performance, making them suitable for open rural spaces. VAWTs operate better in slower or turbulent winds and can be placed closer together, making them ideal for urban or suburban environments.

- Energy harvesting is expected to play a major role in future wireless generations as an aid to overall ES. The major obstruction to the wide adoption of energy harvesting is the reliability of each of the green energy sources. In many scenarios, it is possible to deploy multiple energy-harvesting techniques in parallel. According to the type of energy that is harvested, the energy sources are classified into mechanical, solar or light sources; and electromagnetic energy sources. Mechanical or kinetic sources include sources such as wind, piezoelectric, and electrostatic sources. Light energy sources can come from the sun or artificial light energy sources. Lastly, the most viable electromagnetic energy source is the radio-frequency source. While many of the energy-harvesting techniques have been around for a very long time, energy-harvesting research remains active.

The future 6G AI is expected to be a key enabling technology for a great increase in flexibility and resource-efficient usage of the network. But the application of AI itself goes not without resource consumption. A green design of AI functionality will not only bring AI benefits to the network but will also ensure that AI implementations will not outweigh their resource-saving benefits. Thus, it should be an integral part of all AI designs for the future 6G network.

13.4.4 Technologies to Enhance EE and Low Power Consumption

Future network devices could become energy neutral. Low EC issues can be considered from both the user device and the network's perspectives. Technological advances in various aspects will contribute to lowering power consumption [72].

- **New Architecture:** Operators have proposed various solutions, expecting to build a new information infrastructure that integrates multiple elements such as AI, blockchain, cloud, data, network, edge, terminal, and security. Specifically, the focuses are as follows in terms of infrastructure, unified cloud platform technology stack, and unified orchestration management.
- **AI Native:** AI native improves the intelligent learning and scenario adaptation capabilities of different layers (physical layer, MAC layer, network layer) in an all-round manner, and supports the continuous evolution of personalized intelligent service capabilities. In addition, the data plane and logical intelligence plane are being developed further and included in the network architecture design to enable self-adapting, self-learning, self-correcting, and self-optimizing intelligent networks in the future. Furthermore, by deeply integrating intelligence, computing, and networks, AI native enables industries to create a ubiquitous intelligent ecosystem and networks to achieve intelligent autonomy. It also continuously ensures the quality of AI services and resources through unified orchestration and scheduling.
- **Power Efficiency:** 6G uses higher bands as signal carriers, with a data rate of terabits per second. At the same time, emerging demands for new services, such as holographic communications, intelligent interaction, sensory interconnection, digital twinning, and communications perception, bring

a surge of hundreds and thousands of times of network traffic. To reduce power consumption as much as possible when traffic increases, a new generation of green equipment should be developed based on highly integrated chipsets, high-efficiency devices, upgraded processes, and strong computing algorithms, so that extremely deep sleep and extremely short response time can be achieved. Many technologies, such as intelligent power-saving technology, power consumption reduction by multi-frequency coordination, and power efficiency enhancement by traffic navigation, are implemented on the basis of low-power hardware.

- **Deep Sleep:** The core idea of power-saving technology for wireless software is to reduce redundant network capacity and network power consumption by shutting down some wireless capabilities to match the traffic load distributed in networks. In the future, modular hardware design, scalable software architecture design, and a delicate hibernation scheme will be used to continuously improve the shut-off depth of devices.
- **Multi-dimensional Synergy:** The deep synergy of multi-frequencies and multi-systems reduces network power consumption, realizes dynamic matching of services and resources, and achieves power saving in multi-layer wireless networks without significantly affecting user experience. By combining multiple-domain data (spectrum utilization, user experience, traffic trends) to assess network power efficiency, choosing power-saving techniques based on the particular situation, and taking into account station type, network configuration, traffic statistics, and forecasts, the intelligent self-configuration of power-saving parameters will be made possible. Config needs, UE mobility, and UE business models can all be useful in fine-tuning settings and conserving power without negatively affecting user experience. Through network synergy, information about beam adaptation and cell state (such as sleep state) can be shared between cells. Power usage can be reduced by adjusting the configuration based on the information gathered.
- **MetaCell Is Developed:** It is a new network architecture that maximizes spectrum value and improves spectral efficiency based on a service-centered spectrum allocation strategy, providing superior QoE. MetaCell supports uplink/downlink decoupling and service-centered full-band integrated networking. With MetaCell, ES and low-carbon targets are achieved through higher EE and operation and management are simplified via multi-network integration.
- **Smooth Virtual Cell (SVC):** In contrast to conventional geographic location-based cells, SVC has overcome the limitations of physical cells by dynamically adding or removing some access points (APs) to create a user-specific cell. SVC can guarantee a very constant user experience without being impacted by a physical cell's boundaries. SVC aims to offer users unified, effective services and ensure a smooth transition when they move, no matter where they are. In contrast to the conventional centralized antenna array, SVC is made up of a large number of inexpensive, dispersed APs with few antennas. SVC is applicable in low-latency and mission-critical circumstances in addition to large-capacity settings. By balancing centralized and

distributed processing, SVC can greatly increase spectral and EE when compared to centralized antenna designs, especially in indoor and hotspot environments. Key technologies in SVC include coordination set selection, power control, user scheduling, and beam training, which can be modeled as an optimization problem. Depending on whether the variables are continuous or discrete, there are two types of optimization: combinatorial optimization and continuous optimization. Power control determines power parameters based on the known coordination sets and scheduled users, and it is continuous optimization.

13.5 CONCLUDING REMARKS AND FURTHER STUDY

Identifying and developing new mobile broadband technologies is most important in achieving the targeted performance of 5G networks. At the same time, it is impossible to ignore the impact on the environment and operational processes. The importance of these aspects is obvious to researchers and industry. The issue of sustainability along with the anticipated significant increase in EC has made green communication a crucial 5G/6G criterion. The objective is to keep the mobile network's electricity usage at its current level while increasing the amount of data transmission along with the number of BSs.

We have discussed the growing need for EE in the 5G mobile networks. EE energy efficiency indicates the efficient use of electricity, and at the same time it is one sustainability metric that can be measured. The focus is on methods and metrics, as well as techniques to reduce consumption at the BS and network level. Metrics retain a significant role in the continuation of the 5G standardization process. To improve ES, optimizing network EC through efficient power management techniques becomes a crucial challenge. The first systematic step is the definition of the KPIs as the collection of use cases and minimum requirements. Next, the focus is on models of EC and metrics to evaluate new aspects of 5G EE. It is in fact necessary to develop a stepwise strategy to achieve enhanced EE of functionalities being developed or standardized, in parallel with the standardization process of 5G. Development, standardization, and implementation of ES features as well as incorporating renewable energy sources and promoting energy-efficient practices in network infrastructure, are potential solutions.

Next-generation 6G networks are expected to achieve greener and socially seamless wireless connections in a global scope. A new transformative technology will enable even faster and more reliable use cases. In the context of Industry 4.0 and beyond, 6G's unprecedented data speeds of terabits per second, ultra-low latency enabling near-instantaneous transmission and processing of vast amounts of data, massive connectivity, and enhanced network intelligence could be potential solutions for the integration of digital technologies, thus reinforcing the sustainable evolution of industries.

As shown in this chapter, EE has gained in the last decade its own role as a performance measure and design constraint for communication networks, but many technical, regulatory, policy, and operational challenges still remain to be addressed.

In conclusion, global technology trends setting the stage for innovative connectivity, sustainable systems, and diverse modes of communication. Dedication to minimizing resource usage, limiting environmental impacts, and increasing resource efficiency underscores our commitment to shaping the future of multimedia communications and contributing to a more sustainable and interconnected world.

REFERENCES

1. W. Tong, P. Zhu, *6G The next horizon: from connected people and things to connected intelligence*, Cambridge University Press, 2021.
2. A.R. Mishra, *Fundamentals of network planning and optimisation 2G/3G/4G: evolution to 5G*, Wiley, 2018.
3. W. Chen, P. Gaal, J. Montojo, H. Zisimopoulos, *Fundamentals of 5G communications: connectivity for enhanced mobile broadband and beyond*, McGraw Hill, 2021.
4. X. Lin, N. Lee (Eds.), *5G and beyond: fundamentals and standards*, Springer, 2021.
5. J.G. Andrews, S. Buzzi, W. Choi, S. Hanly, A. Lozano, A.C.K. Soong, J.C. Zhang, "What will 5G be?" *IEEE Journal on Selected Areas in Communications*, vol. 32, no. 6, pp. 1065–1082, June 2014.
6. M. Shaf et al., "5G: a tutorial overview of standards, trials, challenges, deployment, and practice", *IEEE Journal on Selected Areas in Communications*, vol. 35, no. 6, pp. 1201–1221, 2017.
7. M.J. Al-Dujaili, M.A. Al-Dulaimi, "Fifth-generation telecommunications technologies: features, architecture, challenges, and solutions", *Wireless Personal Communications*, vol. 128, no. 1, pp. 447–469, 2023.
8. R. Li, Z. Zhao, X. Zhou, G. Ding, Y. Chen, Z. Wang, H. Zhang, "Intelligent 5G: when cellular networks meet artificial intelligence", *IEEE Wireless Communications*, vol. 24, no. 5, pp. 175–183, 2017.
9. B. Bertenyi, "5G evolution: what's next?", *IEEE Wireless Communications*, vol. 28, no. 1, pp. 4–8, 2021.
10. S. Buzzi, I. Chih-Lin, T.E. Klein, H.V. Poor, C. Yang, A. Zappone, "A survey of energy-efficient techniques for 5G networks and challenges ahead," *IEEE Journal on Selected Areas in Communications*, vol. 34, no. 4, pp. 697–709, 2016.
11. A. Zappone, L. Sanguinetti, G. Bacci, E. Jorswieck, M. Debbah, "Energy-efficient power control: a look at 5G wireless technologies", *IEEE Transactions on Signal Processing*, vol. 64, no. 7, pp. 1668–1683, 2015.
12. R. Tan, Y. Shi, Y. Fan, W. Zhu, T. Wu, "Energy saving technologies and best practices for 5G radio access network", *IEEE Access,* vol. 10, pp. 51747–51756, 2022.
13. Y.-N.R. Li, M. Chen, J. Xu, L. Tian, K. Huang, "Power saving techniques for 5G and beyond", *IEEE Access,* vol. 8, pp. 108675–108690, 2020.
14. N. Piovesan, N. Piovesan, D. López-Pérez, A. De Domenico, X. Geng, H. Bao, M. Debbah, "Machine learning and analytical power consumption models for 5G base stations", *Communications Magazine*, vol. 60, no. 10, pp. 56–62, 2022.
15. W. Chen, X. Lin, J. Lee, A. Toskala, S. Sun, C.F. Chiasserini, L. Liu, "5G-Advanced toward 6G: past, present, and future", *IEEE Journal on Selected Areas in Communications*, vol. 41, no. 6, pp. 1592–1619, June 2023.
16. W. Chen, J. Montojo, J. Lee, M. Shafi, Y. Kim, "The standardization of 5G-Advanced in 3GPP", *IEEE Communications Magazine*, vol. 60, no. 11, pp. 98–104, Nov. 2022.
17. X. Lin, "An overview of 5G-Advanced evolution in 3GPP Release 18", *IEEE Communications Standards Magazine*, vol. 6, no. 3, pp. 77–83, Sept. 2022.
18. Y. Kim et al., "New radio (NR) and its evolution toward 5G-Advanced", *IEEE Wireless Communications*, vol. 26, no. 3, pp. 2–7, June 2019.

19. R. Bassoli, F.H.P. Fitzek, E.C. Strinati, "Why do we need 6G?", *ITU Journal on Future and Evolving Technologies*, vol. 2, no. 6, pp. 1–31, 2021.
20. S. Cavallero, N. Decarli, G. Cuozzo, C. Buratti, D. Dardari, R. Verdone, "Terahertz networks for future industrial Internet of Things", *ITU Journal on Future and Evolving Technologies*, vol. 4, no. 1, pp. 196–208, 2023.
21. A. Fehske, J. Malmodin, G. Biczok, G. Fettweis, "The global footprint of mobile communications – the ecological and economic perspective", *IEEE Communications Magazine*, vol. 49, no. 8, pp. 55–62, 2011.
22. G. Auer et al., "How much energy is needed to run a wireless network?" *IEEE Wireless Communications*, vol. 18, no. 5, pp. 40–49, 2011.
23. D.J. Goodman, N.B. Mandayam, "Power control for wireless data," *IEEE Personal Communications*, vol. 7, pp. 48–54, 2000.
24. 3GPP TS 28.541 *5G Network Resource Model (NRM)*, 2024.
25. 3GPP TS 28.310 Management and orchestration; *Energy Efficiency of 5G*, 2024.
26. 3GPP TS 28.552 Management and orchestration; *5G Performance Measurements*, 2024.
27. 3GPP TS 28.554 Management and orchestration; *5G End to End Key Performance Indicators (KPI)*, 2024.
28. 3GPP TS 28.622 Telecommunication management; *Generic Network Resource Model (NRM) Integration Reference Point (IRP); Information Service (IS)*, 2024.
29. 3GPP TS 28.813 *Study of New Aspects of Energy Efficiency for 5G*, 2021.
30. 3GPP TR 38.864 *Study on Network Energy Savings for NR*, 2023.
31. 3GPP TR 38.840 *Study on User Equipment (UE) Power Saving in NR*, 2019.
32. 3GPP TR 22.840, *Study on Ambient Power-Enabled Internet of Things*, 2023.
33. 3GPP SA TR21.915, Release 15 Description; *Summary of Rel-15 Work Items*, Oct. 2019.
34. 3GPP SA TR21.916, Release 16 Description; *Summary of Rel-16 Work Items*, June 2021.
35. 3GPP SA TR21.917, Release 17 Description; *Summary of Rel-17 Work Items*, Sep. 2022.
36. 3GPP SA TR21.918, Release 18 Description; *Summary of Rel-18 Work Items*, March 2024.
37. 3GPP SA TR21.919, Release 19 Description; *Summary of Rel-19 Work Items*, March 2024.
38. D. Sabella, D. Rapone, M. Fodrini, C. Cavdar, M. Olsson, P. Frenger, S. Tombaz, Energy management in mobile networks towards 5G, in *Energy management in wireless cellular and ad-hoc networks* (M.Z. Shakir, M.A. Imran, K.A. Qaraqe, M.-S. Alouini, A.V. Vasilakos, Eds), Springer, 2016.
39. M. Usama, M. Erol-Kantarci, "A survey on recent trends and open issues in energy efficiency of 5G", *Sensors*, vol. 19, no. 14, p. 3126, 2019.
40. 3GPP TR 32.972 *Study on System and Functional Aspects of Energy Efficiency in 5G Networks* (Release 16), Sept. 2019.
41. 3GPP TR 22.866 *Enhanced Relays for Energy Efficiency and Extensive Coverage* (Release 16), Dec. 2019.
42. 3GPP TS 28.310 *Energy Efficiency of 5G* (Release 16), May 2020.
43. 3GPP TS 28.813 *Study on New Aspects of Energy Efficiency (EE) for 5G* (Release 17), June 2020.
44. R. Schwartz, J. Dodge, N.A. Smith, O. Etzioni, "Green AI", *Communications of the ACM*, vol. 63, no. 12, pp. 54–63, 2020.
45. C.E. Tripp, J. Perr-Sauer, J. Gafur, A. Nag, A. Purkayastha, S. Zisman, E.A. Bensen, Measuring the energy consumption and efficiency of deep neural networks: an empirical analysis and design recommendations, arXiv:2403.08151, 2024.
46. D. Hernandez, T.B. Brown, Measuring the algorithmic efficiency of neural networks, arXiv:2005.04305, 2020.

47. ETSI TR 103 542, *Study on Methods and Metrics to Evaluate Energy Efficiency for Future 5G Systems*, June 2018.
48. ETSI ES 202 706-1, *Metrics and Measurement Method for Energy Efficiency of Wireless Access Network Equipment, Part 1: Power Consumption - Static Measurement Method*, Oct. 2016.
49. ETSI ES 202 706-2, *Metrics and Measurement Method for Energy Efficiency of Wireless Access Network Equipment, Part 2: Energy Efficiency - Dynamic Measurement Method*, Nov. 2018.
50. ETSI ES 203 228, Environmental Engineering (EE); *Assessment of Mobile Network Energy Efficiency*, Jan.2015.
51. ITU-T L.1330, *Energy Efficiency Measurement and Metrics for Telecommunication Networks*, Mar. 2015.
52. ITU-T L.1310 Supplement 36, *Energy Efficiency Metrics and Measurement Methods for Telecommunication Equipment*, July 2017.
53. ITU-T L.1331, *Assessment of Mobile Network Energy Efficiency*, Apr. 2017.
54. 3GPP TR 38.913 *Study on Scenarios and Requirements for Next Generation Access Technologies* (Release 15), June 2018.
55. 3GPP TR 21.866 *Study on Energy Efficiency Aspects of 3GPP Standards* (Release 15), June 2017.
56. ITU-R IMT-2020 M.2083, *Vision – Framework and Overall Objectives of the Future Development of IMT for 2020 and Beyond*, Oct. 2015.
57. 3GPP RAN WG1 TR38.843, Release 18; *Study on Artificial Intelligence (AI)/Machine Learning (ML) for NR Air Interface*, June 2022.
58. 3GPP SA WG4 TR26.802, Release 17; *Multicast Architecture Enhancement for 5G Media Streaming*, June 2021.
59. 3GPP SA WG4 TR26.998, Release 17; *Support of 5G Glass-Type Augmented Reality/ Mixed Reality (AR/MR) Devices*, March 2022.
60. K. Samdanis, P. Rost, A. Maeder, M. Meo, C. Verikoukis (Eds), *Green communications: Theoretical fundamentals, algorithms, and applications*, Wiley, 2015.
61. I. Chih-Lin, S. Han, S. Bian, "Energy-efficient 5G for a greener future", *Nature Electronics*, vol. 3, pp. 182–184, April 2020.
62. Z. Al-Dunainawi, R.S. Alhumaima, H.S. Al-Raweshidy, "Green network costs of 5G and beyond: expectations vs reality", *IEEE Access*, vol. 6, pp. 60206–60213, 2018.
63. S. Chen, Y.-C. Liang, S. Sun, S. Kang, W. Cheng, M. Peng, "Vision, requirements, and technology trend of 6G: how to tackle the challenges of system coverage, capacity, user data-rate and movement speed", *IEEE Wireless Communications*, vol. 27, no. 2, pp. 218–228, 2020.
64. E. Oh, K. Son, B. Krishnamachari, "Dynamic base station switching-on/off strategies for green cellular networks," *IEEE Transactions on Wireless communications*, vol. 12, no. 5, pp. 2126–2136, 2013.
65. ITU-R Report M.2516, *Future Technology Trends of Terrestrial International Mobile Telecommunications Systems Towards 2030 and Beyond*, 2022.
66. ITU-R Recommendation M.2160, *Framework and Overall Objectives of the Future Development of IMT for 2030 and Beyond*, 2023.
67. ITU-R Report M.2410, *Minimum Requirements Related to Technical Performance for IMT-2020 Radio Interface(s)*, Nov. 2017.
68. ITU-R Report M.2411, *Requirements, Evaluation Criteria and Submission Templates for the Development of IMT-2020*, Nov. 2017
69. ITU-R Report M.2412, *Guidelines for Evaluation of Radio Interface Technologies for IMT-2020*, Oct. 2017
70. ITU-R Report M.2150, *Detailed Specifications of the Terrestrial Radio Interfaces of International Mobile Telecommunications-2020 (IMT-2020)*, Feb. 2022.

71. M.W. Akhtar, S.A. Hassan, R. Ghaffar, H. Jung, S. Garg, M.S. Hossain, "The shift to 6G communications: vision and requirements", *Human-Centric Computing and Information Sciences*, vol. 10, article:53, Springer, 2020.
72. T. Huang, W. Yang, J. Wu, J. Ma, X. Zhang, D. Zhang, "A survey on green 6G network: architecture and technologies", *IEEE Access*, vol. 7, pp. 175758–175768, Dec. 2019.

14 AI-Powered Agile Project Management in Sustainable Development of Industry 4.0

Vladan Pantovic, Milan Smigic, Milan Djordjevic, and Dragorad A. Milovanovic

14.1 INTRODUCTION

Project management as a human activity has a tradition of thousands of years. The best historical examples are the construction of the pyramids in Egypt and the Great Wall of China. As a scientific discipline, project management emerged in the second half of the 20th century – which coincides with the Third Industrial Revolution.

Positive feedback from the Third and Fourth Industrial Revolutions with the development of project management is evident. The rapid advancements of Industry 4.0 have significantly influenced the landscape of project management. This industrial revolution, characterized by the integration of cyber-physical systems (CPSs), the Internet of Things (IoT), and advanced data analytics, demands project managers to adapt to the new paradigm.

The integration of advanced digital technologies in Industry 4.0 has significantly shifted the focus toward creating sustainable and Agile industrial environments. A key element in achieving this is the enhancement of project management maturity, which plays a crucial role in driving business excellence. The study by Fajsi et al. (2022) emphasizes the significant impact of project management maturity on organizational business excellence within the context of Industry 4.0. By utilizing the Project Management Maturity Model (ProMMM), the research identified a strong correlation between higher levels of project management maturity and superior business performance.

The Scaled Agile Framework (SAFe) is an effective project management solution for Industry 4.0, ensuring that the benefits of Agile methodologies are realized on a larger scale. SAFe promotes iterative development, continuous feedback, and collaboration among cross-functional teams.

This chapter provides a general overview of project management methodologies in the context of Industry 4.0, explores the transition from traditional methods to Agile practices, and discusses the role of artificial intelligence (AI) in shaping the future of project management.

DOI: 10.1201/9781003511298-14

14.2 INDUSTRIAL REVOLUTIONS AND PROJECT MANAGEMENT

14.2.1 Historical Overview

Over the years, society has experienced four major industrial revolutions, each significantly transforming manufacturing and production processes. The First Industrial Revolution in the late 18th century introduced mechanization through water and steam power, revolutionizing factories and shifting economies from agrarian bases to industrialized systems. The Second Industrial Revolution in the late 19th and early 20th centuries brought about mass production facilitated by electricity and the assembly line, greatly enhancing production efficiency and capabilities.

The Third Industrial Revolution, which began in the late 20th century, introduced automation and computerization, integrating electronics and information technology to automate manufacturing processes. This era saw the rise of programmable logic controllers and robotics, which improved precision and productivity while reducing reliance on human labor.

Currently, we are witnessing the Fourth Industrial Revolution or Industry 4.0, characterized by the fusion of digital, physical, and biological worlds through advanced technologies such as CPSs, IoT, big data analytics, and cloud computing (Esteves et al., 2020). Industry 4.0 represents a paradigm shift in industrial innovation, enabling the development of smart factories where machines, devices, and systems are interconnected and communicate autonomously. This revolution leverages real-time data to drive intelligent decision-making and optimize manufacturing processes.

14.2.2 Key Components

The core components of Industry 4.0 include CPSs, IoT, cloud computing, and big data analytics, which collectively enhance operational efficiency, flexibility, and product quality.

- **Cyber-Physical Systems (CPS)** integrate computation with physical processes, allowing for real-time monitoring and control through interconnected networks. CPS enables the interaction between the physical and digital worlds, facilitating automated and optimized production processes.
- **Internet of Things (IoT)** involves the interconnection of physical devices, such as sensors and machinery, via the Internet, enabling them to collect and exchange data. This technology enhances visibility and control over manufacturing processes, allowing for predictive maintenance and improved resource management.
- **Internet of Services (IoS)** extends the capabilities of IoT by offering services over the Internet. This includes cloud-based solutions that provide access to data and applications from anywhere, enabling more efficient collaboration and resource allocation.
- **Smart Factories** utilize CPS, IoT, and IoS to create a highly automated and interconnected production environment. These factories can adapt to changing conditions in real time, optimize production schedules, and reduce waste, leading to greater efficiency and sustainability.

14.2.3 Transforming Project Management

Crucial technologies for Industry 4.0 simultaneously have a great impact on the transformation of project management as a professional and scientific discipline (Milovanović and Pantović, 2024):

- **Big Data** is fundamental to Industry 4.0, offering extensive support for automation and intelligent systems. By analyzing vast amounts of data, project managers can gain real-time insights into production conditions and market trends. This allows for the optimization of processes, as managers can use comprehensive reports to improve production strategies, anticipate potential issues, and make informed decisions that support sustainable project growth.
- **Automation and Sensors** contribute to the optimization of project execution. Automation leverages data to streamline repetitive and conditional tasks through machine learning, enhancing production speed and minimizing errors. Sensors play a critical role by enabling continuous monitoring of processes without the need for shutdowns or manual interventions. This capability allows project managers to maintain high levels of efficiency and operational accuracy throughout the project lifecycle.
- **IoT** introduces intelligence into the production process by facilitating autonomous operations. IoT enables devices and machines to communicate and interact with each other, significantly shortening production times and enhancing product quality. This interconnectedness allows for real-time adjustments and proactive maintenance, ensuring that the entire production process is more responsive and adaptive to changing conditions.
- **3D Printing** offers significant flexibility in processing custom resources, reducing waste, and allowing for more precise production schedules. This technology enables project managers to produce parts and components on demand, adapting quickly to design changes and customer requirements. The ability to manufacture items in a flexible and efficient manner enhances the overall agility and responsiveness of the production process, making it a valuable asset in project management.

14.3 AGILITY IN THE CONTEXT OF INDUSTRY 4.0

Agile project management, characterized by its iterative approach and focus on customer feedback, has transformed how projects are executed across various sectors (PMI, 2017). Agile project management involves shorter project cycles, constant testing and adaptation, and overlapping work by multiple teams or contributors (Ambler & Lines, 2022). One of the aims of an Agile or iterative approach is to release benefits throughout the process rather than only at the end. At the core, Agile projects should exhibit central values and behaviors of trust, flexibility, empowerment, and collaboration. An incremental approach breaks the development process down into small, manageable portions known as increments. Each increment builds on the previous version so that improvements are made step by step. Even though the Agile

Manifesto was published some time ago (2001), there are still many organizations that did not incorporate Agile principles into their project management practices.

The rapid advancements of Industry 4.0 have significantly influenced the landscape of project management. Marnewick and Marnewick (2020) explore how the implementation of Industry 4.0 technologies necessitates a shift in leadership styles within project management. Their study highlights that traditional "command-and-control" leadership approaches are less effective in this new environment, advocating instead for servant leadership as a more suitable style to manage the complexities and dynamic nature of Industry 4.0 projects.

Agile methodologies have become increasingly vital in the context of Industry 4.0 due to their ability to adapt to rapid changes and foster continuous improvement. Marnewick and Marnewick (2020) discuss how Agile approaches, including frameworks like Scrum and Extreme Programming (XP), align well with the needs of Industry 4.0 by emphasizing iterative development, customer collaboration, and flexibility. The Agile mindset promotes a shift away from rigid, sequential project management methods toward more dynamic and responsive practices. This shift is essential for managing the complex and evolving technological projects characteristic of Industry 4.0.

14.3.1 Project Management Maturity and Business Excellence

The integration of advanced digital technologies in Industry 4.0 has significantly shifted the focus toward creating sustainable and Agile industrial environments. A key element in achieving this is the enhancement of project management maturity, which plays a crucial role in driving business excellence. The study by Fajsi et al. (2022) emphasizes the significant impact of project management maturity on organizational business excellence within the context of Industry 4.0. By utilizing the Project Management Maturity Model (ProMMM), the research identified a strong correlation between higher levels of project management maturity and superior business performance.

The emergence of Industry 4.0 has significantly transformed the operational dynamics of firms, driving the necessity for agility and adaptability in business processes. According to Şen & İrge study (2020), the integration of advanced technologies such as CPSs, IoT, and cloud computing is pivotal in shaping the modern industrial landscape. These technologies facilitate the creation of smart factories, where digital and physical processes are seamlessly integrated, enabling real-time data exchange and automated decision-making.

In the context of Industry 4.0, agility is a critical strategic approach that determines the success and sustainability of firms. The study emphasizes that Agile firms are characterized by their ability to rapidly respond to changes, innovate continuously, and adapt their processes to meet evolving market demands. Agile methodologies, such as Scrum and XP, are increasingly being implemented to enhance flexibility and responsiveness. These methodologies support iterative development and continuous feedback, which are essential for managing the complexities and uncertainties inherent in Industry 4.0 projects.

14.3.2 Agile Manufacturing and Industry 4.0

The integration of information and communication technologies (ICTs) with Industry 4.0 and Agile manufacturing is essential for achieving competitive advantage in today's dynamic business environment. Mohammed and Trzcielinski (2021) emphasize that ICT serves as a critical enabler for both Industry 4.0 and Agile manufacturing systems, enhancing the agility of manufacturing processes by facilitating real-time data exchange and decision-making.

Agile manufacturing, which focuses on adaptability and rapid response to market demands, is significantly bolstered by the implementation of Industry 4.0 technologies. These technologies include CPS, IoT, 5G and cloud computing, which collectively create smart manufacturing environments (Milovanović & Pantović, 2024). The use of ICT in these systems enables seamless communication between machines, devices, and humans, allowing for automated and optimized production processes.

The integration of Agile manufacturing with Industry 4.0 technologies presents a transformative opportunity for manufacturing enterprises to enhance flexibility, responsiveness, and efficiency. According to Yaqoob & Azam study (2018), Agile manufacturing emphasizes adaptive capabilities to meet future changes and uncertainties by integrating organizational information comprehensively.

One of the key enablers of Agile manufacturing within the Industry 4.0 framework is the formation of virtual enterprises (VE). VEs allow various member companies to collaborate effectively, combining their competencies to develop new or enhanced products tailored to customer requirements. This collaborative approach not only enhances product quality but also reduces time-to-market and increases operational efficiency.

Perez (2022) in the study posits that each technological revolution follows a pattern of installation, characterized by financial bubbles and speculative investment, followed by a deployment phase where the technology is widely adopted and leads to a "golden age" of economic growth. The transition between these phases is often marked by a financial crisis, as speculative bubbles burst and capital is reallocated to more productive uses.

14.3.3 SAFe as a Project Management Solution for Industry 4.0

The Scaled Agile Framework (SAFe) is an effective project management solution for Industry 4.0 and the emerging Industry 5.0, providing a structured approach to achieving business agility. SAFe enables organizations to scale Agile practices across multiple teams, departments, and even entire enterprises, ensuring that the benefits of Agile methodologies are realized on a larger scale.

In Industry 4.0, the need for rapid response to market changes and technological advancements is paramount. SAFe addresses this need by promoting iterative development, continuous feedback, and collaboration among cross-functional teams. These principles are critical in managing the uncertainties and fast-paced changes inherent in Industry 4.0 projects.

14.3.3.1 Principles and Core Values of SAFe

The SAFe is underpinned by a set of core values and principles that guide its implementation and operation. These core values – alignment, built-in quality, transparency, and program execution – serve as the foundation for successful Agile practices at scale. Alignment ensures that all teams and departments work cohesively toward common goals and strategic objectives, thereby enhancing organizational coherence and focus. Built-in quality emphasizes the necessity of maintaining high standards throughout the development process, ensuring that all products meet or exceed customer expectations. Transparency fosters open communication and visibility into all aspects of the project, enabling stakeholders to make informed decisions and maintain trust. Program execution focuses on the effective planning, coordination, and execution of tasks across multiple teams, ensuring the consistent delivery of value.

14.3.3.2 SAFe Implementation Roadmap

The implementation of SAFe within an organization follows a structured roadmap designed to facilitate smooth adoption and integration. This roadmap begins with recognizing the need for change, often referred to as reaching a tipping point, followed by training Lean-Agile change agents who will drive the transformation process. The next phase involves establishing a Lean-Agile Center of Excellence to support and sustain the transformation. Subsequent steps include training executives, managers, and leaders in Lean-Agile principles and identifying value streams and Agile Release Trains (ARTs). ARTs are central to the SAFe framework, consisting of multiple Agile teams that work collaboratively toward common objectives. The roadmap culminates in the launch of ARTs, coaching ART execution, and fostering a culture of relentless improvement through continuous learning and adaptation (Scaled Agile, Inc., 2024).

SAFe is designed to be scalable and adaptable to organizations of all sizes. It provides different levels of implementation, including essential SAFe for small teams, Large-solution SAFe for medium-sized enterprises and portfolio SAFe and full SAFe for large organizations. This flexibility ensures that SAFe can be effectively applied to various organizational contexts, facilitating growth and innovation at every scale.

14.3.3.3 SAFe Roles and Responsibilities

SAFe delineates specific roles and responsibilities to ensure effective implementation and operation. Key roles within SAFe include the Release Train Engineer (RTE), Product Owner, Scrum Master, and Agile Team. The RTE acts as a servant leader and coach for the ART, facilitating events and processes to ensure continuous delivery of value. Product Owners are tasked with defining and prioritizing the product backlog, ensuring that the team works on the most valuable features. Scrum masters serve as coaches and facilitators for Agile teams, assisting them in adhering to Agile practices and removing impediments. Agile teams, comprising developers, testers, and other specialists, collaborate to deliver high-quality products in an iterative and incremental manner (Leffingwell at al., 2017).

14.3.3.4 SAFe Practices and Tools

SAFe incorporates a variety of practices and tools to support Agile development at scale. One of the key practices is program increment (PI) planning, a cadence-based event that brings together all teams within the ART to plan and synchronize their work for the upcoming increment. PI planning ensures alignment and commitment to shared goals and objectives. Additionally, SAFe integrates the use of Kanban and Scrum frameworks to manage work, visualize workflows, and enhance process efficiency (Scaled Agile, Inc., 2024). DevOps practices are also integral to SAFe, promoting collaboration between development and operations teams, automating deployment processes, and ensuring continuous delivery of value. Tools such as the SAFe Program Board provide a visual representation of features and dependencies, enabling teams to manage and track their progress effectively (Knaster & Leffingwell, 2020).

SAFe's structured yet flexible approach positions it as an ideal framework for organizations integrating AI into project management. Its emphasis on alignment, transparency, and continuous improvement aligns seamlessly with the demands of AI-powered Agile project management. For virtual organizations and those experiencing rapid scaling, SAFe provides a scalable and adaptable structure that fosters innovation and agility (Leffingwell at al., 2017). By promoting collaboration across cross-functional teams and facilitating the integration of advanced technologies like AI, SAFe enables organizations to manage complex projects effectively, maintain high standards, and respond swiftly to market changes. This capability is especially critical in the context of sustainable development in Industry 4.0, where continuous adaptation and innovation are paramount for long-term success.

14.4 AI IN PROJECT MANAGEMENT

The integration of AI into project management has gained significant attention, revolutionizing traditional methodologies and enhancing various aspects of project execution indicates Pantović (2024). Bharati and Sandbrink (2024) emphasize that AI's ability to provide real-time data analysis, risk assessment, and predictive insights has transformed planning, scheduling, and decision-making processes in project management.

Moreover, the study highlights the dual impact of AI on both productivity and ethical considerations within project management. AI-driven analytics offer substantial benefits, including workload reduction, enhanced decision-making, and access to insightful data that supports proactive project management.

14.4.1 Integration and Impact

The integration of AI into project management has become a critical area of focus for organizations aiming to enhance productivity, optimize resource allocation, and minimize risks. Jannat, Ahmed, and Rajput (2024) provide a comprehensive analysis of how AI technologies such as predictive analytics, natural language processing, and machine learning are being utilized to improve project outcomes.

Despite the evident benefits, the study also identifies several challenges that hinder the seamless integration of AI in project management. Organizational reluctance, expertise shortages, and data privacy concerns are prominent barriers that need to be addressed.

The advent of Industry 5.0 marks a significant shift from the automation-focused Industry 4.0 toward a more human-centric approach, integrating advanced technologies with human creativity and collaboration. One of the key aspects of Industry 5.0 is the use of collaborative robots, or cobots, which work alongside humans to enhance productivity and safety in manufacturing environments. Unlike traditional industrial robots that operate in isolation, cobots are designed to interact directly with human workers, providing assistance and augmenting human capabilities (Doyle-Kent, & Kopacek, 2021).

Agile methodologies play a crucial role in the successful implementation of Industry 5.0 by fostering a responsive and adaptive management approach. Agile principles, which emphasize iterative development, continuous feedback, and cross-functional collaboration, align well with the dynamic and interconnected nature of modern industrial environments (Kılıç & Balcıoğlu, 2023).

Adding to this dynamic landscape, the integration of AI promises to further revolutionize these methodologies. AI, with its capability to process large volumes of data, learn from outcomes, and predict future trends, aligns well with the Agile philosophy of rapid adaptation and continuous improvement.

As AI technologies mature, they offer tools for enhancing decision-making processes, automating routine tasks, and personalizing stakeholder interactions. This fusion of AI with Agile methodologies not only augments the efficiency of processes but also helps in navigating the complexities of modern project demands more effectively.

AI can support Agile professionals, from automating administrative tasks to providing predictive insights that enhance project planning and execution. The benefits of employing AI in Agile projects are vast, including increased accuracy in tasks and better resource management. However, integrating AI also presents significant challenges, not only for Agile but overall use of AI, such as data privacy concerns, potential bias in AI algorithms, and the need for teams to adapt to new technologies.

Currently, the integration of AI in Agile project management is an emerging trend rather than a standard practice. Agile methodologies, such as Scrum, Kanban, Disciplined Agile, SAFe, and Lean (there is still discussion if we should name Scrum, Kanban, Lean, Disciplined Agile, SAFe, etc. as methodologies, frameworks, or approaches), have traditionally relied heavily on human expertise and adaptability. However, as projects become increasingly complex and data-driven, the limitations of human capabilities in data processing and decision speed become apparent.

AI tools are gradually being adopted to enhance these capabilities. For instance, AI-driven traditional/waterfall project management software can automate the scheduling and tracking of tasks, predict project outcomes based on historical data, and even recommend resource reallocation to optimize productivity. Tools equipped with machine learning algorithms can analyze past project performance to identify successful patterns or strategies, which can be replicated in future projects.

Moreover, AI applications in Agile environments include chatbots and virtual assistants that facilitate communication and collaboration, reducing the time teams spend on meetings and emails. Predictive analytics are used to foresee risks and provide quantitative insights into probable project bottlenecks before they become critical issues.

Despite these advances, the full integration of AI in Agile projects is still developing. Many organizations are in the experimental phase, determining the best practices for combining AI with their existing Agile frameworks. The transition involves not only technological upgrades but also a cultural shift within teams to embrace AI tools as partners in project management.

14.4.2 How AI Can Help Agile Professionals?

AI has the potential to significantly assist Agile professionals by enhancing efficiency and accuracy in their tasks. One of the key contributions of AI is in the realm of data analysis. Agile projects generate vast amounts of data, and AI can provide deep insights from this data, which are beyond the scope of manual analysis. For instance, AI can analyze user feedback to identify patterns or trends that might inform product development, helping teams prioritize features that offer the most value to customers.

Additionally, AI can automate routine and repetitive tasks, such as updating progress on tasks, sending follow-up emails, or managing project backlogs. This automation frees up Scrum Masters, Product Owners, and team members to focus on more strategic activities that require human intervention, such as brainstorming sessions, critical decision-making, or complex problem-solving.

Here are a few examples of how AI can be integrated into these Agile activities:

- **Managing the Backlog:** AI can automate the prioritization of backlog items based on historical data about project outcomes, team preferences, and deadlines. Machine learning algorithms can analyze past sprint performances to suggest which items should be tackled next, optimizing the flow of work, and ensuring that the most critical tasks are addressed promptly. Additionally, AI can provide real-time updates and recommendations, helping to keep the backlog organized and up-to-date with minimal manual intervention.
- **Writing User Stories and Epics:** Large language models chatbots like ChatGPT, Gemini, and CoPilot, Claude can be applied to improve the creation and refinement of user stories and epics. AI-powered tools can help in drafting these elements by suggesting improvements in language for clarity and consistency based on best practices. They can also automatically tag and categorize stories into epics based on their content, facilitating better organization and tracking. Furthermore, AI can analyze user feedback from various channels to generate new user stories that accurately reflect customer needs and preferences.
- **Estimating Tasks:** AI can significantly help with task estimation, which is crucial for planning and resource allocation in Agile projects. By using historical data on how similar tasks were estimated and executed, AI models can provide more accurate estimates for new tasks. Techniques like "shirt sizing" or "planning poker" can be enhanced with AI by offering

data-driven reference points, reducing biases that typically occur in group estimation sessions. For instance, AI can suggest a "size" for a task based on its characteristics and historical similarities, helping teams reach a consensus more quickly and with greater confidence.

- **Stand-Up and Retrospective Meetings:** AI can also play a role in stand-up meetings by automating the collection and reporting of what each team member has done, plans to do, and any obstacles they face. Voice recognition and natural language processing can be used to transcribe meetings and highlight action items, decisions, and blockers. This information can be automatically updated in the supporting tools, ensuring that all team members have real-time access to the latest project updates without manual data entry.

During retrospective meetings, AI can help by analyzing data from past sprints to identify trends and patterns in performance, obstacles, and team dynamics. This analysis can provide insights into what went well and what didn't, helping teams to make more informed decisions about changes to implement in future cycles. AI tools can also facilitate the collection of anonymous feedback, ensuring that all voices are heard and that the feedback is compiled and presented efficiently and effectively.

Across all these activities, AI can provide predictive analytics to foresee and mitigate risks, enhance communication through automated summarization and translation, and foster a deeper understanding of team dynamics through sentiment analysis. By automating administrative tasks and offering data-driven insights, AI frees up Agile teams to focus more on creative problem-solving and innovation.

14.4.3 Challenges of Using AI in Agile Projects

While the benefits are considerable, integrating AI into Agile project management brings several challenges that teams must navigate to leverage this technology effectively. Below you will find a few of them.

- **Cultural Resistance and Change Management:** One of the most significant barriers to integrating AI within Agile teams is cultural resistance. Agile methodologies thrive on human interaction, collaboration, and decision-making. Introducing AI can be perceived as a threat to established norms and roles within the team. Team members might be skeptical about relying on AI for decision-making or fear that AI could replace their roles. Overcoming this resistance involves careful change management, education, and demonstrations of AI as a supportive tool rather than a replacement.
- **Training and Skill Gaps:** AI tools require a certain level of technical proficiency. Teams may find themselves ill-equipped in terms of skills to handle sophisticated AI technologies. Training becomes a crucial factor in such scenarios. Organizations need to invest in upskilling their workforce, not only in the technical use of AI tools but also in understanding the

underlying principles and capabilities of AI. This challenge is compounded by the fast pace of technological advancement, which requires continual learning and adaptation.

- **Integration with Existing Tools and Workflow:** Agile teams often use a variety of tools and systems tailored to support specific aspects of Agile methodologies, such as JIRA for task tracking or Slack for communication. Integrating AI solutions with these existing tools without disrupting the current workflows can be challenging. Technical compatibility, data synchronization, and maintaining workflow continuity all pose potential hurdles that require thoughtful planning and execution.
- **Data Issues: Quality, Quantity, and Privacy:** AI systems are heavily dependent on data to train algorithms and provide insights. The quality and quantity of historical data available can significantly impact the effectiveness of AI. In Agile environments, where projects are diverse and data may be unstructured, collecting and preparing high-quality data for AI use can be challenging. Furthermore, concerns about data privacy and security are paramount, especially when sensitive information is involved. Ensuring compliance with data protection regulations (like GDPR – General Data Protection Regulation in Europe) while using AI tools is critical.
- **Over-Reliance on AI:** Relying too heavily on AI for decision-making can lead to a decrease in critical thinking and problem-solving skills among team members. AI models, despite their capabilities, may not always account for nuanced or unprecedented project scenarios. There is a risk that teams might accept AI-generated solutions without sufficient scrutiny or fail to intervene when unique or complex problems arise that require human judgment.
- **Bias in AI Algorithms:** Bias in AI is a well-documented challenge. AI systems learn from historical data, which can itself be biased. This can lead to AI models perpetuating existing prejudices in project decisions, risk assessments, or task allocations. Identifying, mitigating, and continuously monitoring for bias in AI implementations is essential to ensure fairness and effectiveness in Agile project management.
- **Maintaining Agile Principles:** Agile methodologies emphasize flexibility, individual interactions, and customer collaboration over rigid processes and tools. Integrating AI should not detract from these core principles. There is a risk that the mechanization brought by AI could overshadow the Agile values of team dynamics and customer-centric approaches, especially if not implemented with sensitivity to these aspects.
- **Scaling AI Solutions:** As Agile teams often work in varied project environments, an AI solution that works for one team or project may not be suitable for another. Scaling AI solutions across different teams or projects in a way that aligns with specific needs, yet maintains efficiency and effectiveness, presents a complex challenge. Customization and configuration of AI tools must be managed carefully to ensure they provide relevant and actionable insights across projects.

Addressing these challenges requires a balanced approach, combining technical solutions with strategic management and continuous learning. Successful integration of AI into Agile projects involves not just deploying technology but also fostering an environment where technology enhances human capabilities and aligns with Agile values.

14.5 CONCLUSIONS

The mutual correlation between the last two industrial revolutions (Industry 3.0 and Industry 4.0) with the development of project management is obvious. These industrial revolutions required better project management, and improved project management had a feedback effect on the acceleration of industrial revolutions. For the complex and evolving challenges that we try to solve, we use technologies that are also complex and constantly changing.

Agile project management can best be described as the approach of building a product by empowering and trusting people, acknowledging change as the norm, and promoting constant feedback. It is not important to only be or do Agile. True business agility comes from freedom, not frameworks, and that is the reason why a winning mindset is a hybrid of the leading Agile, Lean, and traditional approaches. It is not possible to have one set of ideal practices which will optimize outcomes for every working context. The real goal is to effectively achieve desired organizational outcomes.

The potential of AI to enhance Agile project management is vast, offering opportunities for improved efficiency, better decision-making, and enhanced project outcomes. Authors recently had a chance to facilitate a session on using generative AI in projects, and we could see in practice how these chatbots can help create user stories, estimate effort, build your own chatbot, and much more. It is amazing what we can do already and can only imagine how much more is ahead.

However, as with any technological integration, it comes with a set of challenges that must be thoughtfully addressed. As AI technology continues to evolve, it will become increasingly important for teams to stay informed about the latest developments and think critically about how best to integrate AI into their workflows. Training and continuous learning will be crucial in helping team members adapt to AI tools and use them effectively.

While the path to integrating AI into Agile project methodologies involves navigating various challenges, the potential benefits make it a worthwhile endeavor. The overall purpose of AI is to create IT tools that can function in an intelligent manner. This requires the use of statistical methods, computational intelligence, and optimization techniques. AI techniques require not only an understanding of technology but also an understanding of psychology, linguistics, neuroscience, and many other knowledge areas. With careful implementation and ongoing adaptation, AI can significantly empower Agile teams to meet the demands of modern project environments more effectively, deliver superior results, and further accelerate the effects of Industry 4.0.

AI integration in project management will simply shift the focus of project managers from tactical to strategic, leaving tactical and repetitive tasks to AI and bots while empowering project and organizational leaders to make strategic decisions. This is a mindset shift from project management to project leadership.

REFERENCES

Ambler, S., Lines, M. (2022). *Choose Your WoW - Second Edition: A Disciplined Agile Approach to Optimizing Your Way of Working*, Project Management Institute.

Bharati, A., & Sandbrink, C. (2024). *The Implementation of Artificial Intelligence in Project Management. KEEP ON PLANNING FOR THE REAL*, https://archive.corp.at/cdrom2024/papers2024/CORP2024_78.pdf

Doyle-Kent, M., & Kopacek, P. (2021). Adoption of Collaborative Robotics in Industry 5.0: An Irish Industry Case Study. *IFAC-PapersOnLine*, 54(13), 90–95. https://www.sciencedirect.com/science/article/pii/S2405896321019200

Esteves, M. C., Rodrigues, T. V., Sanjulião, L. K. A. F., & Santos Filho, V. H. (2020). Project Management in Industry 4.0: Technologies and Skills Supporting Project Managers. *Revista Mundi Engenharia, Tecnologia e Gestão*, 5(8), 301–315. https://doi.org/10.21575/25254782rmetg2020vol5n81426

Fajsi, A., Morača, S., Milosavljević, M., & Medić, N. (2022). Project Management Maturity and Business Excellence in the Context of Industry 4.0. *Processes*, 10(6), 1155. https://doi.org/10.3390/pr10061155

Jannat, S. F., Ahmed, M. S., & Rajput, S. A. (2024). AI-Powered Project Management: Myth or Reality? Analyzing the Integration and Impact of Artificial Intelligence in Contemporary Project Environments. *International Journal of Applied Engineering & Technology*, 6(1), 1810–1820. https://romanpub.com/resources/ijaet20v6-1-2024-169.pdf

Kılıç, A., & Balcıoğlu, Y. S. (2023). *Embracing the Future: A Comprehensive Review of Industry 5.0 Research Trends.* ResearchGate. https://www.researchgate.net/publication/370073339_EMBRACING_THE_FUTURE_A_COMPREHENSIVE_REVIEW_OF_INDUSTRY_50_RESEARCH_TRENDS

Knaster, R., & Leffingwell, D. (2020). *SAFe® 5.0 Distilled: Achieving Business Agility with the Scaled Agile Framework.* Addison-Wesley Professional.

Leffingwell, D., Yakyma, A., Knaster, R., Kemelman, D., & Oren, I. (2017). *SAFe® 4.5 Reference Guide: Scaled Agile Framework® for Lean Enterprises.* Addison-Wesley Professional.

Marnewick, A. L., & Marnewick, C. (2020). The Ability of Project Managers to Implement Industry 4.0-Related Projects. *IEEE Access*, 8, 314–324. https://doi.org/10.1109/ACCESS.2019.2961678

Milovanović, D., & Pantović, V. (2024). *5G Mobile Networks and Smart Connection of AIoT Internet Objects* (in Serbian), Faculty of Information technology and Engineerinng, University "Union – Nikola Tesla", Beograd, Serbia.

Mohammed, I. K., & Trzcielinski, S. (2021). The Interconnections between ICT, Industry 4.0, and Agile Manufacturing. *Management and Production Engineering Review*, 12(4), 99–110. https://doi.org/10.24425/mper.2021.139998

Pantović, V. (2022). Emerging Trends in Project Management in Engineering (Keynote), *Proceedings of the 2022 4th IEEE International Conference on Emerging Trends in Electrical, Electronic and Communications Engineering (ELECOM),* Mauritius, p. IV–V, ISBN 978-1-6654–6696–7.

Pantović, V., Vidojević, D., Vujičić, S., Sofijanić, S., & Jovanović-Milenković, M. (2024). Data-Driven Decision Making for Sustainable IT Project Management Excellence, *MDPI Sustainability*, 16(7), 3014. https://www.mdpi.com/2071-1050/16/7/3014, https://doi.org/10.3390/su16073014

Perez, C. (2002). *Technological Revolutions and Financial Capital: The Dynamics of Bubbles and Golden Ages*. Edward Elgar Publishing.

PMI. (2017). *Agile Practice Guide*. Project Management Institute. ISBN 978-1628251999

Scaled Agile, Inc. (2024). Business Agility. Retrieved from Scaled Agile Framework https://scaledagileframework.com/business-agility/

Şen, E., & İrge, N. T. (2020). Industry 4.0 and Agile Firms. In Akkaya, B. (ed.), *Agile Business Leadership Methods for Industry 4.0* (pp. 209–232). Emerald Publishing Limited. https://doi.org/10.1108/978-1-80043-380-920201013

Yaqoob, S., & Azam, M. K. (2018). Industry 4.0 and Agile Manufacturing. In *The European Proceedings of Social & Behavioural Sciences*. Future Academy. 1st Edition https://doi.org/10.15405/epsbs.2019.05.02.17

15 Industry 4.0 Intelligence on the Edge

Visham Hurbungs, Tulsi Pawan Fowdur, and Vandana Bassoo

15.1 INTRODUCTION

Industry 3.0 was centred on automation in individual factories with the objective of saving money by improving efficiency and productivity with the substitution of human labour for machines and computers. Managing the flow of materials through production processes has long been a critical component of business success for many manufacturing industries. Continuously improving manufacturing processes through automation was the key to operational success in the industry. Driven by the same objectives, a number of new concepts and practices are now emerging: Industrial Internet of Things (IIoT), digital twins, Edge artificial intelligence (AI) and Industry 4.0. IIoT connects industrial devices with the aim of gathering and exchanging information along with remotely controlled operations. Digital twins provide virtual environments and models that offer real-life representations of physical assets and objects. Edge AI adds a layer of intelligence to industrial devices while Industry 4.0 integrates digital technologies into automated manufacturing processes. These emerging concepts and practices can optimise manufacturing processes, improve productivity and eventually lead to new business models [1].

Edge computing is a major player in smart environments. With the advancements in networking infrastructures such as 5G and Edge computing, Internet of Things (IoT) devices can now process and analyse data faster at the periphery of the network. Only the most important data is transmitted to the server, thereby reducing latency and network bottlenecks. In today's competitive business environment, Edge computing is crucial since it ensures more efficient operations, faster response times, functionality in remote locations, increased security, data privacy and reduced networking costs among others [2,3]. IoT devices, Edge computing and wireless technologies are increasingly transforming the way businesses operate. Edge computing use cases span multiple industries such as autonomous vehicles, smart healthcare, equipment monitoring, energy efficiency, inventory management, predictive maintenance, patient monitoring and virtualised radio networks [4,5]. Cloud providers such as Amazon Web Services, Microsoft Azure and Google Cloud are now targeting industrial sectors with Edge computing solutions.

Edge AI combines the benefits of Edge computing and AI. AI algorithms are executed directly on interconnected devices at the network edge. Edge AI adds a layer of intelligence at the network boundary where Edge devices not only

DOI: 10.1201/9781003511298-15

collect data from sensors but also process and analyse the data in real time and provide faster insights locally [6,7]. Edge AI is attracting many industries as they discover new ways to optimise workflows, computerise business processes and find innovative prospects while tackling challenges such as cost reduction, resource optimisation and security. Edge AI products include IBM Edge Application Manager, Microsoft Edge, Edge Impulse and Intel Edge Insights for Industrial among others. Gartner's top ten strategic technology trends for 2024 [8] put more emphasis on AI for businesses which drive business goals.

The fourth industrial revolution integrates digital technologies into the manufacturing process. The digital element aims at providing innovative and smart solutions to the manufacturing sector. Digitisation is being powered by IoT, Edge computing, AI and 5G networks. In 2019, Gartner predicted a rise in embedded sensors and advanced AI capabilities in Edge devices through 2028. Gartner also predicted that intelligence will move towards the Edge in a variety of industrial applications [9]. Various industries are therefore adopting Edge and AI services to cut down costs, automate processes, improve decision-making and optimise manufacturing operations. Industry 4.0 offers greater benefits in terms of decentralisation, interoperability and analytics. Future advances in intelligence on the Edge point to a bright future for handling data closer to its source.

15.2 EDGE COMPUTING IN INDUSTRY 4.0

Edge computing and Industry 4.0 create unlimited possibilities for industries which are no longer centred on the physical world. With the latest advancements in network connectivity, IoT and sensor technologies, possibilities range from providing better visibility into manufacturing operations and workflows, improving worker safety, enhancing asset management, performing predictive maintenance and reducing equipment downtime [10]. We are now moving towards an industry with millions of industrial Edge devices, and manufacturers have a vast set of data generated from these devices and connected systems. However, in its raw form, industrial data is meaningless. With Edge computing, local data on industrial devices can be processed and analysed immediately thereby optimising production and manufacturing processes.

Industrial Edge computing now enables new use cases for smart manufacturing. For example, Edge brings the following advantages to the industry [11]:

1. **Lower Latency**: The main potential of Edge is to reduce latency. If a connected industrial device encounters a malfunction, it would be too costly to send the data to the Cloud.
2. **Increased Security**: Additional connected devices in the network mean more points of attack. The decentralised nature of Edge makes smart factories more secure as there won't be a central point of failure. Since data processing happens close to its source, there is less risk of attacks.
3. **More Manageable Data Analytics**: Data processing at the network edge reduces the stress on central systems. Each device on the Edge manages its own data and AI models.

4. **Expanded Interoperability**: The effectiveness of an IoT network depends on its interoperability. Heterogeneous devices can hinder the expansion of smart manufacturing, and Edge computing eliminates to some extent the need for a standard protocol.
5. **Reduced Storage Costs**: Each industrial device transmits only important data to the Cloud, and there is no need for large data storage. Some data can be stored temporarily on the Edge for processing and analysis.

15.2.1 Industrial Edge Computing Infrastructures

Industrial Edge computing infrastructures provide a common framework with a high level of abstraction for connecting all assets used in manufacturing, oil, gas, energy, retail and other industrial processes. Table 15.1 summarises the main features of the Far-Edge project, the Edge Computing Consortium (ECC) and the Industrial Internet Consortium (IIC) reference architectures (RA) for Industry 4.0. These references are guided by international standards defined by ISO/IEC/IEEE 42010:2022 [12]. The architectures are explained in more detail in the following sub-sections in terms of functions, domains and viewpoints.

15.2.1.1 Far-Edge RA [14]

Far-Edge is one of the first initiatives to investigate factory automation with Edge computing. The main objective of Far-Edge is to provide virtualisation of decentralised factory automation. The Far-Edge RA is based on two main components: Edge Computing and Distributed Ledger Technologies. The platform explores the use of Edge Computing and Blockchain technology for industrial automation that supports real-time interactions and large-scale analytics. Far-Edge RA decomposes factory automation into the following components as shown in Figure 15.1:

- Three high-level functional domains
  1. **Automation:** Automate control and configuration of physical control processes to enable plug-and-play of factory equipment.

TABLE 15.1
Edge Computing RA Features [13]

Characteristics	Far-Edge	ECC	IIC
Connectivity	■	■	■
Device management	■	■	■
Data collection	■	■	■
Scalability	■	■	■
Standards		■	■
Security	■	■	■
Blockchain	■		

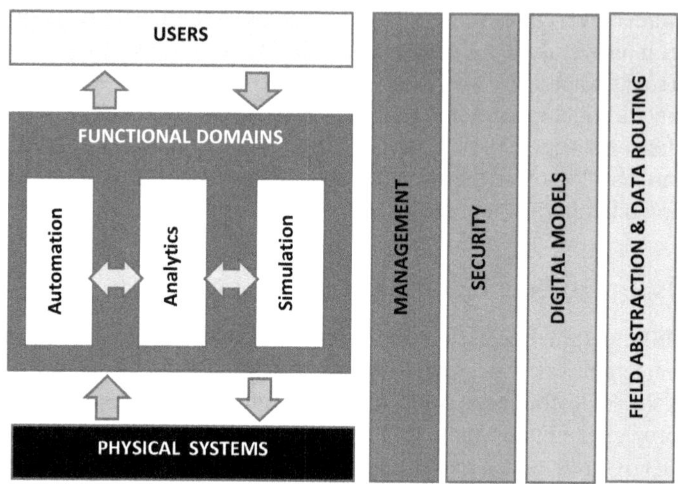

FIGURE 15.1 Far-Edge RA functional domains and functions [14].

2. **Analytics:** Process data and provide intelligence to users.
3. **Simulation:** Simulate the behaviour of physical production processes using digital models.

- Four cross-cutting functions
  1. **Management:** Low-level functions for monitoring factory equipment and IT components.
  2. **Security:** Identity management, authentication, access control and encryption.
  3. **Digital Models:** Modelling of real-world objects.
  4. **Field Abstraction and Data Routing:** Device discovery, communication protocols and unidirectional data channels.

15.2.1.2 Edge Computing RA 2.0 [15]

The Edge Computing RA is a model-driven architecture based on the Model-Driven Engineering principles. It models both the physical and the digital world to achieve coordination between the physical and digital worlds, cross-industry collaboration, reduced system heterogeneity, simplified cross-platform migration and effective support for system lifecycle activities. From the horizontal perspective, the architecture provides smart services, intelligent coordination between development and deployment service framework, smart service orchestration, simplified distributed Edge intelligence architecture for services and intelligent Edge computing nodes. From a vertical perspective, the architecture provides management services, lifecycle data services and security services to deliver smart services in the entire production process. The Edge computing RA offers the following viewpoints as illustrated in Figure 15.2:

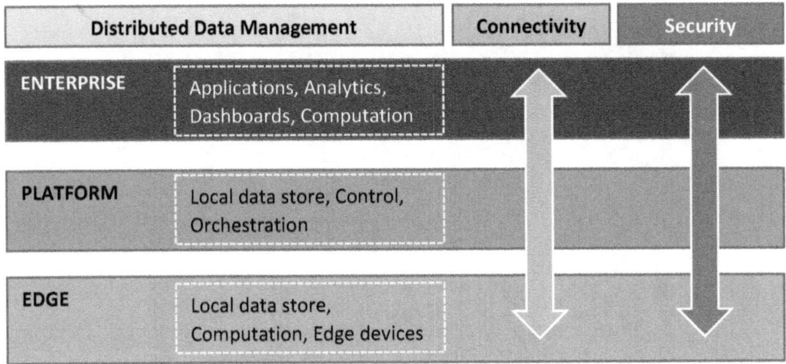

FIGURE 15.2 Edge computing RA 2.0 [13].

1. **Concept View:** Domain models and key concepts of Edge computing. Four types of intelligent Edge computing node development frameworks are defined: real-time computing, lightweight computing, smart gateway and smart distributed systems. Based on these development frameworks, six types of products can be manufactured: embedded, controller, independent controller, terminal perception, smart gateway, edge distributed gateway and edge distributed server.
2. **Function View:** Functions and design concepts of the service, deployment and operation frameworks. Three main layers are defined: edge virtualisation function (industry-oriented services), virtualisation layer and basic resources layer (network, computing and storage).
3. **Deployment View:** System deployment process and deployment scenarios. Support for the following key services is provided: model-based development, emulation and integration release services. The deployment operation service framework provides service orchestration, application deployment and application market.

15.2.1.3 IIC RA [16]

The Industrial Internet RA (IIRA) is a standards-based architecture for IIoT systems. It addresses the requirement for a common framework for interoperable IIoT systems to suit various industry use cases. IIRA 1.9 provides information about wireless communications for industrial automation systems, while IIRA 2.0 improves stakeholder experience by addressing user concerns. Stakeholders who have IIoT concerns include product managers, system engineers, system architects, component architects, developers, integrators and operators. IIRA addresses the concerns of these stakeholders in terms of:

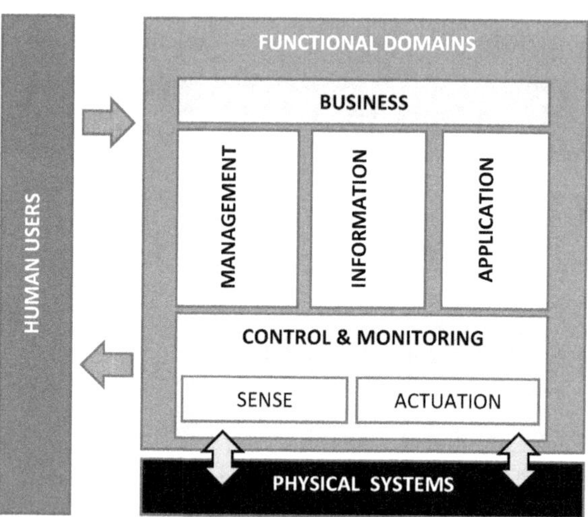

FIGURE 15.3 Industrial Internet Consortium RA [16].

1. **Business Viewpoint:** Business-oriented objectives in setting up an IIoT system. The value-driven model consists of visions, values, key objectives and fundamental capabilities.
2. **Usage Viewpoint:** Expectations of system users related to the use of the IIoT system. Usage concerns include efficiency of operations, usability, ergonomics, configuration, service agreements, training needs, security and safety among others.
3. **Functional Viewpoint:** Functional components and system-specific requirements of the IIoT system. Functional domains are decomposed into control and monitoring, information, application, business and system management. Figure 15.3 illustrates the data and control flow between the functional domains.
4. **Implementation Viewpoint:** Technologies required to implement functional and system components. IIoT system implementation follows well-established architectural patterns such as IoT component capability model pattern, three-tier architecture, digital twin core and layered databus among others.

15.3 INDUSTRIAL INTELLIGENCE ON THE EDGE

Intelligence has now become one of the main drivers of innovation in certain industrial sectors. IoT devices, actuators, sensors, wireless networks and AI make industrial machines smarter than ever before. Manufacturers are integrating IoT, Edge computing and AI into their production processes and all over their manufacturing operations. Machine-learning algorithms improve product quality and flexibility of the production process [6]. Intelligent industrial concepts can be applied across various industrial sectors including manufacturing, oil, gas, mining and other industrial segments. Two important use cases for industrial intelligence are discussed in this section: smart manufacturing and greenhouse gas (GHG) emissions.

15.3.1 Smart Manufacturing in Industry 4.0

Smart manufacturing and Industry 4.0 represents the future of manufacturing. Smart manufacturing is not about breaking down traditional factories and replacing them with machines, robots or computer systems. The main objective is to make the factory smarter by using intelligent devices embedded with software, network connectivity and sensors to improve the manufacturing process. Enterprises are now looking beyond profit as a measure of success. Instead, they seek efficiency, productivity, responsiveness, sustainability, quality control, increased visibility, transparency, adaptability and end-to-end savings [17]. This new business model equips businesses with intelligent factories, intelligent products, intelligent assets and empowered stakeholders. Real-world applications of smart manufacturing include predictive maintenance to reduce equipment downtime, real-time monitoring for improved product or service quality control and simulation in a risk-free virtual environment. Table 15.2 highlights the latest trends in smart manufacturing.

TABLE 15.2
Latest Trends in Smart Manufacturing

Research Work	Description
The role of IoT in smart manufacturing for enhanced household product quality [18]	This chapter explores the role of IoT in transforming manufacturing processes to improve household product quality. It covers data-driven insights, predictive maintenance, product customisation and sustainability. The network of interconnected devices and sensors can help manufacturers modernise operations, decrease costs and eventually deliver superior quality household products that meet the changing requirements of consumers.
Survey of multimodal human–robot interaction (HRI) for smart manufacturing [19]	The authors present a survey of multimodal HRI with emphasis on vision, auditory, touch, motion and bodily parts sensing in human-centric smart manufacturing. HRI prioritises human requirements in the collaboration between humans and robots. The main objective is to increase productivity while reducing human strain in intelligent manufacturing systems.
A smart manufacturing process in the automation of the textile industry [20]	This work investigates automated cotton fabric inspection along with resource allocation in smart manufacturing. The case study of Jagatjit Cotton Textiles manufacturing company in Punjab is presented as a proof of concept for error-free fabric inspection that involves human and machine interaction. As part of the experiments, four different textile supply chain models with industrial automation were developed and tested with triangular and trapezoidal fuzzy uncertainties.

(*Continued*)

TABLE 15.2 (*Continued*)
Latest Trends in Smart Manufacturing

Research Work	Description
Review of Deep Reinforcement Learning (DRL) in smart manufacturing [21]	This systematic review attempts to bridge the gap between Deep Neural Networks and Reinforcement Learning in the field of smart manufacturing. DRL applications are examined in the four engineering lifecycle stages namely design, manufacturing, logistics and maintenance. The review also highlights emerging DRL-related technologies and solutions that pave the path towards smart manufacturing.
Time-series techniques and applications in the manufacturing domain [22]	In this literature review, the authors present time-series classification algorithms in manufacturing as well as guidelines for manufacturers. They analyses time-series techniques and applications in the manufacturing domain. The proposed ontology provides a visual roadmap to simplify manufacturing problems using time-series analytics.
Real-Time Fault Diagnosis (RTFD) technologies techniques for industrial smart manufacturing [23]	This work reviews RTFD technologies in industrial process and machine condition monitoring. RTFD methods are categorised from the smart manufacturing perspective in terms of implementation methods and their industrial applications. RTFD processes are described in detail starting from data acquisition, pre-processing and selection of the RTFD method based on feature extraction, end-to-end neural networks and qualitative method reasoning. This work also highlights challenges and trends in RTFD.
Data-driven decision-making system using smart and Lean manufacturing concepts in Industry 4.0 [24]	This work investigates the control of production activities on the shop floor using a data-driven decision-making system. Critical conditions affecting production time and resource utilisation that could lead to inefficient operating conditions were investigated using the Lean and smart manufacturing approaches. The results showed that productivity was improved by controlling shop floor production activities.
Automated material selection in smart manufacturing [25]	The composition and properties of construction materials were studied in this work. An automated material selection approach using linear regression analysis was used to determine the quality of carbon steels from AISI 1010 to AISI 1060. Mechanical properties were determined by analysing parameters such as strength, stiffness and stability.
Smart manufacturing framework for small and medium enterprises (SMEs) [26]	This work highlights the struggle of SMEs is moving towards smart manufacturing. The authors proposed a smart manufacturing adoption framework based on the needs and challenges faced by SMEs. The framework consists of five stages: identify manufacturing data, readiness assessment of financial and technical conditions, develop awareness about smart manufacturing, develop a smart manufacturing vision and identify tools and practices to achieve the tailored smart manufacturing vision.

(*Continued*)

TABLE 15.2 (*Continued*)
Latest Trends in Smart Manufacturing

Research Work	Description
Real-time wine monitoring system [27]	The authors took advantage of low-cost sensors and wireless communication technology to investigate a real-time monitoring system to control wine barrel status. They focused on a special type of white wine (Fino) manufactured in southern Spain. A smart cork prototype was developed to monitor the health, evaporation, room temperature and light intensity inside the barrel. This setup allowed the wine manufacturer to identify cracks in the wine barrel and abnormal environmental conditions before the wine loses its quality.

15.3.2 GHG Emissions in Industry 4.0

The fourth industrial revolution now overlaps with the urgent need for environmental sustainability since the manufacturing sector plays a major role in global GHG emissions. Industry 4.0 presents many opportunities in the fight against climate change. Industrial energy consumption is a major contributor to GHG emissions, which can be significantly reduced by adopting Industry 4.0 applications. Integration of Industry 4.0 not only contributes to the transition towards a circular economy but also promotes a more sustainable industrial ecosystem. Sensors and real-time data analytics can provide insights into energy usage patterns and therefore make factories more environmentally friendly. For example, an automotive supplier, Gestamp, reduced its energy consumption by 15% using an efficient energy management strategy co-developed with Siemens. The supplier had above 100 production plants and manufactured vehicle body parts, chassis and car parts for major car manufacturers. This helped Gestamp to move closer to sustainable connected plants with Industry 4.0 solutions. With its reduced energy consumption, the company was also able to reduce its carbon emissions by 15% and easily achieved its environmental protection goals [28]. With the expected continuous rise in industrial production, it is a matter of responsibility for industries to invest in digitisation to reduce GHG emissions. Table 15.3 highlights the latest trends in industrial GHG emissions.

15.4 FRAMEWORK FOR PRODUCT CLASSIFICATION IN INDUSTRY 4.0

The Industry 4.0 approach to product classification can boost production rate, decrease manufacturing costs, enhance product quality and reduce time to market. The new generation of intelligent and connected industrial devices equipped with sensors and intelligence make smart factories a reality. The latter can respond faster to production requirements and demands than conventional production lines.

TABLE 15.3
Latest Trends in Industrial GHG Emissions

Research Work	Description
Smart building energy management using Industry 4.0 [29]	This work investigates interior climate conditions in a Heating Ventilation and Air Conditioning system with the main objective of achieving energy efficiency along with occupant comfort and indoor air quality through AI and IoT sensor-based adjustments. By employing Artificial Neural Networks (ANNs) and the Response Surface Methodology, this investigation demonstrated that Industry 4.0 can achieve an increase of 16% in efficiency thereby reducing GHG emissions. A 15% reduction in costs for controlled rooms was also observed.
Role of AI in methane emission estimation in livestock [30]	This work explores methane (CH_4) emissions in the livestock industry with emphasis on cattle. The study highlights the importance of monitoring methane emissions since it is a powerful GHG. AI techniques have been used to control CH_4 produced by cattle since this gas is an important element in climate change. AI techniques such as regression, supervised learning models and ANNs in monitoring CH_4 emission have the potential to considerably improve accuracy.
The role of industrial robots in reducing carbon dioxide emission [31]	The impact of digitisation on carbon dioxide (CO_2) emissions was studied by using data from 40 countries between 1993 and 2015. The authors presented three possible factors that influence CO_2 emissions by industrial robots: economic growth, technological progress and structural transformation. The findings of this empirical analysis showed that the digital factory can significantly reduce CO_2 emissions across many industries.
Enhanced energy management systems for Industry 4.0 [32]	This paper analysed how Industry 4.0 can manage the energy requirements of smart factories by designing, monitoring and controlling industrial plants. The proposed framework takes into consideration key aspects of the ISO 50001 standard [33], which is geared towards organisational efficient energy management. The framework defines five implementation steps that offer practical business applications for energy management and reduction of harmful gas emissions.
Achieving environmental sustainability with carbon emissions management and Industry 4.0 [34]	In view of the environmental pollution issues in the textile industry, this paper aims at optimising profit, carbon and resource efficiency by employing mathematical models for carbon emission. One of the focuses was the use of an Industry 4.0 digital monitoring system to manage production processes in the textile industry.
A digital lifecycle management framework for energy-intensive smart industries [35]	The authors investigated the interaction between process mining and simulation modelling for a sustainable industry. A framework was proposed to support sustainable smart manufacturing in energy-intensive industries. This study concluded that intelligence is important to improve energy and material efficiency in the realisation of sustainable smart manufacturing.

(*Continued*)

TABLE 15.3 (*Continued*)
Latest Trends in Industrial GHG Emissions

Research Work	Description
Monitoring and control of smart greenhouses in Industry 4.0 [36]	This work studied the application of Industry 4.0 concepts in a greenhouse environment with the view of efficiently controlling indoor parameters such as light, ventilation, humidity, temperature and carbon dioxide level. Industry 4.0 brings new opportunities and challenges for intelligent greenhouse crop cultivation with a focus on reduced energy consumption and CO_2 emissions.
Improving environmental sustainability by integrating Industry 4.0 [37]	Twenty Industry 4.0 applications that create a sustainable environment are discussed in this study ranging from production environments, supply chains, delivery and marketing. One main consideration for sustainability is reduced carbon footprint and pollution control in the digital automation of operations in sectors such as mining, oil and gas.
Environmental impact of Industry 4.0 [38]	The objective of this report was to assess the opportunities of Industry 4.0 to reduce the environmental impact of manufacturing. The focus is on sustaining circular economy strategies to minimise emissions. Case studies discussed in this report show how Industry 4.0 could be applied to reduce environmental impacts in the manufacturing industry.
Green IIoT for the smart industry [39]	This work presents enabling technologies for Green Industrial Internet of Things applications for the smart industry while highlighting the importance of renewable energy sources to lower CO_2 emission. Applications range from smart factory, smart agriculture, smart grid, smart city and smart transportation to smart health.

This new business model supports both business and industrial goals in today's competitive manufacturing environment. The following sub-section defines a framework for smart wine classification in Industry 4.0. The proposed framework extends the real-time monitoring system to supervise the status of wine casks or barrels designed by Cañete et al. [27].

15.4.1 Smart Wine Classification in Industry 4.0

The smart cork prototype allowed winemakers to monitor wine casks in real time and take immediate actions. For example, parameters such as cask structure, evaporation level, light intensity, temperature and cracks could be determined before the wine loses its quality [27]. In the proposed wine classification framework for Industry 4.0, intelligence has been added to the wine manufacturing and monitoring process to classify the quality of the Fino white wine directly on the Edge, i.e. inside the wine cask as shown in Figure 15.4.

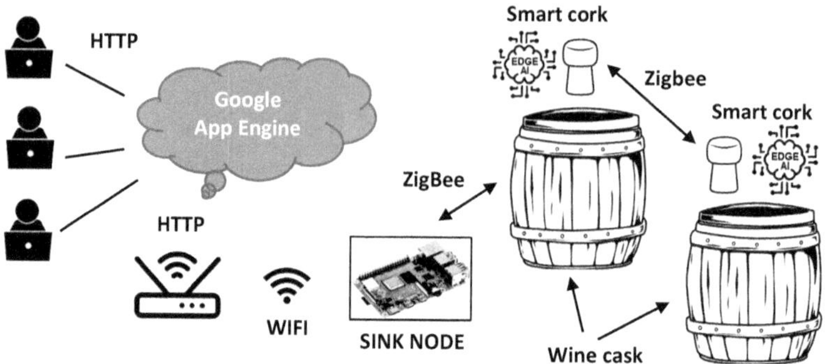

FIGURE 15.4 Proposed wine classification framework for Industry 4.0 [27].

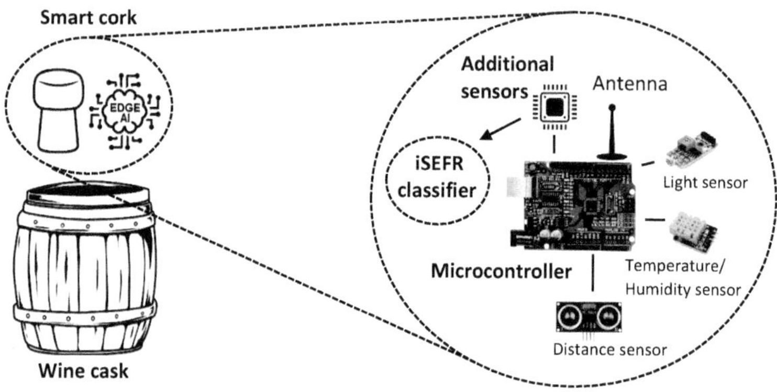

FIGURE 15.5 Intelligent smart cork with additional sensors [27].

Additional sensors can be connected to the microcontroller to capture acidity, sugar, chlorides, sulphur dioxide, density, PH, sulphates and alcohol levels of the Fino wine, as shown in Figure 15.5. These parameters can then be input to the improved Scalable, Efficient and Fast classifieR (iSEFR) [40], which is a resource and energy-efficient binary classifier designed for Edge devices. The classifier can perform both training and testing on a resource-constrained device with a limited memory footprint.

As a proof of concept, three classification experiments were performed using the Vinho Verde white wine dataset [41] on an eight-bit Arduino Uno microcontroller with 32 kb of flash memory and 2 kb of Static Random Access Memory (SRAM). The Arduino Uno with limited computing resources was used to simulate an industrial device on the Edge. The multiclass wine dataset consists of 11 features, and the target label is the wine quality, which is between 0 (very bad) and 10 (very excellent). For this wine classification experiment, the dataset was converted to binary by categorising wine quality less or equal to 5 as negative samples and above 5 as positive samples. A balanced dataset with an equal number of negative and positive wine

TABLE 15.4
Training and Testing Metrics

	Experiment 1		Experiment 2		Experiment 3	
Metrics	Training (100 Records)	Testing (150 Records)	Training (150 Records)	Testing (100 Records)	Training (200 Records)	Testing (50 Records)
TP	47	69	63	47	84	24
TN	45	57	67	41	86	23
FP	5	18	8	9	14	2
FN	3	6	12	3	16	1
Recall	0.94	0.92	0.84	0.94	0.84	0.96
Precision	0.90	0.79	0.89	0.84	0.86	0.92
Accuracy	0.92	0.84	0.87	0.88	0.85	0.94
MCC	0.84	0.69	0.73	0.77	0.70	0.88
Error	0.08	0.16	0.13	0.12	0.15	0.06

samples was input to the iSEFR binary classifier. Table 15.4 illustrates the results of the experiments in terms of the following classification metrics: true positive (TP), true negative (TN), false positive (FP), false negative (FN), recall, precision, accuracy, Matthews correlation coefficient (MCC) and error rate [40].

The average training and testing metrics for each experiment are shown in Figure 15.6. The experimental results demonstrate a high level of accuracy, precision and recall with a low error rate in classifying the Vinho Verde white wine. These metrics can be used to make faster decisions on the wine quality determined directly

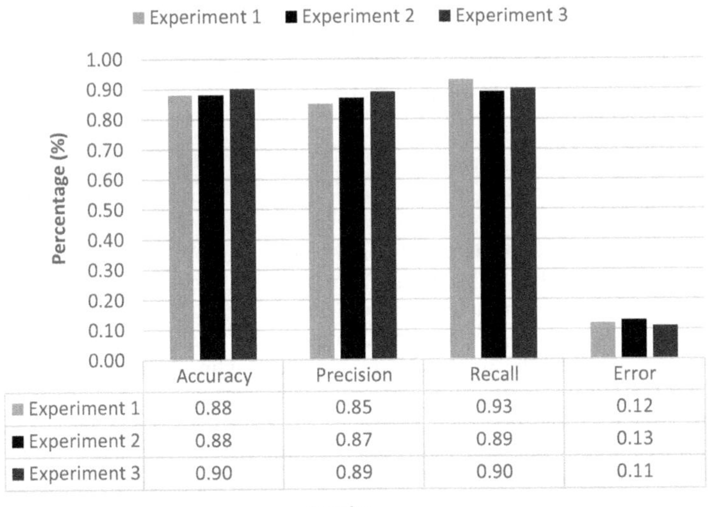

FIGURE 15.6 Average training and testing metrics.

from the cask. The proposed framework is therefore a possible solution to intelligent wine cask monitoring and wine quality classification in Industry 4.0.

15.5 OPEN ISSUES AND CONCLUDING REMARKS

Industry 4.0, also called the fourth industrial revolution, is the era of automation, connectivity, intelligence, digitisation and advanced-manufacturing technologies that are creating a wave of change in the manufacturing sector. In other words, Industry 4.0 is re-inventing how businesses design, manufacture and distribute their products by integrating intelligent digital technologies into manufacturing and industrial processes. Industry 4.0 solutions range from collaborative design platforms, predictive maintenance, supply chain optimisation, agile manufacturing, quality control, defect detection and circular economy practices to carbon emission monitoring [42]. Industry 4.0 intelligence on the Edge enables real-time decision-making without continuous reliance on Fog or Cloud infrastructures. Edge computing and AI algorithms empower industrial devices without any connectivity located at the network periphery, thereby enabling innovative Industry 4.0 applications such as predictive maintenance, smart manufacturing and intelligent factories among others.

Existing literature puts a lot of emphasis on the benefits derived from Industry 4.0 and the digitisation of manufacturing assets and processes. However, some research works have underlined potential areas of concern that have not been addressed so far. The adoption of Industry 4.0 technologies may have unintended negative impacts on the environment with higher levels of energy consumption; millions of IoT devices communicating with each other can generate large data volumes which can overpower computing infrastructures; and replacement of obsolete devices can escalate electronic waste and higher investment costs and increased complexity in replacing the entire manufacturing processes. Moreover, the implementation of sensors, networks and AI requires higher consumption of natural resources than traditional manufacturing technologies. Social implications such as reduced employees' autonomy, loss of employment, unhealthy work–life balance and health and safety issues when interacting with automated systems have also been associated with Industry 4.0 [43].

REFERENCES

1. SLT Partners, "Elisa smart factory: how to win over industry leaders in two years," March 2019. [Online]. Available: https://stlpartners.com/research/elisas-smart-factory-how-to-win-over-industry-leaders-in-just-two-years/. [Accessed 17 June 2024].
2. Microsoft, "What is edge computing?," August 2022 [Online]. Available: https://azure.microsoft.com/en-us/resources/cloud-computing-dictionary/what-is-edge-computing. [Accessed 10 June 2024].
3. V. Hurbungs, V. Bassoo and T. P. Fowdur, "Fog and edge computing: concepts, tools and focus areas," *International Journal of Information Technology,* vol. 13, no. 2, pp. 511–522, 2021.
4. N. Siersted, "10 Edge computing use case examples," STL Partners, 16 December 2021. [Online]. Available: https://stlpartners.com/articles/edge-computing/10-edge-computing-use-case-examples/. [Accessed 15 June 2024].

5. A. C. Baktir, A. Ozgovde and C. Ersoy, "How can edge computing benefit from software-defined networking: a survey, use cases, and future directions," *IEEE Communications Surveys & Tutorials,* vol. 19, no. 4, pp. 2359–2391, 2017.
6. BARC, "Machine learning in Industry 4.0: five use cases," 30 January 2018. [Online]. Available: https://barc.com/de/machine-learning-in-industry-4-0/. [Accessed 15 June 2024].
7. R. Singh and S. S. Gill, "Edge AI: a survey," *Internet of Things and Cyber-Physical Systems,* vol. 3, pp. 71–92, 2023.
8. M. Ava, "Gartner top 10 strategic technology trends for 2024," Gartner, 16 October 2023. [Online]. Available: https://www.gartner.com/en/articles/gartner-top-10-strategic-technology-trends-for-2024. [Accessed 17 June 2024].
9. Gartner, "Gartner top 10 strategic technology trends for 2019," Gartner, 15 October 2018. [Online]. Available: https://www.gartner.com/smarterwithgartner/gartner-top-10-strategic-technology-trends-for-2019. [Accessed 17 June 2024].
10. J. Elliot, "How edge AI is fueling Industry 4.0 outcomes," Forbes, 20 June 2023. [Online]. Available: https://www.forbes.com/sites/forbestechcouncil/2023/06/20/how-edge-ai-is-fueling-industry-40-outcomes/. [Accessed 17 June 2024].
11. M. R. Nichols, "Edge computing is essential for smart manufacturing success," i–SCOOP, 1 May 2020. [Online]. Available: https://www.i-scoop.eu/edge-computing-explained/edge-manufacturing-industry/. [Accessed 17 June 2024].
12. ISO, "Software, systems and enterprise — architecture description," ISO, 2022. [Online]. Available: https://www.iso.org/standard/74393.html. [Accessed 18 June 2024].
13. M. Isaja, R. S. Alonso, S. Rodríguez-González, J. A. García Coria and F. De La Prieta, "Edge Computing Architectures in Industry 4.0: A General Survey and Comparison," In *14th International Conference on Soft Computing Models in Industrial and Environmental Applications*, Seville, Spain, 2019.
14. M. Isaja, "Reference Architecture for Factory Automation using Edge Computing and Blockchain Technologies," In J. Soldatos, O. Lazaro and F. Cavadini (eds.), *The Digital Shopfloor: Industrial Automation in the Industry 4.0 Era*, Taylor and Francis, pp. 71–101, 2018.
15. ECC/AII, "Edge computing reference architecture 2.0," 2017. [Online]. Available: http://en.ecconsortium.net/Uploads/file/20180328/1522232376480704.pdf. [Accessed 18 June 2024].
16. Industrial Internet Consortium, "The industrial internet reference architecture," 7 November 2022. [Online]. Available: https://www.iiconsortium.org/wp-content/uploads/sites/2/2022/11/IIRA-v1.10.pdf. [Accessed 18 June 2024].
17. SAP, "What is smart manufacturing?," SAP, July 2024. [Online]. Available: https://www.sap.com/uk/insights/smart-manufacturing-in-the-cloud.html. [Accessed 19 June 2024].
18. M. Nasir Ali, T. S. Senthil, T. Ilakkiya, D. S. Hasan, N. Bala and S. Boopathi, "IoT's Role in Smart Manufacturing Transformation for Enhanced Household Product Quality," In G. Revathy (ed.), *Advanced Applications in Osmotic Computing*, IGI Global, 2034, pp. 255–289.
19. W. Tian, P. Zheng, S. Li and L. Wang, "Multimodal human–robot interaction for human-centric smart manufacturing: a survey," *Advanced Intelligent Systems,* vol. 6, no. 3, 2023.
20. G. Kaur, B. K. Dey, P. Pandey, A. Majumder and S. Gupta, "A smart manufacturing process for textile industry automation under uncertainties," *Processes,* vol. 12, no. 4, pp. 778–778, 2024.
21. C. Li, P. Zheng, Y. Yin, B. Wang and L. Wang, "Deep reinforcement learning in smart manufacturing: a review and prospects," *CIRP Journal of Manufacturing Science and Technology,* vol. 40, pp. 75–101, 2023.

22. M. A. Farahani, M. R. McCormick, R. Gianinny, F. Hudacheck, R. Harik, Z. Liu and T. Wuest, "Time-series pattern recognition in smart manufacturing systems: a literature review and ontology," *Journal of Manufacturing Systems,* vol. 69, p. 208–241, 2023.
23. W. Yan, J. Wang, S. Lu, M. Zhou and X. Peng, "A review of real-time fault diagnosis methods for industrial smart manufacturing," *Processes,* vol. 11, no. 2, pp. 369–369, 2023.
24. V. Tripathi, S. Chattopadhyaya, A. K. Mukhopadhyay, S. Saraswat, S. Sharma, C. Li, S. Rajkumar and K.-H. Chang, "Development of a data-driven decision-making system using lean and smart manufacturing concept in Industry 4.0: a case study," *Mathematical Problems in Engineering,* vol. 2022, pp. 1–20, 2022.
25. I. Pavlenko, J. Pitel, V. Ivanov, K. Berladir, J. Mižáková, V. Kolos and J. Trojanowska, "Using regression analysis for automated material selection in smart manufacturing," *Mathematics,* vol. 10, no. 11, pp. 1888–1888, 2022.
26. S. Mittal, M. A. Khan, J. K. Purohit, K. Menon, D. Romero and T. Wuest, "A smart manufacturing adoption framework for SMEs," *International Journal of Production Research,* vol. 58, no. 5, pp. 1555–1573, 2019.
27. E. Cañete, J. Chen, C. Martín and B. Rubio, "Smart winery: a real-time monitoring system for structural health and ullage in fino style wine casks," *Sensors,* vol. 18, no. 3, p. 803, 2018.
28. M. Meyer, "Saving energy with big data," Siemens, 21 May 2018. [Online]. Available: https://www.siemens.com/global/en/company/stories/infrastructure/2018/energy-efficiency-excelling-with-800-million-data-points.html. [Accessed 21 June 2024].
29. M. Seraj, M. Parvez, O. Khan and Z. Yahya, "Optimizing smart building energy management systems through Industry 4.0: a response surface methodology approach," *Green Technologies and Sustainability,* vol. 2, no. 2, pp. 100079–100079, 2024.
30. J. G. Nejad, M.-S. Ju, J.-H. Jo, K.-H. Oh, Y.-S. Lee, S.-D. Lee, E.-J. Kim, S. Roh and H.-G. Lee, "Advances in methane emission estimation in Livestock: a review of data collection methods, model development and the role of AI technologies," *Animals,* vol. 14, no. 3, pp. 435–435, 2024.
31. W. Yao, L. Liu, H. Fujii and L. Li, "Digitalization and net-zero carbon: the role of industrial robots towards carbon dioxide emission reduction," *Journal of Cleaner Production,* vol. 450, pp. 141820–141820, 2024.
32. V. Introna, A. Santolamazza and V. Cesarotti, "Integrating Industry 4.0 and 5.0 innovations for enhanced energy management systems," *Energies,* vol. 17, no. 5, pp. 1222–1222, 2024.
33. ISO, "ISO 50001 energy management," ISO, [Online]. Available: https://www.iso.org/iso-50001-energy-management.html. [Accessed 21 June 2024].
34. W.-H. Tsai, "Balancing profit and environmental sustainability with carbon emissions management and Industry 4.0 technologies," *Energies,* vol. 17, no. 17, p. 6175, 2023.
35. M. Kaniappan Chinnathai and B. Alkan, "A digital life-cycle management framework for sustainable smart manufacturing in energy intensive industries," *Journal of Cleaner Production,* vol. 419, p. 138259, 2023.
36. C. Bersani, C. Ruggiero, R. Sacile, A. Soussi and E. Zero, "Internet of things approaches for monitoring and control of smart greenhouses in Industry 4.0," *Energies,* vol. 15, no. 10, p. 3834, 2022.
37. M. Javaid, A. Haleem, R. P. Singh, R. Suman and E. S. Gonzalez, "Understanding the adoption of Industry 4.0 technologies in improving environmental sustainability," *Sustainable Operations and Computers,* vol. 3, no. 1, pp. 203–217, 2022.
38. H. Berg, P. Bendix, M. Jansen, K. Le Blévennec, P. Bottermann, M. Magnus-Melgar, E. Pohjalainen and M. Wahlström, "Unlocking the potential of Industry 4.0 to reduce the environmental impact of production," European Topic Centre Waste and Materials in a Green Economy, 2021.

39. M. Tabaa, F. Monteiro, H. Bensag and A. Dandache, "Green industrial Internet of Things from a smart industry perspectives," *Energy Reports,* vol. 6, pp. 430–446, 2020.
40. V. Hurbungs, V. Bassoo and T. P. Fowdur, "An enhanced binary classifier for edge devices," *Microprocessors and Microsystems,* vol. 93, p. 104596, 2022.
41. UC Irvine, "Wine quality," 2009 June 2009. [Online]. Available: https://archive.ics.uci.edu/dataset/186/wine+quality. [Accessed 05 March 2024].
42. SAP, "What is Industry 4.0?," SAP, May 2023 [Online]. Available: https://www.sap.com/products/scm/industry-4-0/what-is-industry-4-0.html. [Accessed 22 June 2024].
43. M. Dieste, G. Orzes, G. Culot, M. Sartor and G. Nassimbeni, "The "dark side" of Industry 4.0: how can technology be made more sustainable?," *International Journal of Operations & Production Management,* vol. 44, no. 5, pp. 900–933, 2023.

Index

Note: **Bold** page numbers refer to tables and *italic* page numbers refer to figures.

For Product Safety Concerns and Information please contact our EU representative GPSR@taylorandfrancis.com Taylor & Francis Verlag GmbH, Kaufingerstraße 24, 80331 München, Germany

Batch number: 10397790

Printed by Printforce, the Netherlands